The Mekong: A Socio-
to River Basin Develop

An international river basin is an ecological system, an economic thoroughfare, a geographical area, a font of life and livelihoods, a geopolitical network and, often, a cultural icon. It is also a socio-legal phenomenon. This book is the first detailed study of an international river basin from a socio-legal perspective. The Mekong River Basin, which sustains approximately 70 million people across Cambodia, China, Laos, Myanmar, Thailand and Vietnam, provides a prime example of the socio-legal complexities of governing a transboundary river and its tributaries.

The book applies its socio-legal analysis to bring a fresh approach to understanding conflicts surrounding water governance in the Mekong River Basin. The authors describe the wide range of uses being made of legal doctrine and legal argument in ongoing disputes surrounding hydropower development in the basin, putting to rest lingering caricatures of a single, 'ASEAN' way of navigating conflict. They call into question some of the common assumptions concerning the relationship between law and development. The book also sheds light on important questions concerning the global hybridization or crossover of public and private power and its ramifications for water governance. With current debates and looming conflicts over water governance globally, and over shared rivers in particular, these issues could not be more pressing.

Ben Boer is Distinguished Professor of Law in the Research Institute of Environmental Law at Wuhan University, People's Republic of China, and Emeritus Professor at the University of Sydney, Australia.

Philip Hirsch is Professor of Human Geography at the School of Geosciences, University of Sydney, Australia.

Fleur Johns is Professor of Law at the University of New South Wales, Australia.

Ben Saul is Professor of International Law and Australian Research Council Future Fellow at the University of Sydney, Australia.

Natalia Scurrah is a Researcher in Human and Environmental Geography at the School of Geosciences, University of Sydney, Australia.

Earthscan Studies in Water Resource Management

Adaptation to Climate Change through Water Resources Management
Capacity, Equity and Sustainability
Edited by Dominic Stucker and Elena Lopez-Gunn

Hydropower Development in the Mekong Region
Political, Socio-economic and Environmental Perspectives
Edited by Nathanial Matthews and Kim Geheb

Governing Transboundary Waters
Canada, the United States and Indigenous Communities
By Emma S. Norman

Transboundary Water Politics in the Developing World
By Naho Mirumachi

International Water Law and the Quest for Common Security
By Bjørn-Oliver Magsig

Water, Power and Identity
The Cultural Politics of Water in the Andes
By Rutgerd Boelens

Water and Cities in Latin America
Challenges for Sustainable Development

Edited by Ismael Aguilar-Barajas, Jürgen Mahlknecht, Jonathan Kaledin and Marianne Kjellén

Catchment and River Basin Management
Integrating Science and Governance
Edited by Laurence Smith, Keith Porter, Kevin M. Hiscock. Mary Jane Porter and David Benson

Transboundary Water Management and the Climate Change Debate
Anton Earle, Ana Elisa Cascão, Stina Hansson, Anders Jägerskog, Ashok Swain and Joakim Öjendal

Rules, Norms and NGO Advocacy Strategies
Hydropower Development on the Mekong River
Yumiko Yasuda

The Mekong: A Socio-legal Approach to River Basin Development
Ben Boer, Philip Hirsch, Fleur Johns, Ben Saul and Natalia Scurrah

For more information and to view forthcoming titles in this series, please visit the Routledge website: http://www.routledge.com/books/series/ECWRM/

The Mekong: A Socio-legal Approach to River Basin Development

Ben Boer, Philip Hirsch, Fleur Johns, Ben Saul and Natalia Scurrah

Routledge
Taylor & Francis Group

LONDON AND NEW YORK

from Routledge

First published 2016
by Routledge
2 Park Square, Milton Park, Abingdon, Oxon OX14 4RN

and by Routledge
711 Third Avenue, New York, NY 10017

Routledge is an imprint of the Taylor & Francis Group, an informa business

© 2016 Ben Boer, Philip Hirsch, Fleur Johns, Ben Saul and Natalia Scurrah

The right of Ben Boer, Philip Hirsch, Fleur Johns, Ben Saul and Natalia Scurrah to be identified as authors of this work has been asserted by them in accordance with sections 77 and 78 of the Copyright, Designs and Patents Act 1988.

British Library Cataloguing in Publication Data
A catalogue record for this book is available from the British Library

Library of Congress Cataloguing in Publication Data
Names: Boer, Ben.
Title: The Mekong : a socio-legal approach to river basin development / Ben Boer, Philip Hirsch, Fleur Johns, Ben Saul and Natalia Scurrah.
Description: New York, NY : Routledge, 2016. | Series: Earthscan studies in water resource management | Includes bibliographical references and index.
Identifiers: LCCN 2015020519
Subjects: LCSH: Water resources development–Social aspects–Mekong River Watershed | Water resources development–Law and legislation–Mekong River Watershed. | Watershed management–Mekong River Watershed. | Mekong River Commission. | Mekong River Watershed–International status.
Classification: LCC HD1698.M4 B64 2016 | DDC 333.91/620959–dc23
LC record available at http://lccn.loc.gov/2015020519

ISBN: 978-1-138-78844-2 (hbk)
ISBN: 978-1-138-78845-9 (pbk)
ISBN: 978-1-315-76555-6 (ebk)

Typeset in Bembo
by Out of House Publishing
Printed and bound in Great Britain by
Ashford Colour Press Ltd, Gosport, Hampshire

Contents

List of figures and tables   viii

List of abbreviations   ix

Acknowledgements   xii

1   Introduction   1

2   River basins as socio-legal arenas   39

3   Making the Mekong River Basin: donors, developers
    and experts   61

4   Governing a river basin: the work of the Mekong River
    Commission   87

5   Assessing a river basin: the politics of the technical   114

6   Disclosing a river basin: transparency and its discontents   137

7   Contesting a river basin: civil society's legal strategies   165

8   Conclusion   187

    Appendix 1: Selected national legislation, regulations, decisions
    and ministerial instructions   200

    Appendix 2: Selected international, regional and bilateral treaties
    pertaining to Mekong states, and selected 'soft' law human rights,
    environmental and financial standards   203

    Bibliography   211

    Index   244

Figures and tables

Figures

1.1   Mekong River Basin   7
1.2   Lower Mekong Basin dams (with Xayaburi detail)   23
1.3   Upper Mekong Basin dams   25
1.4a 3S (Sesan, Srepok and Sekong Rivers) dams   27
1.4b Lower Sesan River dams detail   29
1.4c A Luoi dam detail   30
1.5   Dams in Laos (with Nam Theun 1, Nam Theun 2 and Theun
      Hinboun detail)   32
1.6   Dams in Thailand (with Pak Mun detail)   34

Tables

1.1   Basin area and runoff in the Mekong's six riparian countries   8
5.1   Main elements of environmental assessment laws and guidelines in
      the Lower Mekong   117

Abbreviations

3SPN	3S Rivers Protection Network
ACC/ISGWR	United Nations Administrative Coordination Committee – Intersecretariat Group for Water Resources
ADB	Asian Development Bank
ADHOC	Cambodian Human Rights and Development Association
AECEN	Asian Environmental Compliance and Enforcement Network
AFD	Agence Française de Développement *or* French Agency for Development
AOP	Assembly of the Poor
ASEAN	Association of Southeast Asian Nations
BDP	Basin Development Plan
BOOT	build–own–operate–transfer
BOT	build–own–transfer
CEO	chief executive officer
CEPA	Culture and Environment Preservation Association
CIA	cumulative impact assessment
CLEC	Community Legal Education Center
CNMC	Cambodian National Mekong Committee
CSR	corporate social responsibility
DOS	development opportunity space
DSF	Decision Support Framework
EA	environmental assessment
ECA Watch	Export Credit Agencies Watch
ECAFE	Economic Commission for Asia and the Far East
EGAT	Electricity Generating Authority of Thailand
EGCO	Electricity Generating Public Company Ltd
EIA	environmental impact assessment
EIS	environmental impact statement
EPF	Electricity Power Forum
ERI	EarthRights International
EVN	Electricity of Vietnam
GMS	Greater Mekong Subregion
GoL	Government of Laos
HIA	health impact assessment
HSAP	Hydropower Sustainability Assessment Protocol
ICEM	International Centre for Environmental Management

IEIA	initial environmental impact assessment
IFAD	International Fund for Agricultural Development
IFC	International Finance Corporation
IFReDI	Inland Fisheries Research and Development Institute
IGA	intergovernmental agreement
IHA	International Hydropower Association
ILC	International Law Commission
IMF	International Monetary Fund
INGO	International Non-Governmental Organization
IPP	independent power producers
IUCN	International Union for Conservation of Nature
IWRM	integrated water resources management
LNF	Lao Front for National Construction
LPRP	Lao People's Revolutionary Party
LWU	Lao Women's Union
M-IWMRP	Mekong IWRM Project
MEE Net	Mekong Energy and Ecology Network
MLAI	Mekong Legal Advocacy Institute
MLN	Mekong Legal Network
MoU	memorandum of understanding
MRC	Mekong River Commission
MWRAS	Mekong Water Resources Assistance Strategy
NARBO	Network of Asian River Basin Organizations
NCP	national contact point
NGO	non-governmental organization
NHRC	National Human Rights Commission
NIEO	New International Economic Order
NMC	National Mekong Committee
OECD	Organisation for Economic Co-operation and Development
PanNature	People and Nature Reconciliation
PAPs	Project Affected Peoples
PCPD	Public Consultation, Participation and Disclosure
PNPCA	Procedures for Notification, Prior Consultation and Agreement
POE	Panel of Experts
PPA	Power Purchase Agreement
PPP	public–private partnership
RBO	river basin organization
RCC	Rivers Coalition in Cambodia
RPTCC	Regional Power Trade Coordination Committee
SEA	strategic environmental assessment
SIA	social impact assessment
Sida	Swedish International Development Agency
SPSI	science–policy–stakeholder interface
TbEIA	transboundary environmental impact assessment
TERRA	Towards Ecological Recovery and Regional Alliance
TVA	Tennessee Valley Authority
UN	United Nations
UNCTAD	United Nations Conference on Trade and Development
UNDP	United Nations Development Programme

UNECE	United Nations Economic Commission for Europe
UNESCAP	United Nations Economic and Social Commission for Asia and the Pacific
UNIDO	United Nations Industrial Development Organization
UNOHCHR	United Nations Office of the High Commissioner for Human Rights
UNTAC	United Nations Transitional Authority on Cambodia
USAID	United States Agency for International Development
VND	Vietnamese Dong
VRN	Vietnamese Rivers Network
WTO	World Trade Organization
WUP	Water Utilization Program
WWC	World Water Council
WWF	World Wide Fund for Nature

Acknowledgements

Cross-border collaboration – in a variety of senses – is a theme of this book and it has also been the lived experience of its creation. Conducting fieldwork across four nations, among people (interviewers and interviewees alike) from a number of different fields, and writing together across disciplinary boundaries and disparate authorial styles: this has been a challenge. Yet it has also been edifying and enjoyable – filled with at least as many moments of humour and camaraderie as with instances of frustration. Through this, we have come to appreciate anew Italo Calvino's call, in his *Six Memos for the Next Millennium*, for writing to 'replace the oneness of a thinking "I" with a multiplicity of subjects, voices, and views of the world', as well as the difficulties of doing so (Calvino, 1988). We hope that our effort in that regard comes through in this book's rendering of the Mekong River Basin as socio-legally polyphonic.

If writing as and for 'a multiplicity of subjects' is appropriate to our subject matter, it has nonetheless proven time-consuming. As such, it has placed demands on many, perhaps few more than Tim Hardwick and Ashley Wright of Earthscan within the Routledge/Taylor & Francis Group. Tim and Ashley had their patience sorely tested, yet never wavered in their support for this project – no matter how many extension requests hit their inbox. For that commitment and flexibility, we are immensely grateful.

Also essential to this project have been the kindness of strangers and the generosity of friends and acquaintances, most of whom go unnamed in the text: to all who agreed to be interviewed and allowed us to partake so freely of their time, wisdom and experience, we are greatly indebted. We are thankful, also, to those who assisted us with mediation and transcription in the planning and aftermath of these interviews and with translation in Vietnam and Cambodia; without this support, our engagements would have been far more limited and our records far less reliable. We gratefully acknowledge, also, that research for this book was supported under the Australian Research Council's Discovery Projects funding scheme (project DP110102987).

As we moved from the field to the writing desk – or rather to the disparate array of desks and transitory surfaces on which this book has been inscribed by its five authors – we called on yet another set of supporters and contributors. Jessie

Connell, Dr Pornsakol Dao Coorey, Naomi Hart and Sarah Schwartz all carried out invaluable research at various stages in the process of writing and otherwise assisted in bringing this book into press. Trina Galido-Isorena created a beautiful set of maps for us and Wandee Chaiyasarn was kind enough to allow us to reproduce a detail from one of his exquisite paintings on the cover of this book. Michelle Nichols did superlative editorial work on the final manuscript. We would like to express our appreciation for all their fine work.

At slightly greater remove – but nonetheless vital – was the support that we have drawn for this project from our respective partners and families. In the form of late-night Skype calls, travel commitments and a series of shared meals, discussion of this book has flowed in and out of our homes and marked our calendars for some years now. For their vicarious embrace of this project and for many a word of encouragement as spirits flagged, we are tremendously thankful. Thanks are due, in particular, to Penelope Johns for accompanying one of our fieldwork trips to assist in caring for the then-eight-month-old Ilka.

1   Introduction

Chance encounters on the Nakai Plateau

The day in late November 2012 was one of contrasts, connections and coincidences.

Some four hours after leaving Vientiane, five Australian academics – the authors of this book – arrived at the Nam Theun 2 hydropower project's visitor centre at the base of the Nakai Plateau, in the Lao People's Democratic Republic. The same trip in 1991 had taken two full days in a Russian jeep. The reduced sense of remoteness was heightened when the affable Australian public relations manager came down the stairs of the stilted exhibition hall to welcome us on friendly first-name terms. He reminded us that several years previously he had given a talk to our research seminar at the University of Sydney.

In an otherwise often bureaucratically closed country, where hydropower can be a very sensitive topic, we were struck by the openness of the manager and of the whole space in which we found ourselves. We had free rein to look around the centre and peruse its intricately constructed models, information posters, maps and timelines. It was cool and airy inside the building. We took our time inspecting the display and watching a promotional video. We were also invited to visit and look inside the nearby powerhouse – out of which water was rushing following its 350-metre descent from the plateau – and any of the resettled villages on the plateau we cared to visit. The sense of transparency was refreshing.

A couple of hours later, we arrived with our minibus driver, but otherwise unchaperoned, at a village on the southern edge of the 450 square kilometre dam reservoir. It was late afternoon, but the village was relatively empty, its residents not yet returned from the daily toil of subsistence living. A few fishing boats were moored; one or two moved about in the distance. We had chosen this site because one of us had visited and stayed in several villages along the Theun River on the Nakai Plateau in 1991, long before it was flooded by the Nam Theun 2 dam. Maps at the visitor centre had indicated that one of the villages previously visited had been resettled to this site. Staying in that community had been a particularly memorable part of the earlier visit. The observations and interviews conducted there served as the basis of an illustrated feature article published a few months later in a Bangkok newspaper (Hirsch, 1991).

We knew of the frequent consultations that had taken up much of villagers' time during the project planning process. We knew, too, of the continuing regular visits that the resettlement sites receive, as Nam Theun 2 has become Lao PDR's and the World Bank's showpiece hydropower project. In light of this, we chose to seek out a casual conversation at the small village shop rather than bother villagers with more formal interviews. After passing an eerily empty marketplace built by the resettlement authorities, we exited the minibus in front of a small ramshackle shop. A shy young woman behind the counter was serving luridly coloured cordials and displaying a few other convenience goods. Our Lao speaker, who had stayed three days in the now-flooded village more than two decades earlier, greeted the young woman's father, who was standing nearby, and asked whether he had in fact moved from that village. He looked up and said, incredulously: 'I recognize you! You stayed with us about 20 years ago, right?' And the extraordinary recognition was mutual.

As we all sat around a rickety wooden table in front of the shop, there followed an hour and a half of outpouring by our host, joined by a steadily growing group of family members and neighbours. Some told us of their satisfaction with access to electricity and nice houses, but the overwhelming story was of the hardships of moving, of losing most of their cattle as well as rice land, of decline in the initially productive reservoir fishery and of the general despair of making a living on the barren soils of the land to which they had been moved. Most poignant and impassioned of all were their tales of the discomfort of having to speak in positive tones and through interpreters to outsiders at scripted meetings about the success of the resettlement.

The same evening, we had our third unexpected encounter of the day. Generously hosted by the hydropower project at its Wooden House guesthouse at Ban Oudomsouk near the road entry to the plateau, we were delighted to find that the Nam Theun 2 Panel of Experts was staying there overnight as well. They were at the start of their 20th mission to Laos since 1997, as part of the World Bank's oversight of the project. The New Zealander and two US nationals who comprise the Panel are all in their 80s and bring with them more than a century and a half of combined experience on dams, resettlement and environmental protection. They have advised three US presidents, the World Bank and numerous governments. They include two former directors-general of the International Union for Conservation of Nature and the leading world authority on the social impacts of dams.

On this three-week mission, they were particularly concerned to address 'handover' issues for the downstream rehabilitation component of the resettlement package. They were worried about the unpreparedness of the local authorities to take on and manage the complex mitigation and compensation tasks that the project had created. Over a lively dinner conversation at a nearby French restaurant in this strangely cosmopolitan backwater of Laos, our research team and the octogenarian experts discussed the challenges of bringing to bear international standards in governing a project such as Nam Theun 2. We talked of these experts' careers and the attachments and insights they had gained from their longitudinal work on this hydropower project. We also marvelled at the ways in which the project had not only brought together the group of people around the dinner table, but also inserted an array of global financial, environmental, international institutional and legal actors into the lives and landscapes of one of the remotest parts of the Mekong.

It was a day replete with odd confluences and striking divergences, story-telling and image-making, abiding hopes and ongoing frustrations – a day that touched upon many of the elements at play in the socio-legal make-up of the Mekong River Basin.

Law and the theatre of an international river basin

Much has been written on the geography, ecology, history, politics, governance and development of the Mekong (Luong, 1996; Kaosa-ard and Dore, 2003; Molle *et al.*, 2009; Myint, 2012; Osborne, 2009, 1975; Gao, 2014). In this body of writing, rivers and their basins are subjects of adventure, exploration, economic endeavour, environmental politics, resource conflict, subsistence and indigenous livelihoods, cultural attachments to land and resources, and other encounters between people and nature. Rather surprisingly, however, there has been little explicitly socio-legal discussion of transnational or other rivers (with the exception of Suhardiman and Giordano, 2014). This is despite the fact that the many ways in which we think about rivers often depend, to some extent, on norms, institutions and instruments associated with law.

Our purpose in this book is to present the Mekong through a socio-legal lens. That is, we aim to show that relations between different actors over the use and management of the river and its basin – indeed, the basin itself – are meaningfully imbued with and constituted through law, understood plurally to include international, regional, national and subnational laws. 'Law' in this context is not restricted to the formal institutions and instruments with which it is most often associated (legislation, court cases, administrative regulations and the agents that create and enforce them). Rather, we place law here on a continuum of social normativity. The legal phenomena we study comprise an array of conduct-shaping and thought-directing rules and standards, many of them informally generated. Popular understandings of 'justice' or 'rights' are, accordingly, as much a feature of 'law' in this account as the dispositions and tendencies of those with legal qualifications. In this pluralist understanding, law is inextricably bound up with and inflected by the political, social, economic and physical ecology of the river. Law does not act exogenously on or emanate from these conditions; it is continuous with and inextricable from them. In Chapter 2 we discuss this approach further, and relate it to some salient historical controversies in legal and social thought.

Conceiving of the Mekong in this way begs the question of how a socio-legal treatment relates to alternative ways of understanding and representing rivers and their basins. We therefore commence by considering different meanings attached to rivers and their basins. These alternative meanings invoke law in a variety of ways: through, for instance, the actual or potential disputes, jurisdictional issues and regulatory requirements at stake within them.

Rivers and river basins are often thought of, in a naturalistic sense, as flows and connections. As a flow, a river channels what is sometimes referred to as a 'fugitive resource', in this case water, from upstream to downstream locations (Smith, 2008, p. 475). It also channels sediment, nutrients and pollutants. The river basin delivers water from land to waterway, either directly through surface flow or via groundwater systems. In some respects, the river also serves as a conduit for flows in the other direction – for example, upstream fish migration or navigation movement. Likewise, the connection of rivers and their floodplains through flood overspill events links aquatic and terrestrial parts of the basin as an ecosystem. In a less naturalistic connection, irrigation takes water out of rivers and onto land surfaces to achieve subsistence or economic ends. As flows and connections, rivers and river basins sustain and link actors across places, so that use of or interference with one part of a river or its basin by some actors has implications for conditions in other parts and affects other actors.

Rivers and their basins are also perceived as they are represented cartographically: in one- or two-dimensional space. Rivers, and the catchment boundaries of the territory that drains into them, are mapped as lines and polygons respectively. The division of territory among sovereign states and subnationally within those states, places, rivers and basins, or parts of them, within particular political and legal jurisdictions. Riverine borders, channels and geographic features may support or undermine territorial claims by nation-states, indigenous peoples and other parties in national and international law. The spatiality of rivers and river basins may overlap, reinforce or exist in tension with human-imposed boundaries on the Earth's surface.

Rivers can be treated, too, as relational phenomena. Between territories, but also within them, rivers and river basins are subject to use, abuse, claims and impacts by a diverse set of interests and actors. These include nation states *qua* states, and different ministries and departments that come under them, together with subnational territories and their related political and administrative agencies, whose boundaries ill-fit the natural spaces of river basins or lines of rivers. To these are added different sorts of economic actors. Private sector corporate entities often lay claim to water or engage in activities or investments that impact on river systems. Myriad household-level activities by small-scale farmers and fishers, including indigenous peoples, impact upon, and are impacted by, the activities of others. Non-consumptive users of water derive economic, aesthetic, spiritual and other values and benefits from rivers and are hence subject to impacts of others' claims on them *in situ* or elsewhere in the river system. Among these, urban electricity consumers and industrial power users stand out as particular beneficiaries of hydropower projects, the latter often constructed on rivers located in remote parts of a country or across national borders. Regional and international actors, from transboundary river commissions to multilateral development banks to transnational coalitions of non-governmental organizations (NGOs), may claim rights to govern or influence the use or management of rivers on the basis of their transboundary significance. In this sense, river systems often come to be defined in terms of sets of relationships. These may be characterized by terms such as 'stakeholder configurations', 'socio-ecological systems', 'basin economies', 'hydro-hegemonic regimes' and so on (Zeitoun and Warner, 2006).

River basins are also represented widely in mythological and literary terms. The cultural meanings of water and rivers have been cast in such terms for the Mekong region as 'Mae Nam Khong' (Mother of Khong Waters), 'Cuu Long' (Nine Dragons) and 'Lancang Jiang' (Turbulent River) in Thai, Vietnamese and Chinese respectively. Even the simple 'Tonle Thom' (Big River) – as the Mekong is sometimes characterized in Khmer – attests to the spirit and significance of the river. The mythic appearance of Naga – a river serpent that inhabits and embodies the river as a protective force – and of images of the river as artery, lifeline or life force, with the Tonle Sap Lake as its beating heart, or otherwise as a mystical entity with an identity of its own, all demonstrate the river's significance. In these representations, the river is more than a volume of water draining an area of land with its particular fluvio-geomorphic and ecological characteristics and material functionality. As such, the customary significance of the Mekong invokes rules and responsibilities, taboos and traditional use rights that attest to the plurality of law applied to, and mobilized by, the river.

In addition, river basins generate and hold competing political and economic visions, such as those associated with the term 'sustainable development' or, more broadly, sustainability. As discussed in Chapters 2 and 3, the contested idea of 'development' is central to contemporary socio-legal discourse around rivers, including the means and ends of development and who has a say in determining the forms it takes. A development-versus-conservation dynamic is often played out with reference to specific rivers and their basins. Modernity (in a developmental register) and sustainability may pitch different values and interests against one another, but the concepts also work in tandem to shape ways of thinking about river basins and to frame alternative ways of using and governing them (Hajer, 1997). These concepts – and the tensions and complicities between them – animate many of the socio-legal analyses throughout this book.

Questions of lawful authority and relation are inherent in each of these versions of a river basin. Yet law also quite explicitly enters the scenes that we have sketched, in a range of ways. Rivers and river basins falling within particular national and subnational jurisdictions will be subject to the water laws, environmental laws and other relevant substantive and procedural laws of those jurisdictions. These national and subnational legal regimes will, in turn, be influenced by, as well as contribute to, the formation and interpretation of international norms and regimes concerned with freshwater resources and riparian rights, as well as those concerned with human rights, dispute resolution, trade and investment, and the like. Contending uses of riverine water, and plans for its use, will also be subject to a wide array of 'softer' – that is, more aspirational, informal, not-yet-in-force or non-judicially enforceable – norms, practices and claims, many of them with some direct or indirect legal provenance, whether local or remote.

Each of the river basin characterizations described presents particular problems for legal governance. To think of a river basin as a phenomenon of flows raises challenges of connection and the containment or mitigation of adverse impacts. Conceiving of a river basin cartographically and territorially brings proprietary interests and sovereign conflicts to the fore. Approaching a river basin as a cultural artefact or mythic 'being' changes the temporality of the governance endeavour and highlights questions of identity and non-quantifiable values at stake. These complications may be compounded in the case of a river basin that is international in its dimensions. More than 264 river basins worldwide drain the territory of more than one country; these cover some 47 per cent of the world's land surface and provide home to more than 40 per cent of the world's population (Varis *et al.*, 2008; Wolf *et al.*, 1999). Yet, for the most part, nation-states reserve the sovereign right to use resources within their own boundaries and to make rules about how those resources are accessed, used and managed, albeit that these rules are partly constituted and otherwise influenced by international norms.

When rivers cross national boundaries, tensions arise between the flows and connections that they produce and the mythic traditions they carry, on the one hand, and their cartographic placement within and between sovereign territories, on the other hand. When those in one part of an international river basin come into conflict with those in a jurisdiction governed by a different legal regime, national legal regimes may prove inadequate to express or address the concerns and conflicts in question. International legal regimes may likewise fall short of doing so, insofar as they rely upon, or seek to call forth, some community of interest. When an international river basin is shared by riparian members with different histories and priorities with respect to its development, the articulation of a common vision of modernity, sustainability, or indeed lawful governance, may prove problematic.

Perhaps in no river basin on Earth are these issues manifest as clearly as in the Mekong River Basin.

The Mekong

The Mekong River Basin is the arena in which our socio-legal drama takes place, with the Mekong River as a central actor in this drama. Our analysis is primarily set within the lower section of the basin, among the four countries of Thailand, Laos, Cambodia and Vietnam, which we refer to as the Lower Mekong. The Mekong is one of the world's larger transboundary river systems in terms of population, area and river length. It is one of 13 river basins draining through the territories of five or more countries. The entire Mekong traverses six countries, over a length of 4,900 kilometres, from its source in Qinghai province in China to the mouths of its delta in several provinces of southern Vietnam (Figure 1.1). Its tributaries drain a basin of some 795,000 square kilometres. While it is only the 25th largest river basin draining to the sea measured by land area, it ranks 12th in terms of length and eighth by volume of discharge to the ocean. It supports a population of around 70 million people from multiple ethnic groups within each of the countries in whose territory it lies (Matthews and Geheb, 2015).

The significance of the Mekong varies from one country to another. Myanmar maintains quite a marginal interest in the river, as a result of the small portion of the country that is situated within the basin. In contrast, the Mekong is politically, economically and socially pivotal in Lao PDR and Cambodia, the national territories and populations of which are overwhelmingly contained within the basin.

Figure 1.1 Mekong River Basin

Table 1.1 Basin area and runoff in the Mekong's six
riparian countries

Country	Basin area % of total	Runoff % of total
Cambodia	20	18
China	21	16
Lao PDR	25	35
Myanmar	3	2
Thailand	23	18
Vietnam	8	11
Total	**100**	**100**

Source: MRC (2005, p. 1)

Table 1.1 shows the contribution of the six riparian countries to land area and runoff in the Mekong River Basin. Each country assumes the right to develop water resources within its own sovereign territory, while more or less accepting principles of 'reasonable and equitable' utilization first enumerated authoritatively (albeit without binding legal force) in Article IV of the 1966 Helsinki Rules on the Uses of the Waters of International Rivers (Salman, 2007) and later in Article 5 of the 1997 UN Convention on the Law of the Non-Navigational Uses of International Watercourses (which, as we discuss in Chapter 4, entered into force in 2014, as a result of Vietnam becoming a party – the only Mekong country to do so to date). Interpretations of what is reasonable and what is equitable vary considerably, however, as between these riparian nations, as well as within each of them. In subsequent chapters, we will revisit this contestability of notions of reasonable and equitable utilization, as these have been spelled out in the 1995 Agreement on the Cooperation for the Sustainable Development of the Mekong River Basin (the Mekong Agreement), to which Thailand, Laos, Cambodia and Vietnam are all parties.

The Mekong River Basin is an iconic basin in a number of respects. It has some outstanding features that make it worthy of study in its own right, as well as socio-legal dimensions that may have parallels in, or significance for, other transboundary basins. These features (explained further below) include its hydrological regime; the phenomenon of the Tonle Sap Lake; the size of its fisheries; the history of riparian cooperation amid geopolitical tension, including during the Cold War; the relatively recent pressure for dam construction on a hitherto relatively free-flowing river; and the intensity of external donor interests in, and NGO attention to, the future of the river. These features, in turn, help shape the Mekong River Basin as an arena of contestation.

As a monsoonal basin supplemented by glacial snowmelt in its upper reaches, the seasonal hydrology of the Mekong River and its tributaries shapes prevailing debates about water resource development. Most significantly, the potential of dams to store water during the wet season and release water during the dry season has driven hydropower and irrigation development in the basin; it has also given a flood control and drought-proofing rationale to such schemes. Conversely, dams on the Mekong, particularly those upstream in China, have sometimes been blamed for floods and droughts. Within individual tributary systems, and to a much more limited extent between the Mekong and neighbouring tributary basins, the diversion of water by some dams and irrigation structures from one river to another raises a complex set of issues between concerned parties.

A particular feature of the Mekong hydrological system is the Tonle Sap Lake (Kummu and Sarkkula, 2008). One of the world's largest freshwater bodies and Southeast Asia's largest freshwater lake, Tonle Sap expands some five times in area and rises on average nine metres in depth between the dry and wet seasons. About 60 per cent of the water that causes this expansion comes from the biannual flow reversal of the Tonle Sap River, which links the lake to the Mekong via the confluence at Phnom Penh. In turn, this backflow accounts for about 10 per cent of the annual Mekong flow, helping to regulate the otherwise even more peaked discharge of water through the Delta in Vietnam. Tonle Sap Lake is of vital importance for farmers and fishers in Cambodia, but also for the natural regulation of flows downstream in Vietnam. Its productivity – approximately half a million tons of fish are caught from the lake each year – depends largely on the seasonal rises and falls that link the water body with the terrestrial source of nutrients and fish breeding habitats such as flooded forests (Hortle, 2007; Lamberts, 2008). This means that flow alteration elsewhere in the basin has the potential to affect the livelihoods of large numbers of people living around the lake, downstream of the lake, and even upstream of the lake, as fish migrate annually as far north as Thailand.

The Mekong's large number of fish species makes it the world's second most biodiverse river basin after the Amazon in terms of ichthyofauna (Baird, 1999; Ziv *et al.*, 2012). An estimated 1,300 species occupy the diverse habitats of the Mekong River (Friend *et al.*, 2009, p. 311), its tributaries and other wetlands. The Mekong River Basin is also the world's largest inland fishery. While estimates vary, studies carried out under the Mekong River Commission's (MRC) fisheries programme suggest that 2.6 million tons of fish and other aquatic organisms are caught per year (Hortle, 2007). This represents around 20 per cent of the global wild freshwater catch. These figures underlie the heavy reliance on fisheries by the mainly rural inhabitants of the basin, between 27 and 78 per cent of whose animal protein is derived from fish (Hortle, 2007). Dependence on wild capture fisheries is even higher for the rural poor who have less access to markets and alternative sources of animal protein (Friend *et al.*, 2009). These figures are highly significant in a basin in which planned development projects would have significant impacts on the mainly migratory fishery resources (resources that are hence also 'fugitive', in the sense that they are not permanently within a single jurisdiction). Most notably, the size and susceptibility of the fisheries complicate the hydrology-based legal regime around shared waters, inviting consideration of a further set of legal norms concerned with species conservation, food security and human rights, biodiversity preservation and heritage protection.

Compared with most of the world's large river systems, the Mekong has until recently remained relatively underdeveloped. Only a small fraction of the estimated

30,000 megawatt hydropower potential on the tributaries has been developed (MRC, 2011b, p. 12). Despite more than half a century of planning for dams on the lower part of the mainstream, it was not until 2012 that construction of the first of these dams commenced. Since the 1990s, however, there has been an accelerating pace of dam construction on the tributaries and on the upper part of the mainstream. China has completed six large dams on the upper reaches of the river's mainstream in Yunnan province (called the Lancang Jiang), with at least another 14 dams planned or under construction in Yunnan and Tibet (International Rivers, 2013a). Meanwhile, off-takes from the river for irrigation and other purposes remain quite small. Figures 1.2 and 1.3 show existing dams, dams under construction and planned dams on the Mekong mainstream and its tributaries in the Lower and Upper Mekong. The MRC's Basin Development Strategy estimates that about 5 per cent of the river's flow is used, with the remainder finding its way to the South China Sea through the Delta's various distributaries. Each of the Mekong countries has ambitious plans for increasing this off-take to support dry season agriculture and supplement wet season farming, but it is unclear how much of this planning will materialize. The MRC represents this unused flow as a 'development opportunity space', a concept to which we return in more detail in Chapter 4.

An important reason for the historical 'underdevelopment' of the Mekong is the geopolitical conflict that prevented large infrastructure plans from being realized during the 1960s and 1970s (Nguyen, 1999), decades which were the heyday of modernist development in many other parts of the world. The Vietnam War, the 'secret' war in Laos, the turmoil and conflict in Cambodia, and the Cold War that divided the Mekong region through to the late 1980s all militated against large-scale collaborative development projects. The end of this period of conflict coincided with an intensification of contestation over appropriate paths to development and, more specifically, a global concern over the impacts of large dams, as discussed in Chapter 3. This translated, in Thailand in particular, into opposition to large dam projects. Just as the regional geopolitics shifted in favour of renewed development of hydropower and irrigation works in the early 1990s, so did the ecopolitics of the region set new, cross-cutting lines of conflict. The emergence of a Mekong-specific legal regime – centred on the 1995 Mekong Agreement – can be understood as both an expression of these overlapping agendas and an attempt to transcend or bypass them to some degree.

Notwithstanding the political and military conflagrations that have afflicted it, the Mekong has a nearly unbroken 60-year history of cooperative arrangements for the use and management of its waters. Much writing on the Mekong as an international river is premised on this fact. Since the setting up of the Mekong Committee in 1957, the governments of four of the six countries of the Mekong have sought a degree of common purpose in the use of the river's water resources. The Committee was revamped in the post-Brundtland era (that is, following the publication of the report by the World Commission on Environment and Development in 1987) and in the light of the post-Rio sustainable development era in the mid-1990s (following the World Summit on Sustainable Development in 1992). The MRC – and the Mekong Agreement by which its creation was authorized – have together produced an institutional and legal basis for the set-ting of rules in pursuit of mutual objectives. The Mekong would seem to be a case study of recourse to international law in the development and management of a transboundary river basin.

Even so, as has already been emphasized, the Mekong remains a highly contested development arena. The contestation is not just, or even mainly, over water per se; therein lies the problem of applying generalized water-sharing approaches and other primarily hydrological bases for governance. Rather, the contested develop-ment of the basin has to do with competing visions of appropriate development and unaddressed concerns over the displacements and complex socio-ecological outcomes of dams and other projects. It has to do, as well, with unresolved ambigu-ities and contradictions between public and private interests inherent in develop-ment planning, and contending parties seeking the capacity to make or influence decisions in this context. It is to the actors in this setting that we now turn.

Protagonists and antagonists

Many actors, interests and agendas have shaped development on the Mekong. Up until the 1990s, water resource development and management were considered more or less within the exclusive domain of states, intergovernmental organizations and multilateral institutions. Early development projects in the 1960s and 1970s were designed and implemented by state agencies with financial and technical support from the United States and the World Bank (in the case of Thailand and Laos) or the Soviet Union (in the case of Vietnam). Technocrats housed in state 'hydraucracies' and international bureaucrats and technical experts contracted to intergovernmental bodies were the early champions of large-scale river basin development in the Mekong (Molle *et al.*, 2009).

Over time, however, a myriad of non-state actors emerged onto the scene as figures of influence and agency. Civil society actors formally unaffiliated with state governments, particularly international, regional and local NGOs, sought to respond to the needs and interests of a growing number of marginalized com-munities negatively impacted by large dams. As discussed in Chapter 7, while the space for civil society is highly constrained by political conditions in the Mekong countries, activism on Mekong issues has grown in all of the coun-tries since the 1990s. More recently, a diverse mix of private sector actors has emerged to become the principal source of funding for dams in the Mekong, overtaking traditional intergovernmental funders and guarantors such as the World Bank and the Asian Development Bank (ADB) (Middleton *et al.*, 2009). The authority of these non-state actors often articulates with state agencies and with each other in different ways. The resulting combinations of public–private power play an important role in shaping the various regulatory processes (legal, institutional, political, knowledge-based) that govern development and water usage in the Mekong, or endeavour to do so.

Alongside these protagonists, the Mekong River itself is, as noted above, a leading actor, whether as a mythico-historical figure and attractor of disparate loyalties and attachments in that sense, or as a bearer of material agitants and activants (water, sediment, fish, pollutants, and power-generation capacity).

The four Mekong riparian states

This multiplicity of actors notwithstanding, the four sovereign states that share the Lower Mekong River Basin – Cambodia, Laos, Thailand and Vietnam – drive 'national agendas' for water resource development and management in the Lower Mekong. By way of introduction, it is worth considering some of the key social, political and economic factors that have shaped – and continue to shape – water resource development agendas in each of the four Lower Mekong countries.

Thailand

Thailand was the first country to embark on a programme of large-scale river basin development, which was supported by the United States and the World Bank to counter the spread of communism in the northeast of the country. In the new geopolitics of the post-Cold War era, dam construction in Thailand was cast as a necessary response to rising energy demands that accompanied the country's economic growth and industrialization process. With most of the best dam sites already developed by the turn of the twenty-first century, and local civil society opposition limiting the construction of large power stations domestically, Thailand turned to its neighbours as sources of energy. Increasingly, the state-owned utility, the Electricity Generating Authority of Thailand (EGAT), has favoured importing hydropower from Laos, where it is difficult for civil society to voice opposition to dams. Thai companies have become key investors in Lao hydropower (as financiers and as primary power purchasers) and are behind some of the more controversial projects in the basin, including the Xayaburi dam on the Mekong mainstream.

While Thailand's priority is to import hydroelectricity to meet the growing energy demands of its people and industries, extracting water from the Mekong River to irrigate its arid and impoverished northeast region has also entered onto the agenda of successive governments. Isan or northeast Thailand makes up 85 per cent of the Thai territory that drains to the Mekong River via the Chi-Mun and Songkhram river systems (Molle and Floch, 2007). A number of grandiose projects have been proposed for Isan over the years, including the Khong-Chi-Mun water diversion scheme, a partially completed and now largely defunct network of dams and canals that promised to irrigate 796,800 hectares of farmland by diverting water from the Mekong River. In a later iteration, the so-called Thai Water Grid Project proposed transferring water from adjacent river basins in Laos and Cambodia to irrigate and modernize the northeast region (Molle and Floch, 2007).

Laos

Due to its mountainous topography, Laos has the largest technically exploitable hydropower potential of the four Lower Mekong countries, estimated at around 18,000 megawatts (Ministry of Energy and Mines, Lao PDR, 2014). Generating revenue from the sale of hydroelectricity to neighbouring Thailand, Vietnam and, to a lesser extent, China is one of the principal means by which Laos hopes to maintain economic growth and reach its defined target of graduating from Least Developed Country status by 2020. As of 2015, Laos has built 17 large hydropower projects with a total capacity of 3,200 megawatts and an additional 40 to 50 projects are under construction or planned (Ministry of Energy and Mines, Lao PDR, 2014). Laos's single-party system and limited space for civil society expression are important contextual factors enabling it to attract investment in potentially risky projects. Central government control, exercised to provide stabilization commitments (that is, contractual promises on the part of the state to compensate developers for loss attributable to some change in law), has mitigated country and regulatory risk to the satisfaction of foreign developers and financiers. As a consequence, Laos is the first country to have pushed ahead with the construction of dams on the Mekong mainstream – despite objections from Vietnam and Cambodia, as well as from vocal sections of Thai civil society.

Vietnam

Vietnam's position in the Mekong River Basin is complicated by the fact that it is both a downstream and an upstream country. On the one hand, Vietnam has the most to lose from the construction of dams on the mainstream and other upstream developments that threaten the country's primary rice-growing area – the Mekong Delta. On the other hand, Vietnam is highly dependent on hydropower, both domestic and imported, to meet its staggering 15 per cent annual growth in energy demand (Nguyen, 2012). Since the 1990s, the state-owned company Electricity of Vietnam has built a number of dams on the Sesan and Srepok rivers in the Central Highlands of the country, two important tributaries that flow into the Mekong River before re-entering Vietnam. These dams have had devastating impacts on communities downstream in Cambodia while also reducing water and sediment flows into Vietnam's own Mekong Delta (Hirsch and Wyatt, 2004; Piman *et al.*, 2013). Vietnam's interests are further complicated by the fact that Vietnamese companies are also proposing to invest in two mainstream dams (Luang Prabang dam in Laos and Stung Treng dam in Cambodia). Vietnamese-owned companies are also minor shareholders in the controversial Lower Sesan 2 dam under construction in Cambodia. At the same time, as noted earlier, Vietnam committed itself to the UN Watercourses Convention in 2014, breaking with its Mekong neighbours, and the prevailing preference for bilateralism and regionalism, in formally subscribing to international river norms.

Cambodia

Due to its lack of infrastructure, Cambodia has the lowest electrification rates of the four Mekong countries. Only about 25 per cent of the population is connected to the electricity grid (Poch and Tuy, 2012). The country is highly dependent on imported diesel and oil for energy, making electricity both costly and unreliable (Poch and Tuy, 2012). In this context, the Cambodian government is prioritizing investments in the energy sector, and developing hydropower for both domestic consumption and export is high on the agenda. While Cambodia initially struggled to attract investment for large hydropower projects, in the past decade the Chinese government has provided high-level support for Cambodia's hydropower development (Middleton, 2008). Of the eight large hydropower dams completed or under construction in Cambodia, seven have been or are being developed by Chinese companies. The eighth dam, the Lower Sesan 2, is a joint venture between Cambodian, Chinese and Vietnamese companies.

The Lower Sesan 2 dam has caused significant controversy within and outside Cambodia because of its predicted impact on fisheries (a staggering 9.3 per cent reduction in fish stocks basin-wide, according to Ziv *et al.*, 2012). Cambodia is also considering building two large mainstream dams at Stung Treng and the Sambor Rapids, which would likely be catastrophic for fisheries and local livelihoods (IFReDI, 2013; Baran, 2010). With so much of the country's population heavily reliant on fisheries for food, nutrition and income, the construction of dams on the Mekong mainstream and tributaries poses a real dilemma for Cambodia (IFReDI, 2013). Despite the push for hydropower, there are signs that some ministries and senior members of the Cambodian government, including Prime Minister Hun Sen himself, understand the significance of fisheries and the Tonle Sap system to the country's food security. This is evident, for example, in the prime ministerial order to ban commercial fishing lots in 2012 (Sokheng and Kunmakara, 2012). Nevertheless, with Laos well on its way to building the second dam on the mainstream, it is conceivable that Cambodia's leaders will argue for their nation's 'right' to get a share of the projected benefits from mainstream developments.

Riparian nation-states' assertion of sovereignty over their respective stretches of the river, and their unilateral pursuit of independent development agendas, have long been identified as key governance challenges in transboundary river basin contexts. In the Mekong, as elsewhere, dam projects are often justified in language that privileges 'national interests' over and above those at subnational or transboundary scales (Hirsch *et al.*, 2006). As the country-level snapshots above suggest, however, state interests are far from clear-cut, unified or territorially bounded. Even assumptions about a state's upstream versus downstream interests turn out to be over simplifications that mask diversity and contestation among actors working across multiple jurisdictions. To understand who is making and influencing decisions at a transboundary level, it is necessary to move beyond simple characterizations of Mekong water governance as a matter of interaction between monolithic states, and examine the wide range of actors and processes at multiple scales that simultaneously support and challenge the state (Suhardiman and Giordano, 2012; Sneddon and Fox, 2006; Hirsch *et al.*, 2006).

The Mekong River Commission, the Greater Mekong Subregion and the Association of Southeast Asian Nations

The question of states' representation and pursuit of their respective national interests is at the core of debates regarding the role, function and legitimacy of the intergovernmental MRC. First established in 1957 under the framework of the old Mekong Committee, the MRC was revamped in 1995 when Cambodia, Laos, Thailand and Vietnam signed the Mekong Agreement. Under the Agreement, the four countries have a mandate and an obligation to 'cooperate in all fields of sustainable development, utilization, management and conservation of the water and related resources of the Mekong River Basin' (Article 1). Beyond the agreement itself – a loosely structured set of articles open to variable interpretation – there is little else that indicates a commonality of purpose or vision for 'sustainable development', 'use' or 'management' of the Lower Mekong River Basin among its four member states. Indeed, there are starkly different understandings and expectations about what kind of organization the MRC is, and whom it serves (Hirsch *et al.*, 2006; Lee and Scurrah, 2009).

On the one hand, the MRC is a donor-funded organization under considerable pressure to act in a manner befitting a river basin organization with a mandate in transboundary governance. The expectation from donors and civil society is that the MRC will balance different interests in the basin through stakeholder consultative processes and promote open, rational decision-making based on good river science and in accordance with principles of 'good governance' (see, for example, Department of Foreign Affairs and Trade, Australia, 2015).

On the other hand, the MRC's charter and operational structure give primacy to its member states and the principle of national sovereignty (that is, to the law-making and decision-making autonomy of nation-states). Decisions over dams with transboundary impacts are often described and understood in terms of a broader transnational political culture of non-interference, wherein governments are reluctant to impose limits on the sovereign rights of neighbouring countries to develop rivers within their territory – a culture often identified with the Association of Southeast Asian Nations (ASEAN) (Middleton, 2014). The transboundary decision-making tools, structures and procedures of the MRC have tended to work at some distance from national decision-making processes (Suhardiman *et al.*, 2012). With limited power to make or enforce decisions likely to decide the Mekong River Basin's fate, the MRC has to some extent been marginalized from core decisions affecting the basin, even as it exerts influence in a range of ways discussed in Chapters 4, 5 and 6 (Dore and Lazarus, 2009).

Despite these obstacles to the MRC exerting decisive authority in the Mekong River Basin, donors continue to support the MRC because it is the only regional body that addresses transboundary water governance (Cronin and Weatherby, 2014). The ADB's Greater Mekong Subregion (GMS) Economic Cooperation Programme, which spans the territory of the four MRC countries plus China and Myanmar, is primarily concerned with regional economic growth and integration through the construction of transborder infrastructure. Similarly, the ten-member ASEAN is primarily geared towards promoting trade and investment, its recent human rights and security focus notwithstanding. Both institutions actively support hydropower development in the region, either through the GMS-funded Mekong region energy grid or ASEAN's 10 per cent renewable energy target (Matthews and Geheb, 2015). As such, they are often in competition with the MRC, to which is left the task of ensuring that water resources are developed 'sustainably'. Compared to the GMS and ASEAN, the MRC has limited political buy-in from member state governments, which do not strongly associate MRC participation with the realization of their respective national interests. Despite China maintaining 'dialogue partner' status with the MRC since 1996 and contributing some data to the organization, the MRC also does not offer its member states a viable way of navigating the most significant relationship currently animating the Mekong River Basin – namely, that with China.

China and Asian regionalism

China's rapid economic growth and rise as a global power has been the most significant twenty-first century geopolitical development influencing the dam-building landscape in the Mekong. Since the turn of the twenty first century, China has nurtured the expansion of Chinese investment in the Mekong region, including in hydropower, by strengthened investment, trade and aid relationships with governments (Middleton, 2008). China's 'Go Out' policy, promoting Chinese investment in resources overseas, and the development of Chinese energy technology and expertise, have been key factors enabling Chinese hydropower companies to take advantage of the Mekong's expanded hydropower market (Heinrich Böll Stiftung *et al.*, 2008). According to Urban *et al.* (2013), more than 50 large hydropower dams have been or are being developed and financed by Chinese companies in the Greater Mekong region.

China's growing influence in the Mekong region is manifest in its growing embrace of multilateralism. China has increasingly engaged with ASEAN, evident in the signing of the China–ASEAN free trade agreement in 2010. It has also provided high-level support for the ADB's GMS programme, which has expanded the scope of economic cooperation between the Mekong region and China two geographically adjacent provinces, Yunnan and Guangxi. China's environmental regulations have also been studied by some Mekong governments, with Chinese encouragement. China's ambition to play a leading role in Asia's future development was made clear with the announcement in October 2014 of its intention to establish a new multilateral institution, the Asian Infrastructure Investment Bank. The move has been interpreted as an attempt to expand China's 'soft power' in the region vis-à-vis the United States and Japan, which still hold most of the voting power in the ADB (*Economist*, 2014).

China has also built strong bilateral relations through aid, trade and investment with resource-abundant countries Cambodia and Laos. A key aspect of this engagement has been the expansion of Chinese investment in hydropower dams. China has a policy of relative non-conditionality in development cooperation, with less oversight of social and environmental impacts and more light-handed management of the projects that it funds in comparison with traditional development partners. This has provided Mekong countries with an attractive alternative source of funding. To some extent this has undermined the safeguards applied by international development banks and donor agencies as developers and financiers of hydropower projects. At the same time, however, Cambodia and Laos have been reluctant to rely entirely on China as their sole foreign investor and aid provider, and have continued to court Western donors as well as attract a range of private investors from other countries – particularly from within the Mekong region itself.

Private sector actors and utilities as financiers in hydropower development

Up until the 1997 Asian financial crisis, most dams in the Lower Mekong River Basin were public investments, based largely on loans from the World Bank and the ADB. As economies recovered, private sector actors began to emerge in Asia, making new sources of international finance available (Middleton *et al.*, 2009). This coincided, roughly speaking, with the emergence of the so-called 'post-Washington Consensus' whereby multilateral investment bodies, including the World Bank and the ADB, adopted a more poverty-focused approach to development. These banks had been chastened by a greater awareness of the detrimental human impact of initiatives they had previously engaged in (including dam projects). Today, as a result of this confluence of factors, all new large dams being constructed and planned in Laos and Cambodia are commercial projects developed as public–private partnerships (PPPs) under build–own–transfer (BOT) or build–own–operate–transfer (BOOT) contractual schemes (Middleton *et al.*, 2015). Most are being developed and financed by companies from China, Thailand, Vietnam, Malaysia, Korea and Japan on a limited recourse basis (that is, on the expectation that the project itself will generate funds sufficient to cover the cost of finance and that financiers will be compelled to look primarily to the liquidation of project assets, in the event of default, in lieu of making claims against the public or private owners of the project).

Chinese companies and Chinese banks are among the biggest builders and financiers of dams in the Mekong. State-owned giants such as Huaneng and Sino-Hydro have built or are in the process of building a large number of Mekong and non-Mekong tributary projects and hold contracts to develop four dams on the Lower Mekong mainstream. Thai energy and construction companies, including Electricity Generating Public Company (EGCO), Ratchaburi Electricity Generating Holding Public Company and CH. Karnchang Public Company, are also actively investing in hydropower projects in Laos, as are two Vietnamese companies, PetroVietnam and Viet-Lao Power Joint Stock Company.

These companies are often supported by public banks or export credit agencies from their own countries. Most prominent among these are the Export-Import Bank of Thailand and the Export-Import Bank of China, which are responsible for indirectly financing a large number of hydropower projects in the Mekong (typically by providing credit and/or insurance to support the purchase of goods and services from their home country for project purposes). Regional commercial banks have also become key financiers of dams. Laos's largest dam to date, the US$3.8 billion Xayaburi dam, is largely financed by five Thai commercial banks (Bangkok Bank, Kasikornbank, Krungthai Bank, Siam Commercial Bank and

TISCO Bank), alongside the contributions of the project's equity investors. While their direct role in dam financing has declined, the World Bank, ADB and bilateral donors still maintain their relevance by funding cross-border power lines and providing technical, legal and institutional support for the Mekong region power grid (see Chapter 3).

State electricity utilities are also key players shaping hydropower development in the Mekong River Basin. Thailand's EGAT is the primary purchaser of hydroelectricity from Laos and the sole distributor of electricity within Thailand. EGAT's reputation as a reliable and creditworthy power purchaser is critical in raising project financing with commercial lenders for dam construction in Laos (Doran and Christensen, 2014). As well as purchasing power, EGAT finances dams in neighbouring Laos either through its subsidiary company, EGAT International, or through independent power producers, EGCO and Ratchaburi, in which EGAT owns majority shares. In Vietnam, around 40 per cent of power generation is through independent power producers (which are mainly state-owned fuel and construction companies, such as PetroVietnam), or BOT-PPP projects (Nguyen, 2012). Électricité du Laos and Électricité du Cambodge have remained under state ownership. While Laos has preferred to take a shareholding through Électricité du Laos within BOT-PPP projects, Cambodia has opted for full private ownership of hydropower dams before selling electricity back to Électricité du Cambodge (Middleton *et al.*, 2015).

Law firms and lawyers

International law firms are influential yet often 'hidden' players actively shaping the hydropower regulatory landscape. The number and size of law firms with Mekong region-specific expertise has grown significantly over the past decade with the opening of legal services markets in ASEAN. A large number of these firms have core expertise in energy, natural resources and infrastructure. DFDL Legal and Tax Services, for example, has multiple offices in Cambodia, Laos, Thailand, Myanmar and Vietnam and extensive experience negotiating and drafting a range of contracts for cross-border hydropower trade in the Mekong. Indeed, DFDL claims to have been involved in pioneering innovative modes of project financing and associated documentation for hydropower projects involving complex cross-border transactions between Laos and Thailand – an experience that the company suggests can be transferred to other countries, such as Myanmar, Bhutan and Nepal (Doran and Christensen, 2014). DFDL clients include the Export-Import Bank of China, Thai lender consortia and the World Bank, among others, and their lawyers have brokered deals associated with 28 hydropower projects in Laos, Cambodia and Vietnam, including Nam Theun 2, Xayaburi and the Lower Sesan 2 (see DFDL Legal and Tax Services, 2014a).

As well as offering advice to governments and client companies investing in or providing services in connection with hydropower, law firms also offer advice to NGOs that solicit their services to contest the legality of projects. For example, International Rivers and the Environmental Defender Law Center solicited the services of the US law firm Perkins Coie to examine whether Laos is violating obligations under international environmental law by constructing the Xayaburi dam (Higgs, 2011; see also Chapter 7). In other instances, NGOs have engaged academic institutions to carry out legal analysis. For example, the Human Rights Centre at Essex University carried out legal analysis for Mekong Watch of the Concession Agreement between the government of Laos and Nam Theun 2 Hydropower Power Company (Can and Leader, 2005).

Civil society actors

In this book, we deal with civil society mainly with regard to NGOs and grassroots community groups affected by hydropower development, particularly in Chapter 7. Some such civil society actors focus specifically on river ecosystems, river-based livelihoods and water resource governance (including hydropower). These groups include the 3S Rivers Protection Network in Cambodia, the Rivers Coalition in Cambodia, the Fisheries Action Coalition Team in Cambodia, the Vietnamese Rivers Network, Living Rivers Siam and the Thai People's Network for Mekong. Other domestic actors have a wider agenda but also address Mekong issues. These groups include People and Nature Reconciliation (PanNature) in Vietnam, CLICK in Laos, the Assembly of the Poor in Thailand and the Culture, and Environmental Preservation Association and the Highlander Association, both in Cambodia.

Thailand is also a regional base for transnational networks working on Mekong hydropower, such as Towards Ecological Recovery and Regional Alliance (TERRA) and the Mekong Energy and Ecology Network (MEE Net). Three key international NGOs (INGOs) that focus extensively on Mekong issues – International Rivers, EarthRights International and Focus on the Global South – have regional offices in Thailand. Increasingly, other regional cities have become nodal points for regional and international organizations. Oxfam's Mekong Water Governance Programme has its main office in Phnom Penh. The Consultative Group for International Agricultural Research's Research Programme on Water, Land and Ecosystems is based in Vientiane, with a dedicated Greater Mekong programme. Japan's Mekong Watch has staff residing in all four Mekong countries but no formal regional office. In all the Mekong countries, human rights and legal NGOs touch on aspects of Mekong hydropower.

Civil society has historical roots in all four Mekong countries in various permutations. Today there are formal constitutional guarantees enabling civil society activity (in principle, at least) in each of the four countries, including guarantees of freedoms of speech, association and assembly. These constitutional rights are additional to the more specific rights provided for in domestic legislation and policies, including laws that relate to hydropower development. Environmental impact assessment laws, for example, include rights to public participation and consultation in development decisions, rights to information and rights for impacted communities to receive just compensation for incurred damages (see Chapter 5). Furthermore, all four countries have acceded to key international human rights treaties that civil society groups use in their advocacy (see Chapter 7 and Appendix 2).

Thailand has the most robust civil society of the four Mekong countries examined, with the capacity to limit the power and reach of the state. An alliance of local, Bangkok-based and international environmental and rural livelihood movements, which emerged in the 1980s, halted the construction of the Nam Choan dam in Kanchanaburi in 1988, led to a logging ban being imposed in 1989, and mounted a forceful campaign against the Pak Mun dam in Ubon Ratchathani from 1990 onwards (Hirsch and Lohmann, 1989; Hirsch, 1997). Region-wide, civil society groups (including groups from Thailand, Cambodia and Vietnam and some head-quartered internationally) have worked collaboratively to oppose the construction of mainstream dams, including through the Save the Mekong coalition, formed in 2009.

In practice, however, civil society is constrained in various ways in each of the four countries. In Thailand, successive military coups have punctuated Thai politics, disrupting and undermining the rule of law and constitutional protections for civil society. The most recent military coup, in May 2014, involved intimidation and violence and placed serious restrictions on civilian freedom of expression, including the suspension of constitutional guarantees. Legal institutions in Thailand have not always upheld formal legal protections, including those established by the 1997 constitution to ensure checks and balances against abuses of executive and legislative power. The judiciary has legitimized coups and adopted a narrow, positivist, technical approach to law, seen by some to thwart social justice (Connors, 2011). While the National Human Rights Commission has 'indigenised' rights and advocated for community rights to natural resources, it was politicized and 'fell into disarray' after the 2006 military coup (Connors, 2011, pp. 107–108). Despite democratization and rights discourse, Thai political culture remains rigid, paternalistic, elitist and centralized, with royalists, the military, the bureaucracy and predatory corporatist interests resisting civil society demands for greater transparency, decentralization and equity (Banpasirichote, 2004, pp. 243–245).

As in Thailand, civil society in Cambodia faces numerous obstacles. Liberal democracy and human rights protections embedded in Cambodia's Constitution mask the reality that Cambodia is effectively a one-party state under the hard-line rule of Prime Minister Hun Sen's Cambodian People's Party. In the 30 years since Hun Sen rose to power, during which he staged a coup in 1997 against the coalition government of which he was a part, elections have been marred by intimidation and violence against opponents, voters, rights activists and the media (Peou, 2011). The government has used the legal system against its opponents and ordinary citizens, particularly by effecting arbitrary arrest or prosecution for vague offences of disinformation, defamation or incitement. The abuse of power by Cambodia's political and business elite, and its military and security forces, is often connected with questionable land deals and natural resource projects. Popular demonstrations against the government around the 2013 elections temporarily widened the space for civil society in Cambodia. However, the possibility of another government crackdown remains ever-present as Hun Sen maintains his unshakable grip on power.

Civil society is most constrained in the communist states of Vietnam and Laos. According to the Marxist-Leninist ideology to which these states nominally subscribe, there is no need for separate civil society because 'the People are the masters' and 'all the state power belongs to the People' (Vietnam Constitution, 2013, Article 2; Laos Constitution, 2003, Article 2). While people in both countries formally enjoy civil freedoms of expression, media, assembly, association and demonstration (Vietnam Constitution, 2013, Article 25; Laos Constitution, 2003, Article 44), there is little room for an independent civil society outside party structures. Formally, 'mass organizations' are the key vehicles for mobilizing sectoral social interests from the village to the central levels of government.

In practice, nevertheless, the space for wider and more diverse civil society has grown in both countries since economic liberalization. NGOs and other forms of association can legally register, although the process is time-consuming and often seen as a means of extending government surveillance and control. Civil society works mainly in partnership with government agencies, emphasizing non-confrontational 'soft advocacy' approaches, and personal connections are important in influencing government. In both countries, there is space for civil society to work on technical, environmental and livelihood issues, but not overt human rights advocacy, legal advisory work or other sensitive political matters. In Laos, the space for discussing sensitive topics such as land and hydropower shrank drastically after the enforced disappearance of well-respected social activist Sombath Somphone in December 2012 (Sombath.org, n.d.). Organized opposition to hydropower in Laos has mainly come from INGOs such as International Rivers or regional networks. In Vietnam, there have been more openings for NGOs, scientists and others to raise social and environmental concerns around hydropower development. Even in Vietnam, however, advocacy is easier where NGO and state interests align, as in Vietnam's opposition to Laos's development of the Xayaburi and Don Sahong hydropower dams.

Five dam suites

Contested development in the Mekong hinges upon a number of processes and projects. By far the most robust of 'plots' in the Mekong development narrative is the construction of dams. Much of the story that we tell in the chapters that follow revolves around this plotline. In order to set the main script lines for each of these stories, we now introduce the dams and the 'suites' of which they are a part.

Xayaburi and Lower Mekong mainstream dams

The longest-standing element of the Mekong dam storyline is the ambition by hydro-engineers to build dams on the mainstream of the Lower Mekong River. In Chapter 3, we trace the history of this ambition as a background to present-day plans and disputes. However, for the most part, we focus in this book on contemporary projects on the mainstream. Specifically, we describe one that is under construction and the proposed cascade of which it is a part.

Xayaburi dam is located approximately 100 kilometres downstream of the former Lao royal capital of Luang Prabang. It is the first mainstream dam under construction in the four Lower Mekong countries that are party to the Mekong Agreement. It is one of 11 dams proposed for the Lower Mekong mainstream; seven of these are in Laos, two straddle the Lao–Thai border and two are in Cambodia. If all of these dams were built, they would comprise the third uppermost rung in the proposed cascade after the Pak Beng and Luang Prabang dams. The second mainstream dam formally proposed for construction is the Don Sahong hydropower project at the Khone Falls in Southern Laos, less than two kilometres from the Lao–Cambodian border. Don Sahong would be the ninth rung of the cascade. Figure 1.2 shows the location of the 11 dams being proposed for the Lower Mekong. A 12th dam proposed at the Khone Falls, Tha Takho, cannot go ahead if Don Sahong proceeds, since there is only sufficient water for one of these projects to function. If all 11 mainstream dams were to be built, some 55 per cent of the length of the Lower Mekong would be reduced to a series of still-water reservoirs. It is anticipated that investors in these dams would primarily come from Thailand, China and Vietnam, but specialized services and debt and equity financing would likely be drawn, to varying degrees, from European companies and banks or export credit agencies.

When completed, Xayaburi dam will be 32.6 metres high and the dam wall will be 810 metres long. Its reservoir will extend 100 kilometres upstream and inundate an area of 49 square kilometres (Ministry of Energy and Mines, Lao PDR, n.d.). It will generate 1,260 megawatts, making it the largest power project in Laos. It is predicted to generate 5,990 gigawatt hours annually, which would give it a 54 per cent factor loading. This percentage of the theoretical maximum output of the dam – were it to operate year-round at full capacity for 24 hours per day – is quite high, increasing the economic feasibility of the dam. This feasibility is predicated on enhanced dry season flows resulting from upstream water storage and releases created as a result of dams operating in China (TEAM Consulting Engineering and Management, 2010).

Ninety-five per cent of the power generated by Xayaburi dam is for export to Thailand. The dam is owned by a consortium led by a Thai company (CH. Karnchang) and including Natee Synergy Company Ltd (a wholly owned subsidiary of Global Power Synergy Company, itself a wholly owned subsidiary of PTT Public Company Ltd, the outcome of the privatization of the Petroleum Authority of Thailand), EGCO, Bangkok Expressway Public Company and a Lao-based company, PT Construction and Irrigation Company Ltd, as well as the Lao state enterprise Électricité du Laos. It is being built by a Thai construction company with support from a Finnish engineering firm (CH. Karnchang and Poyry, respectively) and its debt financing has largely been sourced from Thai banks (Krungthai Bank, Siam Commercial Bank, Kasikornbank, Bangkok Bank and TISCO Bank).

Xayaburi dam is highly controversial and much contested for a number of reasons. The fact that it is the first dam on the lower section of the mainstream that is subject to the governance procedures associated with the MRC means that it is precedent-setting in several ways. National, regional and international NGOs, donors to the MRC and downstream governments are concerned that it will become progressively more difficult to avoid the construction of further mainstream dams once one is in place. Xayaburi also has significant potential

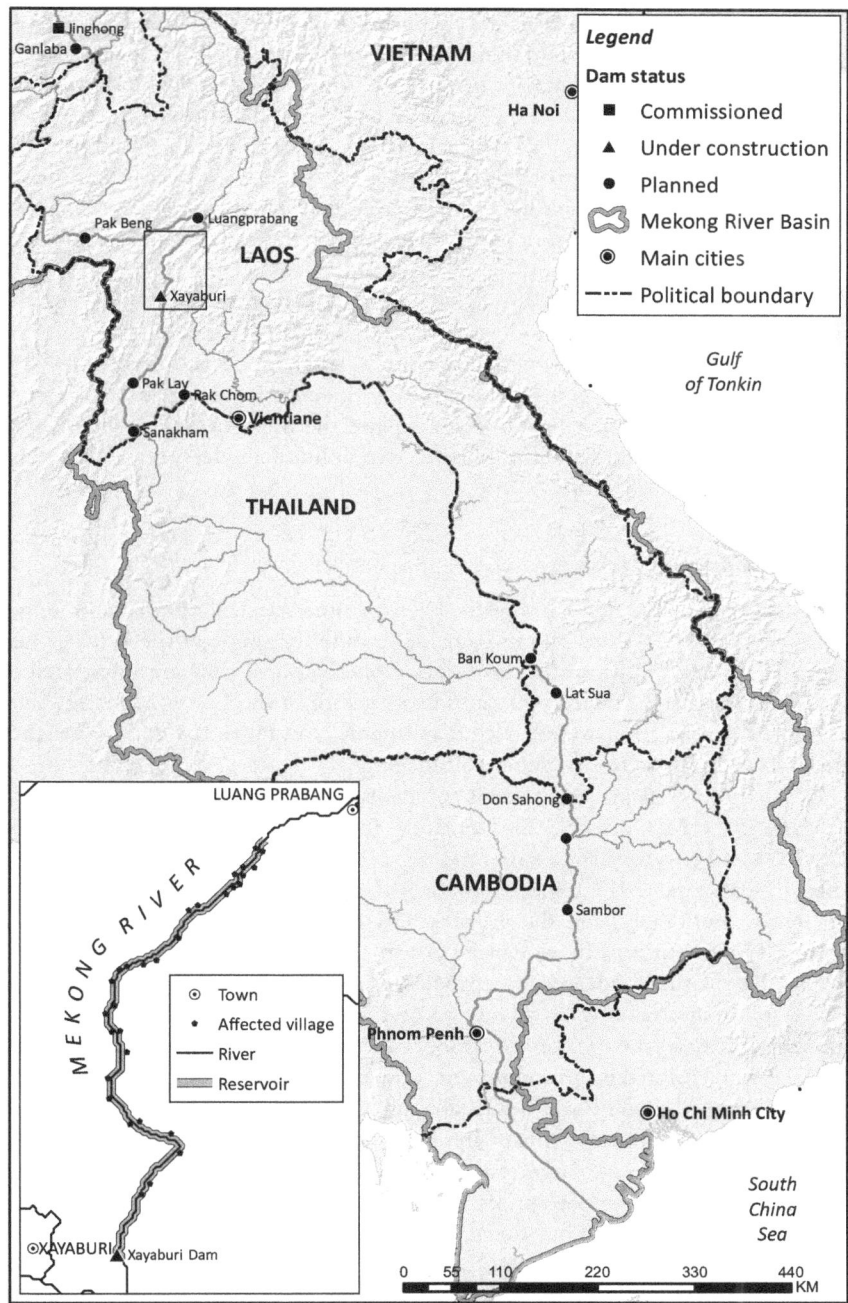

Figure 1.2 Lower Mekong Basin dams (with Xayaburi detail)

impacts in its own right. In particular, the blockage effect on fisheries and the displacement of at least 424 households, or more than 2,000 villagers (TEAM Consulting Engineering and Management, 2010), away from their homes and river-based livelihoods has attracted attention and critique from NGOs and others (International Rivers, 2011;Vaidyanathan, 2011). As discussed in Chapter 5, the environmental impact assessment for Xayaburi only studied impacts 10 kilometres downstream (TEAM Consulting Engineering and Management, 2010), raising concerns that potential transboundary impacts were being neglected (International Rivers, 2011).

Xayaburi also sets a precedent in becoming the first project to trigger the prior consultation procedure under the MRC Procedures for Notification, Prior Consultation and Agreement, which are prominent among the procedures that the MRC has established as the most explicitly directive regulatory component of its governance regime. In Chapter 4 we explore the messy ways in which Laos's recourse to this procedure became embedded in political and scientific debates and confrontations.

Chinese mainstream dams

The Mekong mainstream has already been dammed several times in its upper reaches (Figure 1.3). China and its dams lie outside the scope of the detailed case study analysis and discussion in this book. Nonetheless, as they are implicated in decision-making and financing of dams downstream in the Lower Mekong River Basin, their construction and operation are highly relevant to law and governance around dams in the MRC member countries.

China's now well-known cascade of dams in Yunnan province (Dore and Xiaogang, 2004; McCormack, 2001) includes four medium-sized dams completed since 1994 and two very large structures, the Xiaowan and Nuozhadu dams, completed in 2010 and 2013 respectively. The storage capacity of the latter dams is sufficient to noticeably alter the seasonal flow of the river all the way down the Mekong system (International Rivers, 2013b). In a highly seasonal flow regime such as that of the monsoon-driven Mekong system, smoothing of the annual hydrograph by upstream water storage in, and release from, reservoirs behind very large dams increases the economic viability of downstream low-head dams (Hirsch, 2011). Less well known is the continuing construction of dams in the upper reaches of the Lancang-Mekong River in Qinghai province and Tibet.

The Lancang dams in Yunnan are built and operated by Huaneng Power International, one of five large state-owned enterprises that builds hydropower projects in China. These enterprises are increasingly following China's 'Go Out' policy of taking expertise and financial capital into investments beyond China's borders. Many of the tributary dams in Laos and Cambodia are being built by these enterprises, under quite different socio-legal arrangements from dams funded by international financial institutions or publicly listed companies. As noted above, we do not examine these arrangements in any detail in this book – a choice driven in part by the difficulty of gaining access to relevant documents and personnel.

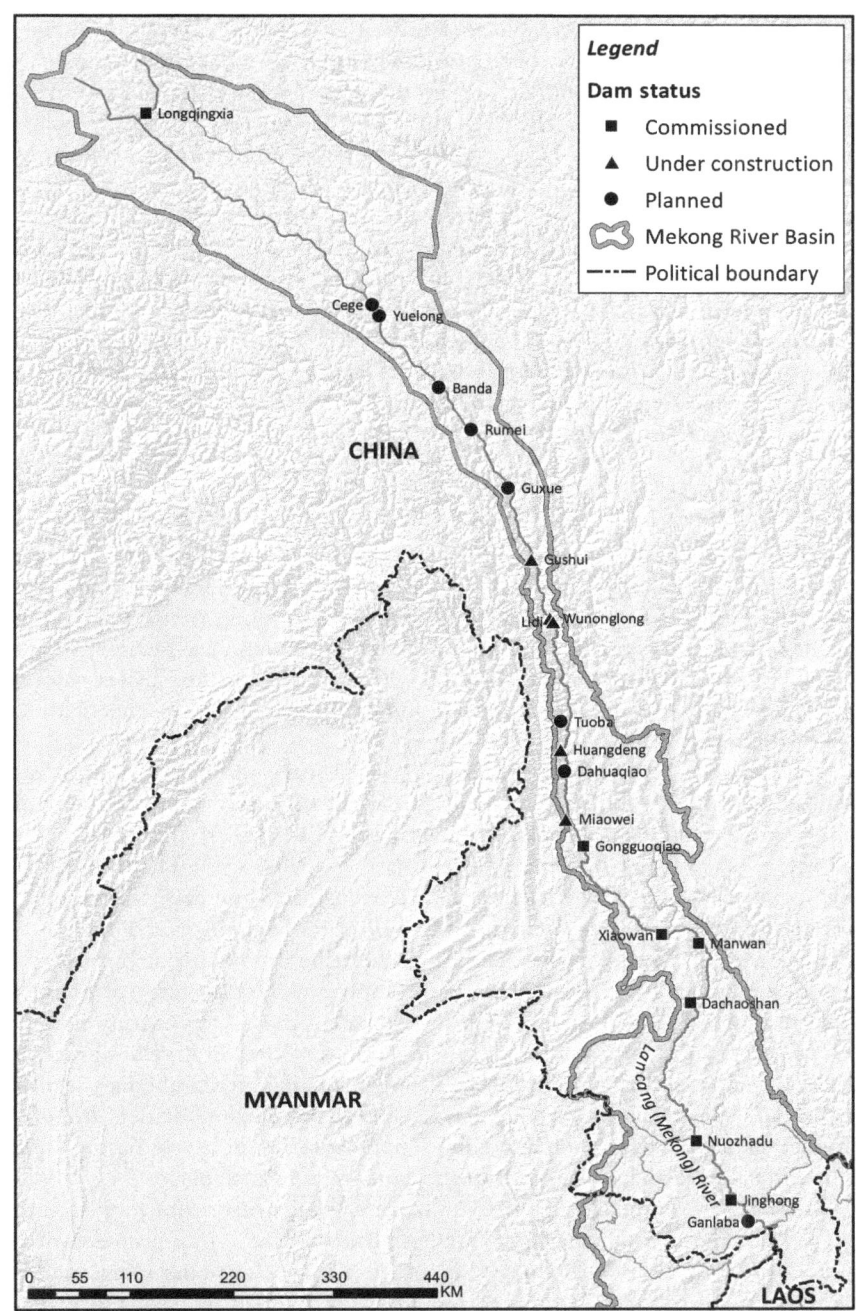

Figure 1.3 Upper Mekong Basin dams

China's mainstream dams have been controversial for a number of reasons. First, since China is neither a signatory to the Mekong Agreement nor a member of the MRC, its projects lie entirely outside the principal international governance arrangements in the Mekong River Basin. China is therefore not legally obliged under any agreement specific to the Mekong River to engage in any negotiations regarding the downstream impacts of its hydropower projects. As a consequence, as both the most powerful and the upstream-most country sharing the Mekong's waters, China has been able to act unilaterally. Second, a series of floods and droughts along the lower part of the Mekong River during the years following completion of the Yunnan dams has led to accusation and counter-accusation regarding the causes of extreme high and low flows. There is significant politics between China and the MRC around information sharing about seasonally specific, real-time hydrological data release, which is intensified by China's observer status at the MRC and ongoing informal negotiations between China and the MRC.

Yali Falls, Sesan 2, A Luoi and the 3S dams

The Sesan, Srepok and Sekong Rivers (also known as the 3S rivers) are three transboundary tributaries that rise in Vietnam's Central Highlands and flow westward, joining the Mekong at a single confluence in Stung Treng in Cambodia. The Sesan and Srepok pass straight from Vietnam into Cambodian territory, joining about 37 kilometres above Stung Treng, while the Sekong first passes through Laos before joining the Sesan just seven kilometres above the confluence with the Mekong (Figure 1.4(a)).

The 3S rivers are significant to the socio-legal make-up of the Mekong in a number of ways. Combined, they form the largest tributary system of the entire basin, contributing 23 per cent of annual flow (MRC, 2009) and between five and 15 per cent of annual sediment (Sarkkula *et al.*, 2010, p. 23). They are also the only major tributaries of the Mekong that cross national boundaries before join-ing the mainstream, raising inter-jurisdictional issues in their own right. The 3S system is the most heavily developed of all the tributary basins for hydropower. This development has been largely in the upstream-most country of Vietnam, while the impacts have been felt disproportionately in the downstream country of Cambodia. Finally, most of the people living in the 3S basin have until recently been indigenous ethnic minorities. In the Vietnamese and Cambodian sections of the basin, in recent years there has been a large influx of settlers from the national majority ethnic Kinh and Khmer populations respectively, raising a host of resource tenure and associated human rights issues in each place.

Yali Falls dam on the Sesan River in Vietnam was the first major project in the Lower Mekong to have significant transboundary impacts. Long planned within the framework of the old Mekong Committee, Yali Falls was built between 1993 and 2000. Vietnam formally notified the MRC of the dam in September 1997. Generating 720 megawatts of power, Yali Falls was at the time by far the largest power project to be built in the Lower Mekong River Basin. It was financed mainly by the Ukrainian and Russian governments, but the World Bank funded

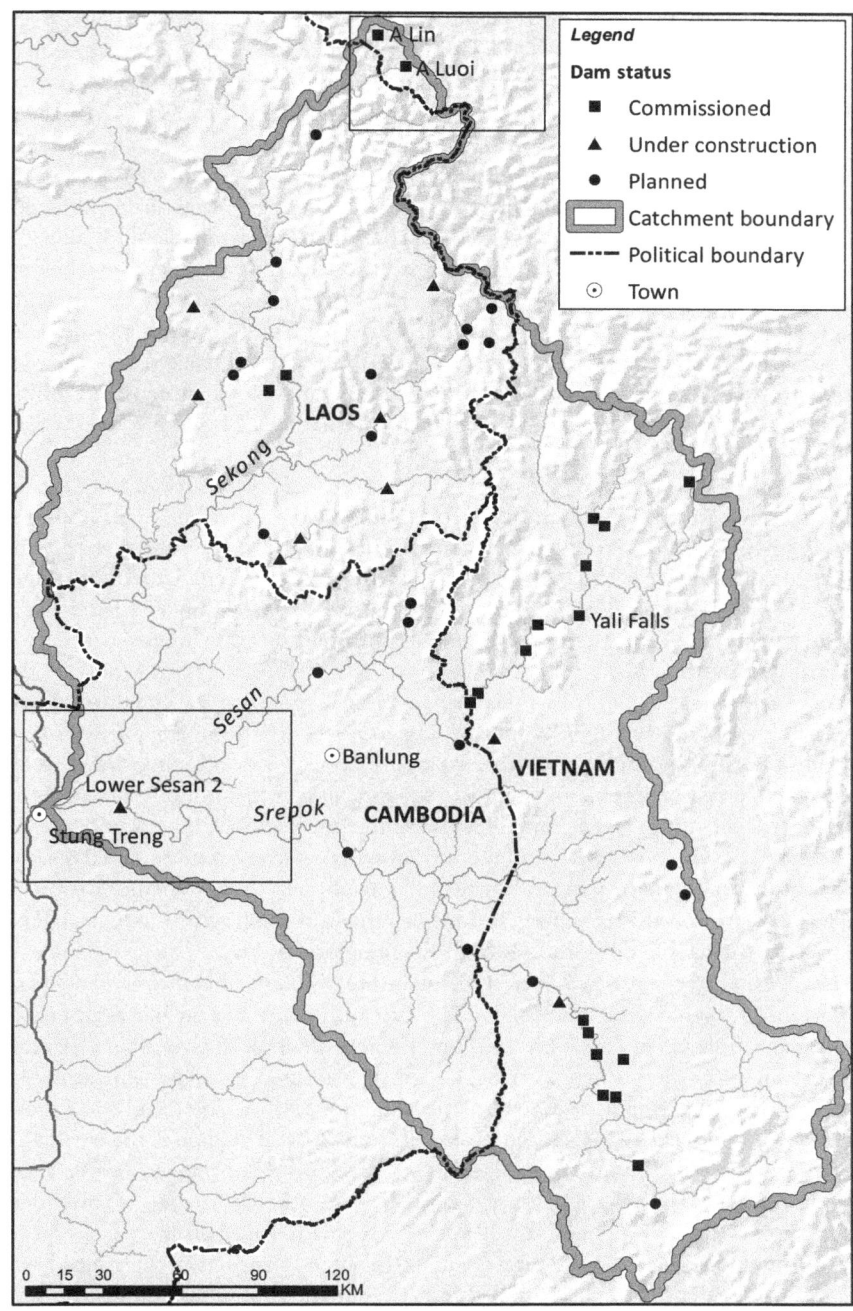

Figure 1.4(a) 3S (Sesan, Srepok and Sekong Rivers) dams

transmission lines that took power from Yali Falls to Vietnam's national electricity grid. The Swiss firm Electrowatt carried out the environmental impact assessment, which was much criticized for only examining impacts eight kilometres downstream and not taking into account the potential downstream impacts in Cambodia.

Yali Falls was also the first major completed dam project to become a *cause célèbre* among local, national, regional and international NGOs. In 1996, and again in 2000, sudden water releases from the dam site resulted in flash flooding for more than 100 kilometres downstream. This resulted in loss of life and property at a time when many NGOs were mobilized concerning prospective tributary and mainstream dams. In addition to these major flooding events, the opening and closing of the power station on a daily basis, to meet the fluctuations in demand by urban electricity users in distant Vietnamese cities, has impacted fisheries and made pre-existing uses of the river for gold panning, riverbank gardens and fishing impractical or risky (Hirsch and Wyatt, 2004; International Rivers Network, 2002; Wyatt and Baird, 2007).

Following the construction of Yali Falls, Vietnam has continued to build dams in its section of the 3S basin, and Laos has an intensive programme of hydropower development on the Sekong and its tributaries. Cambodia, meanwhile, has commenced construction of the 400-megawatt Lower Sesan 2 dam. Six main barrages have been built on each of the Sesan and Srepok Rivers within Vietnamese territory, mostly by Vietnamese state-owned companies. Vietnam followed the procedures set by the MRC for tributary dams, which involve notification of the other three riparian countries signatory to the Mekong Agreement through the Vietnam National Mekong Committee. These procedures do not require or allow for any official response by the other riparians. However, continuing downstream impacts in Cambodia remain a subject of complaint by community networks in Ratanakiri and Stung Treng provinces. Oxfam, International Rivers and the 3S Rivers Protection Network continue to campaign actively but. As the dams are now in place, they have had to shift their demands away from whether or not to build the dams to calls for redress and management of the flows to reduce adverse impact on the lives and livelihoods of those living downstream.

Issues associated with two current dams in the 3S basin inform our socio-legal discussion in the chapters that follow. The first of these is a large and high profile project that receives attention in Chapter 7, where we look at civil society contestations. The 400-megawatt Lower Sesan 2 dam is located just downstream of the confluence of the Srepok and Sesan Rivers (Figure 1.4(b)). The Cambodian government notified the MRC of the project in June 2010. The dam is the first to be built inside Cambodia within the 3S basin. Its developer, Hydropower Lower Sesan 2 Company Ltd, is a joint venture between China's Hydrolancang, Cambodia's influential Royal Group and Electricity of Vietnam International Joint Stock Company. Construction commenced in 2014 after a great deal of controversy. The 75-metre-high dam will flood about 336 square kilometres, inundating several communities of mainly ethnic minority peoples living close to the Sesan and Srepok Rivers and displacing about 5,000 people. Because it is located so close to the Mekong confluence, the dam will isolate a large part of the tributary basin from the mainstream. As a consequence, the dam is predicted to have a larger impact on fish than any other Mekong hydropower project, leading to a loss of about nine per cent of the entire basin-wide catch (Ziv *et al.*, 2012). Lower Sesan 2 was notified to the MRC in 2010.

The second dam that we explore, particularly in Chapter 5 on environmental assessment, is a relatively minor project that has had no limelight to date. The 170-megawatt A Luoi dam is in the small area of the upper Sekong catchment that lies within Vietnamese territory and is situated on the A Sap tributary of the Sekong River (Figure 1.4(c)). A key feature of A Luoi is that it was the first dam situated within the Mekong River Basin to divert water through an inter-basin diversion. Together with a second and smaller inter-basin diversion in the upper Sekong, A Lin hydropower project, this in effect means that most of the water that would otherwise flow into the upper Sekong in Laos is lost to the system. This makes the upper Sekong dams in Vietnam unique, as other tributary dams in the Mekong River Basin maintain water in the overall river system. In the case of A Luoi, it is instead channelled from the small reservoir through a 14-kilometre-long tunnel to the other side of the escarpment, then via a penstock chute through a vertical distance of 490 metres to the power station, following which the water exits to the Bo tributary of the Huong (Perfume) River that flows through Hue into the South China Sea (Power-technology.com, n.d.). Since the A Sap is an upper tributary, the volume of water lost is not large, but nevertheless represents a transboundary

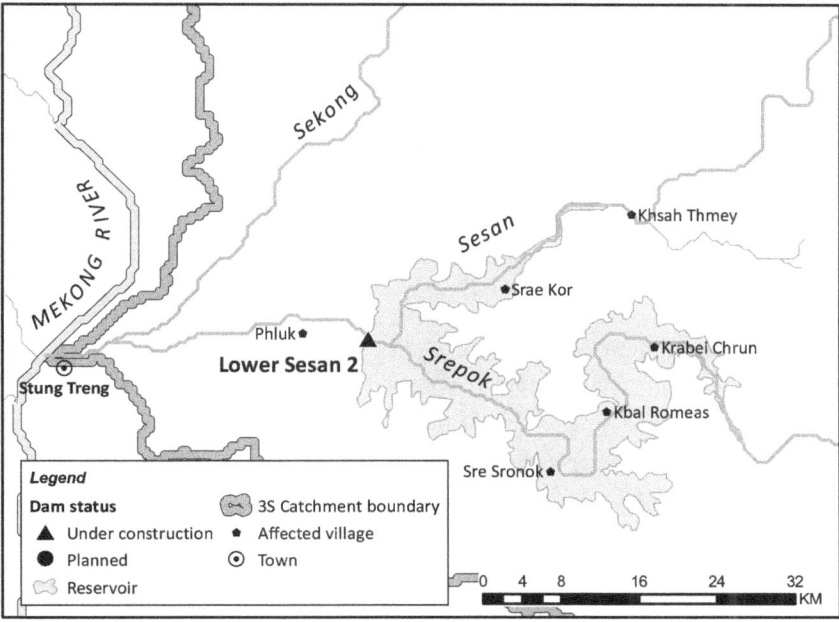

Figure 1.4(b) Lower Sesan River dams detail

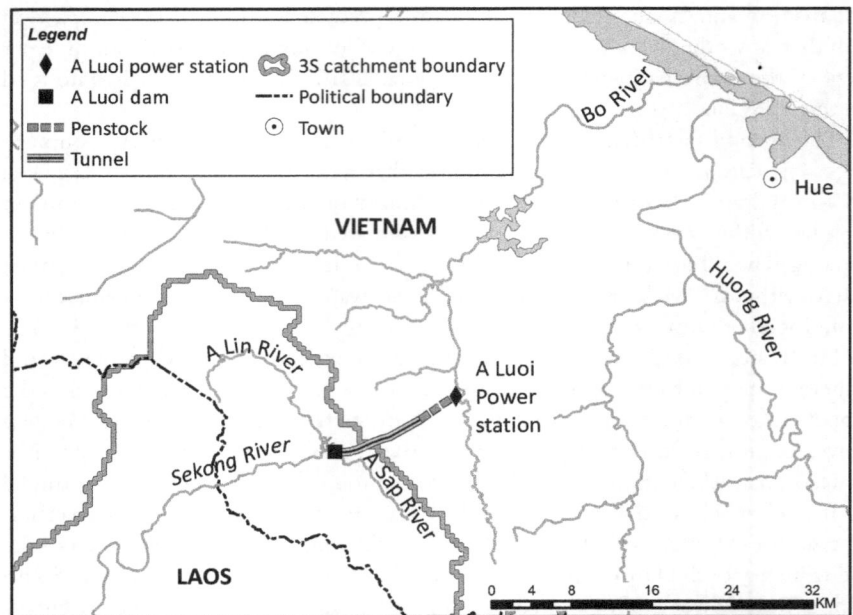

Figure 1.4(c) A Luoi dam detail

impact that remains unassessed. Several villages also faced relocation from the area inundated by the reservoir.

The dam was completed in late 2011 and was never notified through the Vietnam National Mekong Committee to the MRC, unlike the A Lin project that was notified in March 2009. A Luoi was, however, submitted to receive a subsidy for carbon credits under the Clean Development Mechanism, a market-based mechanism under the Kyoto Protocol allowing wealthy countries to offset their carbon emissions through subsidies of alternatives to carbon-intensive projects in less developed countries. The project was built and is owned by the Central Hydropower Joint Stock Company and Vattenfall Energy Trading Netherlands NV. It is financed by the Vietnam Bank of Agriculture and Rural Development together with the Vietnam–Thailand Joint Venture Bank, the An Binh Commercial Joint Stock Bank and the Rubber Financing Company (Power-technology.com, n.d.).

Nam Theun 2 and tributary dams in Laos

The Nam Theun 2 hydropower project in Laos is the largest tributary project in the Lower Mekong River Basin. The 1,075-megawatt dam diverts water off the Nakai Plateau into the Xe Bang Fai tributary of the Mekong, generating power through the 350-metre vertical distance that the water travels (Figure 1.5). The dam itself is quite a modest 50-metre tall structure that nevertheless floods a 450 square kilometre tract of land on the relatively flat Nakai Plateau. It required the resettlement of about 6,200 people, and a much larger number of people are affected by the greatly increased volume of water taken down the Xe Bang Fai and by reduced flows downstream on the Nam Theun River system. The project has the highest profile of any non-mainstream dam in the Mekong River Basin, for a number of reasons. Funded in part by the World Bank, the dam went through a protracted planning and consultation process prior to its approval for financial close in 2004 and construction between 2004 and 2008. It marked the re-entry of the World Bank into large hydropower project financing, following a hiatus after critiques of its earlier engagements and the bank's high-profile withdrawals from projects during the 1980s and early 1990s. Nam Theun 2 thus has significance far beyond its locale, Laos or even the Lower Mekong River Basin. A key feature of the project, as intimated in the vignette at the head of this chapter, is the subjection of the project to various supervisory mechanisms and impact mitigation requirements that were, in principle, to set new standards for hydropower in Laos and the region. In Chapter 6, we explore the socio-legal implications of these arrangements and the modes of governance installed or fostered thereby. The project is a joint venture between EDF International (a wholly owned subsidiary of the predominantly state-owned French company Électricité de France), the Lao government's Lao Holding State Enterprise, and Thailand's partially state-owned, publicly listed company EGCO. Debt financing and guarantees for the project were secured under a complex arrangement involving some 27 financial institutions, led by the World Bank and the ADB.

Nam Theun 2 is one of dozens of tributary hydropower projects in Laos. The first such project, the small five-megawatt Selabam dam on the Sedone River in Champasak province, was built with French assistance in the 1960s. More significant was the 150-megawatt Nam Ngum 1 dam built during the late 1960s under the auspices of the Mekong Committee and completed in 1971, with assistance from ten countries. Until the early 1990s, the power exported to Thailand from Nam Ngum 1 made hydropower the second largest source of foreign exchange for Laos after timber. Since the 1990s, there has been a proliferation of power projects, partly to supply the rapid rural electrification and urban expansion of the country, but mainly for export to Thailand. Electricity exports remain the second largest source of foreign exchange for Laos, after minerals (Maierbrugger, 2014). Vietnam, China and potentially Cambodia are also markets for Lao electricity exports.

Under current plans, most of Laos's tributaries will be heavily dammed by the year 2030 (Ministry of Energy and Mines, Lao PDR, 2014). The country has pinned its economic future on serving as a regional power hub. In governance terms one of the most significant developments in Laos has been the shift away from multilateral financing towards privately owned and financed projects. This is highly significant for the socio-legal configuration of stakeholders involved in and impacted by hydropower development. The growing role of China's hydropower companies in medium-scale hydropower projects in Laos also shapes the ways in which project assessment is carried out: transparency is constrained; mitigation and compensation are internally managed for affected communities and appear to be

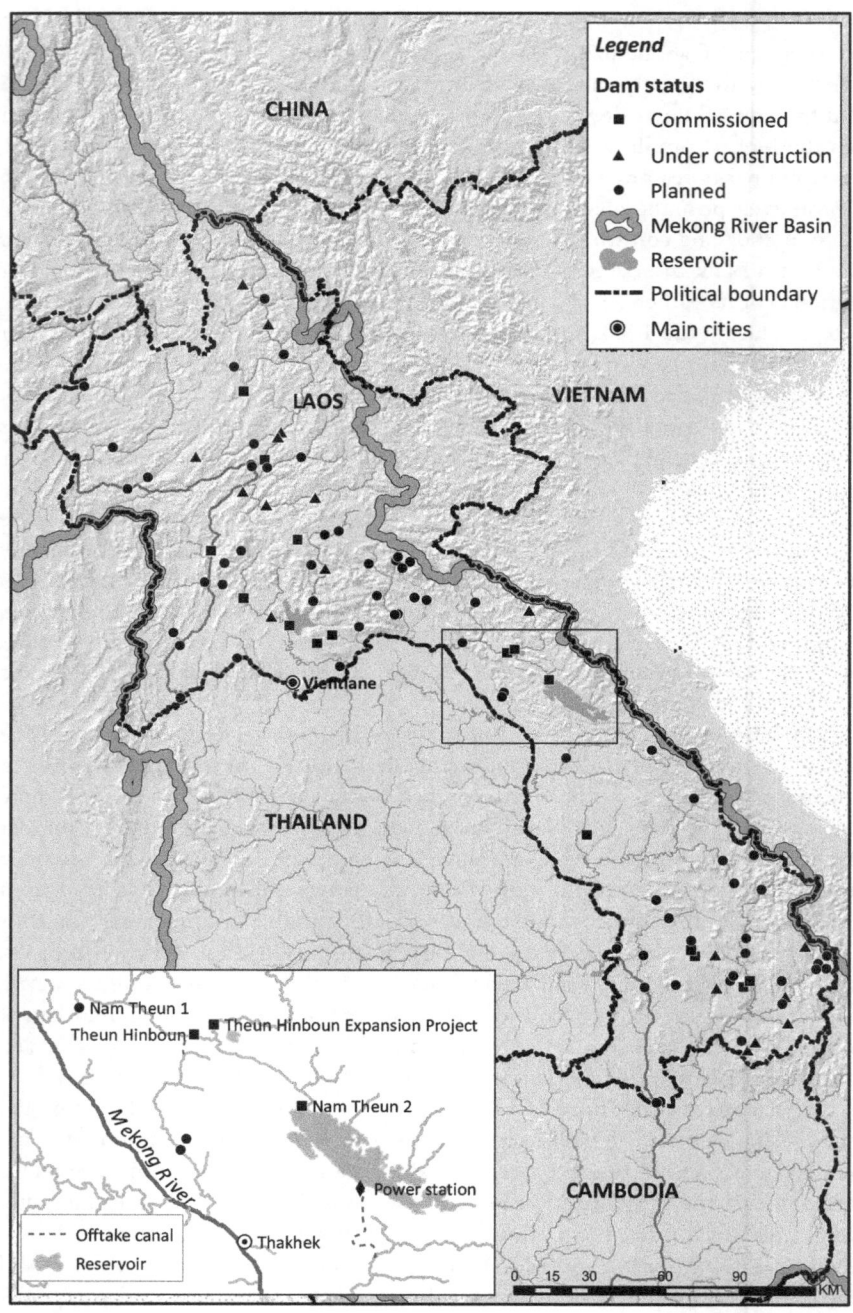

Figure 1.5 Dams in Laos (with Nam Theun 1, Nam Theun 2 and Theun Hinboun detail)

limited; and international donors and civil society have relatively minimal engagement with the governance of such projects.

Pak Mun and Thailand's dam politics

Pak Mun dam is located near the confluence of the Mun and Mekong Rivers in northeastern Thailand (Figure 1.6). Pak Mun was built between 1991 and 1994 amid considerable controversy. It was a public utility owned by EGAT and was part-funded by the World Bank. The dam is a relatively modest structure standing 17 metres high and generating 136 megawatts of power. Its reservoir banks up about 80 kilometres upstream when the dam is full. The major point of controversy surrounding this project has been the impact of the dam on fisheries. Many of the communities affected by Pak Mun had previously been heavily dependent on fishing as a main source of livelihood. Since the Mun River and its tributaries drain most of northeastern Thailand, isolation of the Mun river basin from the Mekong mainstream blocks fish migration and hence interrupts the breeding and feeding lifecycles of fish. The dam became one of the focal projects examined by the World Commission on Dams, which was established jointly by the World Bank and the International Union for Conservation of Nature in 1998 to review the development effectiveness of large dams, establish standards for future projects and identify alternatives to achieve the benefits provided by dams (World Commission on Dams, 2000a).

Controversy over Pak Mun needs to be understood in the wider context of dam construction in Thailand, and more specifically the plans and projects for that section of the country that lies within the Mekong River Basin. Thailand has been damming its rivers since the 1960s, starting with two large hydropower dams on the Ping and Nan tributaries of the Chaophraya River. Further dams were built on other rivers and their tributaries, including the Mae Klong/Khwae systems, until the 1980s, when the proposed World Bank funding of the Nam Choan dam became a *cause célèbre* for its impact on a now World Heritage-listed wildlife sanctuary. The dam was cancelled in 1988 after widespread protests. Pak Mun was the last hydropower dam to be built in Thailand.

Part of the reason that Pak Mun went ahead was that its reservoir did not flood large areas of forest, and the fisheries-based livelihoods of the affected communities were under-evaluated during the assessment process (Foran and Manorom, 2009). Another reason was that it was a key piece of infrastructure in Thailand's plans to water, develop and green the dry and poor northeast through a grand design of 16 dams and channels that would link up the Mekong, Chi and Mun river systems (the so-called Khong-Chi-Mun programme). Several other dams had been built during the 1960s and 1970s. However, the civil society challenges to such infrastructure, combined with opportunities for Thailand to source power from neighbouring countries following regional rapprochement after the late 1980s, meant that this grand scheme has not eventuated – although it has been proposed under new guises several times since.

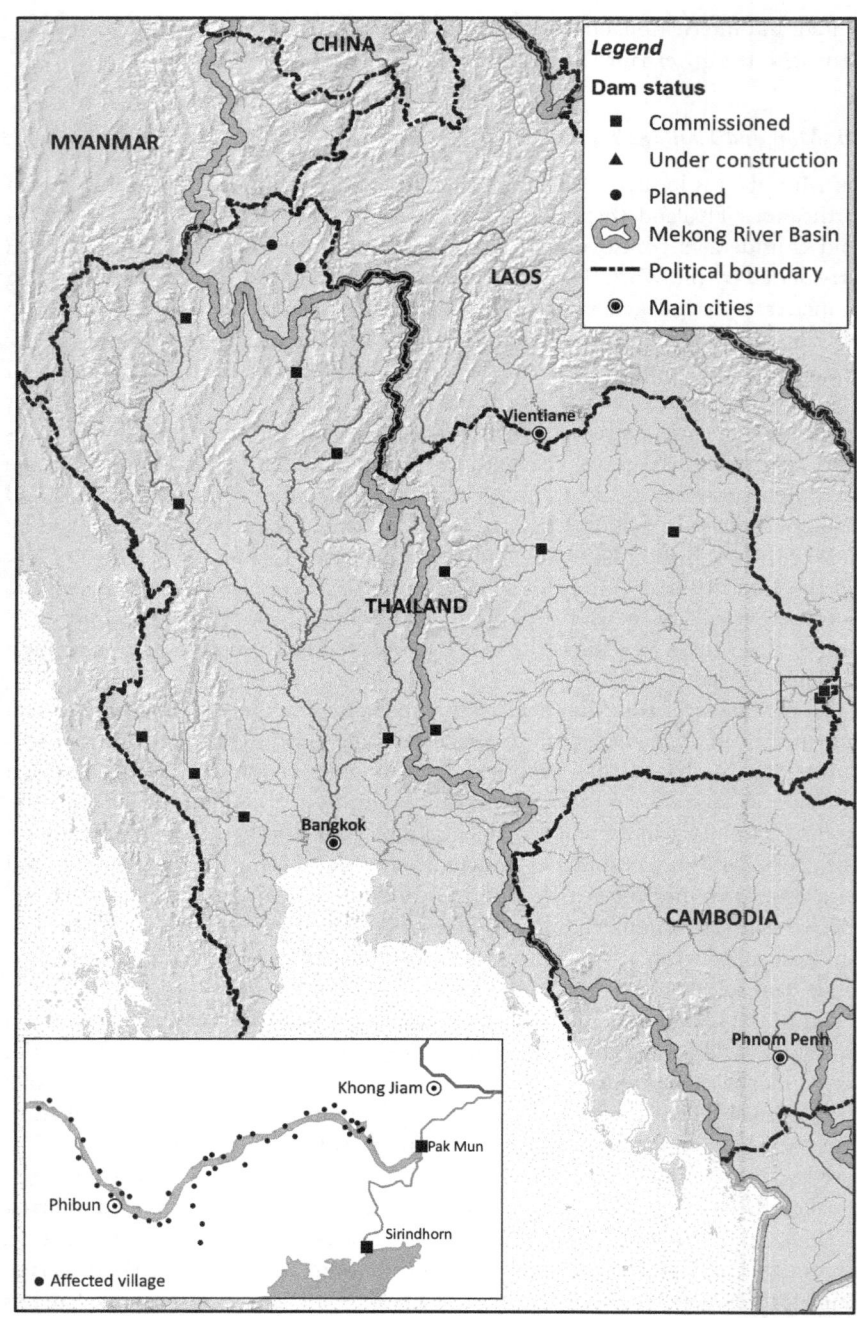

Figure 1.6 Dams in Thailand (with Pak Mun detail)

The other significant, but abortive, proposal for river linkage in Thailand that is relevant to the Mekong system is the so-called Kok-Ing-Nan project (TERRA, 1998). This would have involved damming of the Kok and Ing tributaries of the Mekong in the northern province of Chiang Rai, each of which would have generated power in its own right, and diverting water out of the Mekong River Basin through a series of tunnels and canals to the headwaters of the Nan River in the Chao Phraya River Basin. In turn, this would have produced more power at the existing Sirikit dam and enhanced irrigated dry season rice farming in Central Thailand. However, as well as taking water out of the Mekong system, the Kok-Ing-Nan scheme would mainly have been viable for wet season diversions that would, in turn, have exacerbated existing floods in Nan and downstream provinces, and protests resulted in the shelving of the project.

Our approach

As explained in greater detail in Chapter 2, we employ an interdisciplinary approach to examine the complex network of transnational, regional and national rules, institutions and arrangements governing the use and development of the Mekong River Basin. Our research team brings together diverse identities and research strengths in law, social science, geography and area studies. These combined strengths enabled us to conduct theoretical inquiry into the circulation of legal ideas, institutions and vocabularies alongside ethnographic mapping of particular socio-legal fields and case study projects, exploring at once their public and private law and 'hard' and 'soft' law dimensions.

In our ethnographic inquiry, we have relied on in-depth interviews to assemble a nuanced account of how laws and legal institutions at different levels operate and shape water governance outcomes, claims and expectations in the Mekong. Our team jointly conducted 53 interviews in Laos, Cambodia, Thailand and Vietnam with a wide range of stakeholders involved in Mekong water governance. This included representatives of government at all levels, international and regional organizations, NGOs, the private sector, academia and people living in rural communities. Our interviewees were selected to achieve diversity rather than for their representativeness of the vast array of actors and stakeholders in the Mekong development drama. We used a combination of opportunistic or convenience sampling (whereby interviewees were selected in part on the basis of their accessibility at relevant times) and snowball sampling (whereby some interviewees referred us to others). We clustered our interviews around particular projects and localities associated with specific institutions and developments. Therefore, the views expressed and quoted in this book are not to be treated as representative in the sense of presenting an objective indication of the characteristics of any particular population depicted. Rather, our interest and intent have been to capture the range of understandings and expectations of law that different actors hold in relation to transboundary water governance in the Mekong. We also aim to explore the factors (historical and contemporary, institutional, regulatory, and careerist) by which those understandings and expectations have been shaped.

Collectively, the interviews provide insights into how diverse actors perceive legal rules and institutions, and how associated legal ideas and vocabularies affect decision-making and distributional outcomes in the Mekong. Interviews also allow us to identify the transnational normative codes and relations, or what have elsewhere been termed 'webs of influence' (Braithwaite and Drahos, 2000), at work in transboundary water governance and to situate key legal and policy decision-makers (institutional, individual or collective) in relation to those codes, as well to each other. We also use case studies to enable a more detailed and cross-cutting analysis than has been available to date of how legal influences and ideas are acquired and deployed, change, travel and operate in the Mekong. As outlined in the section above, the 'dam suites' we have selected present distinct hybridizations of law and legal influences – hard and soft, public and private, national, regional and international.

Although our research focuses on regional and transboundary dimensions of law and water governance, a key challenge encountered by the team was mastering the translation of legal concepts across different political, cultural and lexical contexts. Legal vocabularies such as 'soft law', 'governance', 'civil society' and 'public participation' do not easily translate beyond the literal. In our discussions and interviews, terms had to be explained and set in context, particularly at the community level.

The very broad and all-encompassing notion of law that we employ in our research was itself a cause for much discussion and even disagreement within our own team. At the same time, the considerable transnational circulation of legal ideas in the Mekong region about water, the environment, natural resources and human rights provided solid points of departure for our dialogue with diverse interlocutors, and for the discussion and analysis in this book.

Chapter synopses

The chapters that follow explore socio-legal arrangements that shape the Mekong River Basin as actively produced rather than given. Following a conceptual guide to the material covered in the book in the next chapter, the remaining main chapter titles reflect the key actions that produce the socio-legal make-up of the Mekong River Basin: making, governing, assessing, disclosing and contesting.

Chapter 2 builds upon the introductory chapter's explanation of the distinctive contribution and approach of this book by focusing on the challenge of socio-legal inquiry. It explains the understanding of 'law' deployed in this book, the different legal orders and regimes with which it engages (including those specific to hydropower), and its relationship to such notions as 'governance' and 'development'. In so doing, it shows how this book stands apart from typical legal analyses of transboundary river basins. It also calls into question many conventions of 'law and development' work in this field, alongside some common tropes of Mekong-specific expert and donor knowledge.

Chapter 3 examines legal reform agendas and interventions of colonial authorities, donors, developers and experts that have shaped the progress and direction of development in the Mekong River Basin since the late nineteenth century. The chapter traces key periods of law reform, including early efforts to reshape legal systems in French colonial Indochina; the importation of Soviet legal models to promote socialist economic development; and the mushrooming of donor-supported legal reforms that accompanied the rise of 'neoliberalism' from the 1990s onwards. Throughout these different periods, legal borrowing and transplants have encountered resistance and reorientation, often producing outcomes unintended or unanticipated by the original architects of change. The chapter focuses on donor and expert input into the creation of complex legal architecture promoting 'sustainable hydropower' in the Mekong River Basin. The legal architecture suggests plurality in the goals and objectives for hydropower development. These include fostering economic growth and private-sector investment, but also poverty alleviation, the protection of rights for affected communities and the protection of the environment. Nevertheless, laws, policies and regulations encouraging private-sector financing in hydropower development built partly on donor edifices have often, we argue, served to strengthen the role of central governments in decision-making rather than to strengthen rights, fostering a kind of (neo)liberal authoritarianism in some Mekong country contexts.

Chapter 4 turns the attention of the book to the theme of governance, a socio-political practice into which law frequently blends. We instantiate our governance focus through a study of the MRC, the main institution with a transboundary mandate under a binding agreement for river basin management. In particular, we explore the role of science-based technical support in shaping the ways in which projects are proposed, assessed and regulated. Focusing on a central set of procedures instituted by the MRC for notification and consultation on dams, we reveal the interplay between harder and softer modes of regulation and influence in decision-making. We show that the technical field has its own politics, helping to shape the terms within which projects are considered, and tending towards particular outcomes that are often counterintuitive from a science-informs-policy-informs-decisions perspective. We also show how unexpected openings for societal input on decisions are fostered by otherwise highly structured and institutionalized procedures.

Chapter 5 continues the analysis of governance by examining how Mekong River Basin development activities and conditions have been evaluated and measured through environmental assessment processes, including both hard law and soft regulatory approaches. These environmental assessment processes, seen as an aspect of the technical mode of governance, represent a distinctive dimension of the 'politics of the technical' introduced in Chapter 4. The chapter investigates issues of civil society access to information and public participation in hydropower decision-making, and casts some doubt on the acceptability and legitimacy of actual decisions taken on the basis of these processes. It examines several hydropower developments, including A Luoi in Vietnam and Xayaburi and Don Sahong in Laos, in terms of these access issues. The chapter asks *who* participates in the assessment processes, and who does not (or not adequately), as well as *what* is assessed and *how*. The chapter also asks whether more participatory approaches in environmental assessment will actually lead to more equitable outcomes in hydropower development for 'stakeholders' (presaging discussion of this term, and of the pitfalls and potential of 'participation', in Chapters 6 and 7). This, in turn, raises the question of how equitable outcomes themselves might be defined and what role the law has in achieving them.

Chapter 6 elaborates upon the attention in prior chapters to designing, funding, governing and evaluating by focusing on the theme of transparency. Expectations of, and calls for, transparency play a burgeoning role in Mekong River Basin development, the chapter shows, in part due to the influence of private sector and intergovernmental financiers. Despite this, and the involvement of 'stakeholders' in complex rituals of consultation (in the Nam Theun 2 project, for instance), relevant processes of decision-making have in many respects become harder to 'see', the chapter contends, amid a growing hybridization of public and private power and law. In the case of Nam Theun 2, Chapter 6 shows, transparency policies helped to invest hydropower development with a sense of inevitability, insulating first-order commitments and public-private trade-offs from question, all while fostering fluency in a language and practice of information sharing. Rival understandings of transparency persist still, but these seem to have coalesced, for the time being, around a sense of Lao 'publicness' for which hydropower development is an unavoidable touchstone.

Turning to the practice of contestation, Chapter 7 explores the varied role of civil society in hydropower development on the Mekong, in light of the social, economic and political conditions in each country. It finds that civil society is generally reluctant to challenge hydropower development in the courts for many reasons, although it has occasionally done so in Thailand. Rather, civil society frequently utilizes law in a myriad of other ways in all of the Mekong countries: when engaging administrative and political authorities; in confronting private actors directly or through international accountability procedures; and through legal education, empowerment and research. Even where legal strategies are not at the forefront of resistance, as with the mass protest campaigns by the Assembly of the Poor, law still structures and enables resistance in key ways.

We conclude the book in Chapter 8 with a recapitulation of the key assumptions about the place of law in shaping the Mekong River Basin that we have sought to challenge, and in so doing summarize our key arguments. We also suggest ways in which our reinterpretation of the socio-legal make-up of the Mekong River Basin may be taken on board by different players in the Mekong development drama.

2 River basins as socio-legal arenas

Approaching the Mekong River Basin as a socio-legal arena, as this book does, embeds analysis of the basin in a long tradition of socio-legal scholarship, otherwise known as law and society research. This tradition has not featured to date among legal approaches to this transboundary watercourse. At the same time, such an approach dis-embeds the Mekong River Basin from the predominance of integrated water resources management in discussions surrounding the governance of river basins (Mukhtarov and Cherp, 2014; see also Chapter 3). It stretches the terrain of socio-legal research beyond its conventional nation-state frame of reference, and away from concern with the 'deep' study of a particular ethnographic site. Yet, in its focus on one river basin, and four of the six riparian nations that share it, this book spans a field more modest in dimensions than the 'expansive global perspective' that socio-legal research has recently embraced (Darian-Smith, 2013, p. 4; Merry, 2007). This chapter explains how this book is located in each of these senses. It also explains the understanding of 'law' deployed in this book, the different legal orders with which it engages, and its relationship to such notions as 'governance' and 'development'.

Conventional legal approaches to transboundary river basins

Lawyers and legal scholars have long been concerned with the use and abuse of transboundary water resources. This section will focus on the way in which that concern has been manifest in international law, rather than in national law. Since the latter half of the nineteenth century, states have been submitting to arbitral or judicial resolution disputes concerning water usage or jurisdictional boundaries with respect to rivers shared with other states (St John *et al.*, 1876, p. 413; de Martens, 1888). The twentieth century witnessed the adoption of numerous bilateral and multilateral agreements concerning particular transboundary rivers under international treaty law (Dinar, 2008). It was also a period during which national, regional and international law came to express and grapple with river basins as such, taking account of their hydrological and ecological indivisibility (Teclaff, 1996).

On the international plane, many legal efforts in this regard have taken the form of 'soft law', a term denoting norms that are not formally legally binding or enforceable, but are nonetheless designed to condition or inform decision-making, as explained below. From the late nineteenth century onwards, for instance, the Institut de droit international, and later the International Law Association, were among the global bodies of legal professionals that adopted formally non-binding resolutions and 'regulations' concerning the navigational and non-navigational use of transboundary watercourses (FAO, 1998). In 1966, the International Law Association (a not-for-profit body of legal professionals dedicated to the study and development of international law) adopted the Helsinki Rules on the Uses of the Waters of International Rivers. These guidelines sought to inform how rivers and their connected groundwaters that cross national boundaries may be used. Shortly thereafter, in 1970, the International Law Commission (a body of legal experts nominated by states who are charged by the United Nations with the development and codification of international law) commenced studying the law of the non-navigational uses of international watercourses. These efforts by the International Law Association and the International Law Commission both contributed to the adoption of the 1997 UN Watercourses Convention and produced the 2008 Draft Articles on the Law of Transboundary Aquifers (Watercourses Convention, 1997; McCaffrey, 2007; United Nations, 2008). The former treaty entered into force for its states parties in August 2014. This was thanks to Vietnam's accession to the treaty 90 days beforehand, a step that made it the first of the Mekong countries to become a party.

Legal attention has likewise been dedicated to the elaboration and interpretation of principles of international environmental law and international human rights law with respect to transboundary river basins, including the Mekong River Basin (for example, Bourquain, 2008; Gao, 2014). These have taken the form of both non-binding yet custom-shaping declarations, and binding treaty regimes. The 1992 Rio Declaration on Environment and Development has particular salience in this regard, as does the 1991 Convention on Environmental Impact Assessment in a Transboundary Context (the Espoo Convention), to which none of the nations of the Mekong River Basin are party (Viñuales, 2015; Espoo Convention, 1991). The United Nations Economic Commission for Europe (UNECE) Convention on Access to Information, Public Participation in Decision-Making and Access to Justice in Environmental Matters (the Aarhus Convention) was adopted in 1998 and entered into force in 2001; once again, none of the states of the Mekong River Basin is a party to this treaty (Aarhus Convention, 1998).

In all these initiatives and associated scholarship, the predominant legal approach towards the river basins concerned may broadly be characterized as one of trying to enmesh some pre-existing phenomenon in a legal net, thereby tempering and redirecting (as well as authorizing) sovereign discretion in relation to that phenomenon. A river basin, or some element of it, has been taken as an object of legal knowledge open to doctrinal redefinition. So recrafted, the river basin is to serve as a launching pad, and a boundary setter, for the modulation and channelling of a particular, object-specific sphere of human conduct, interaction and decision-making.

Consider, by way of illustration, the following passage from an article by Alistair Rieu-Clarke, a legal scholar who specializes in transboundary watercourse treaty regimes and has written extensively on this subject, including on the Mekong River Basin (for example, Rieu-Clarke and Gooch, 2010):

> Good governance will not by itself deliver what [integrated water resources management] demands unless certain actions are taken in the legal arena. Firstly, disadvantaged and under-represented groups must be explicitly protected in law if equity is to be credibly achieved. In addition, other silent interests such as those of the environment will demand specific legal provision with respect to both prescriptive regulation and market-based instruments, and mechanisms for ensuring that experience is fed back into the policy development sphere must rely heavily on suitable legal provisions in relation to participation, legislative and constitutional processes and scientific evidentiary standards.
>
> (Rieu-Clarke, 2010, p. 247)

In this passage, 'disadvantaged and under-represented groups' and 'the environment' alike are posited as naturally occurring phenomena. They await legal 'provision' through a combination of explicit protection and 'processes and … standards'. In the meantime, they stand exposed and vulnerable to misrepresentation – 'silent interests' amid the noise of 'policy development'. Only 'suitable legal provisions' can ensure their due representation and that conduct and decision-making are modified to deliver on the 'demands' of integrated water resources management, the prospect of which promises achievement over prevailing alternatives. Account is taken, in Rieu-Clarke's writing, of the significance and vagaries of legal process and the importance to legal work of scientific insights. Nonetheless, his scholarly writing is indicative of an approach focused on expanding the reach, completeness, understanding and efficacy of legal doctrine across worldly terrain and subjects understood to pre-exist that doctrine.

The distinctiveness of a socio-legal approach

The adoption of a socio-legal approach to the Mekong River Basin signals an impulse to 'break out' from the strictures of such a doctrinal approach to transboundary river basins, while it also calls aspects of that approach into question (Cotterrell, 2002, p. 633). At the same time, socio-legal inquiry tends to maintain (sometimes latent) aspirations to build upon and refine conventional doctrinal legal endeavour, as this book also does.

Insertion of a river basin into the tradition of socio-legal thought (or law and society scholarship, in North American parlance) takes as a starting point that river basins comprise phenomena of the 'social'. The 'social' in this context does not merely correspond to 'society' or people in the collective. Rather, the 'social' is at once an epistemology – a form or theory of knowledge – and the outcome of a particular type of knowledge-making work that has been underway in a range of places since the nineteenth century. In other words, the 'social' may be characterized as a set of conditions or 'terms' for 'the way in which human intellectual, political and moral authorities, in certain places and contexts, thought about and acted upon their collective experience' (Rose, 1996, p. 329; Simon, 2000, p. 144).

Lawyers have been among those who participated in this work, generating a specifically legal version of the 'social' alongside that generated by other disciplines. Socio-legal scholarship's turn towards the 'social' draws, accordingly, upon two distinct yet related formulations of that term: both a sociological social and a legal social. In other words, focusing on the 'social' through socio-legal scholarship implies more than linking law and social science. It also entails taking sides in a rivalry between different modes of legal consciousness ongoing since the late nineteenth century. Let us deal with each of these versions of the 'social' within socio-legal thought in turn, beginning with law's reading through or with the sociological social.

The work of socio-legal scholarship is often described as a matter of illuminating the significance for law of phenomena made comprehensible through social scientific work, and the role of law in shaping and impacting those phenomena. Socio-legal scholars have long argued that conditions comprising the 'social' are most successfully studied through legal scholars' engagement with the social sciences, although the pre-eminence afforded social scientific methodologies within the field has declined over time (Simon, 2000). As an expression of the social in this sense, a river basin may be cast both as a site and a precondition for the production of certain laws (as social ground upon which the law may be observed working) and as an outcome of the constitutive effects of those laws (as social ground amenable to legal cultivation and remaking). Socio-legal scholarship has envisaged itself repeatedly touching down on such ground – making contact with 'law as it is lived in society' (or, for our purposes, law as is it is lived in a river basin) – even as it has emphasized the permeability of boundaries between 'law' and the 'social' (Calavita, 2010, pp. 1, 3–4; Sarat, 2014).

Yet, as noted above, socio-legal scholarship also draws on a sense of 'the social' specific to legal thought. 'The social' may be understood, as Duncan Kennedy has argued, as a mode of *legal* consciousness traceable to German-speaking writers of the late nineteenth and early twentieth centuries and 'globalized' thereafter in a number of different variants. That which this consciousness opposed was so-called 'classical legal thought': a phenomenon that 'had no essence' but exhibited among its 'important traits' a propensity to think about law as 'a system of spheres of autonomy for public and private actors with the boundaries of [those] spheres defined by legal reasoning' (Kennedy, 2006b, p. 21). Against classical legal thought so understood, legal scholars across a range of jurisdictions deployed 'an abstraction' – one only later dubbed 'the social' – foregrounding 'modern conditions of interdependence' (Kennedy, 2006b, pp. 37–38). Introduction of 'the social' to law would, it was hoped, remedy 'the failure of coherently individualist law' that, in the international arena, signified 'an international order based on the logic of sovereignty' (Kennedy, 2006b, p. 38; see also Koskenniemi, 2001, pp. 266–352).

'A crucial part of the social critique of classical legal thought,' Kennedy has written, 'was the claim that [classical legal thought] maintained an appearance of objectivity in legal interpretation only through the abuse of deduction.' By purporting to deduce particular decisions from a disinterested system of general principles, classical legal thought embedded individualism and the sanctity of private property in the very structure of legal thought: or so the social critique maintained. Legal recourse to the 'social' worked against such a deductive approach and towards purposive reform. Contrary to classical legal thought, law could not operate objectively and should not be expected to do so; it should, rather, advance a particular understanding of the 'social "is"' (that is, the social condition or selected social facts) as the basis for 'an adaptive "ought"'. Legal consciousness of 'the social', in these terms, characterized law as 'a means to accomplishment of social purposes' (Kennedy, 2006b, pp. 39–49).

Social critique more or less in this vein is something that legal scholars have continued to do to this day, with a wide range of political inflections (from 'right' to 'left'). This is the case even as, in Kennedy's assessment, the 'hope of the [s]ocial, that an institutional mechanism based on the recognition of organized groups … can correctly achieve accommodation, has disappeared' (Kennedy, 2006b, p. 67).

The pursuit of socio-legal research in a river basin, then, casts that basin as much more than a locus of observable social phenomena with which law and lawyers must be brought into closer contact. It casts that basin, too, as a battleground for contending modes of legal consciousness – 'the social' against classical legal thought's rendering of individualism as doctrinal objectivity, roughly speaking.

In summary, the task that socio-legal scholarship sets those who would pursue it, and by which its endeavours may be distinguished, has been pithily digested by Jonathan Simon as follows: 'to overthrow [the] paradigm [identifiable as the "normal science of law"]: first, by revealing the work of the conventional paradigm in shaping the understanding of law held by both the public and many participants; second, by identifying the gaps … between the legal world as described by the prevailing legal paradigm and the world as described by empirical research; third, by producing an alternative paradigm for producing legal knowledge', combining legal techniques with social scientific techniques and, increasingly, techniques drawn from the humanities (Simon, 2000, p. 170).

How to understand 'law' in this book

In aspiring to 'overthrow' (or, in some instances, elaborate and improve upon) conventional doctrinal paradigms of legal knowledge, socio-legal research has contributed to the propagation of pluralist understandings of law. These may not correspond to popular understandings of 'law' or 'the legal' and may, accordingly, be unfamiliar to some readers of this book. For this reason, the understandings of law deployed here merit explanation.

Legal pluralism

In venturing an explanation of 'law' in this book, it is important to acknowledge that, in most instances, socio-legal scholarship has tried to 'escape from legal philosophy's "What is law?" conundrums', making it 'possible to locate whatever we choose to designate as law in a continuum, network or web of regulatory practices and techniques' (Cotterrell, 2002, p. 637). Socio-legal research and writing do not, therefore, yield any one answer to the question what is and is not law; this is simply not among the questions by which scholars working in this vein have been most preoccupied. Rather, socio-legal scholars have typically subscribed to a plural understanding of law.

Briefly, legal pluralism renders law in 'multiple instantiations' by attention to a wide range of practices and agents (Merry, 2013, p. 2). It identifies law with 'routine structuring', and the deployment or invocation of authority of a certain tenor, rather than with a particular institutional or professional provenance (Cotterrell, 2002, p. 639). In pluralist terms, many sorts of regulatory forms may be approached and analysed as law, beyond legislation, executive decrees, the rulings of state-sponsored tribunals or doctrinally sanctioned sources of public international law (treaties, customary international law, general principles of law).

Among plural renderings of law, terminological pluralism is also discernible, as 'law' gets folded into 'regulation', for instance. As Christine Parker has argued, understandings of 'regulation' are, themselves, 'highly pluralist', with regulation signifying anything from 'a type of legal instrument' to 'any intentional "process of controlling behavior with reference to some standard or purpose"' (Parker, 2008, p. 350).

Much ink and effort have been expended in trying to pin down the dimensions of this pluralism, its beginnings and endings especially. Brian Tamanaha has, for instance, lamented that legal pluralists of a 'social scientific' bent have been unable 'to locate an agreed definition of "law"' (he refrains from worrying about the frayed edges of 'regulation') (Tamanaha, 1993, p. 192). Differentiation and precision remain important to socio-legal scholars but, as noted above, these are not typically pursued through investment in a definition of 'law' for all times, places and peoples. The instability of legal pluralism does not, moreover, appear to have undermined its cogency. On the contrary, this very yet-to-be-agreed quality has equipped socio-legal scholars with a capacity to 'see' or 'find' law 'operating ... in innumerable social sites and settings' (Cotterrell, 2002, p. 639).

Occasional handwringing notwithstanding, the enthusiasm of socio-legal scholars for this elastic notion of law does not seem to have abated over time. In a 2013 address, legal anthropologist Sally Engle Merry – a key purveyor of what Austin Sarat calls the law and society 'perspective' since the late 1960s – celebrated the continuing usefulness of legal pluralism as an analytical framework (Sarat, 2014, p. 217). Legal pluralism, Merry remarked, 'shows that law affects social life in many ways, both inside and outside formal legal institutions', including by 'defin[ing] identity'. In a pluralist rendering, law comprises 'a bricolage built up from practice' and 'shaped through interactions among multiple legal orders'. Some of these 'legal orders' will be 'informal' in character, invested with various degrees and types of legal authority, and involving a range of actors beyond those with law school training (Merry, 2013, pp. 2–3). Legal pluralism has, moreover, been shown to have particular salience for understanding law's role in conflict over hydropower development outside the context of the Mekong River Basin (Rajagopal, 2005). In this book, we hold to the view that law in and of the Mekong River Basin is best approached through a legal pluralist lens.

Hard and soft laws, and downplaying the distinction

It is in this socio-legal, pluralist tradition that this book roams voraciously across fields, vocabularies and forms of law, understood in the plural. Legal norms, the operation of which will be explored in this book, vary widely in their quality and origins. In particular, they range in ways that legal scholars have often diagnosed in terms of 'hardness' and 'softness'.

'Hard' law denotes legal norms that are relatively clear and binding, typically expressed in legislation or codified as treaties (a relevant selection of which we outline in an appendix). It is a descriptor reserved especially for legal norms, the breach of which may trigger recourse to formal enforcement mechanisms such as courts within national legal orders, and tribunals or other comparable enforcement mechanisms on the international plane. Section 97 of Thailand's Enhancement and Conservation of National Quality Act of 1992 is one possible example:

> Any person who commits an unlawful act or omission by whatever means resulting in the destruction, loss or damage to natural resources owned by the State or belonging to the public domain shall be liable to make compensation to the State representing the total value of natural resources so destroyed, lost or damaged by such an unlawful act or omission.
> (see generally Kititasnasorchai and Tasneeyanond, 2000)

In contrast, the term 'soft' law refers to legal materials that are weak in the obligations imposed and/or loose in the degree of decision-maker discretion for which they allow (Bilder, 2000, p. 71; Abbott and Snidal, 2000, p. 422). The 1995 Mekong Agreement is, for example, often characterized and criticized with regard to its softness, in view of much of its language being hortatory in tone (see, for example, Bearden, 2010; Hirsch *et al.*, 2006).

Soft law may also designate norms that are not legally binding at all, but none-theless exert quasi-legal force in the sense that they shape conduct and induce some 'compliance pull', often as 'standards' or 'principles' (Chinkin, 1989, p. 851). The World Bank's environmental and social safeguard policies comprise one example of such norms, and another is the Equator Principles, voluntarily adopted by a range of commercial banks. The former have been influential in hydropower develop-ment in the Mekong River Basin, although they have become less so since the Nam Theun 2 project (as discussed in Chapter 3). The latter have been the subject of particular focus and lobbying in connection with Mekong hydropower, even as they are yet to be adopted by banks in Southeast Asia (see, for example, IFC, 2012).

In some instances, soft law may comprise a 'self-contained regime' designed to flexibly and responsively embody the intentions and interests of the parties in a given field (Hillgenberg, 1999). One example is the set of dam development guidelines produced by the World Commission on Dams, 'a global multi-stakeholder body initiated in 1997 by the World Bank and the [International Union for Conservation of Nature] in response to growing opposition to large dam projects' (International Rivers, 2014). A sense of their decisive yet aspirational tone may be discerned from the following passage – one of the 'key principles and actions that the Commission propose[d] all actors should adopt and implement' (World Commission on Dams, 2000a, p. xxxiv):

> Public acceptance of key decisions is essential for equitable and sustainable water and energy resources development. Acceptance emerges from recognising rights, addressing risks, and safeguarding the entitlements of all groups of affected people, particularly indigenous and tribal peoples, women and other vulnerable groups.
>
> (World Commission on Dams, 2000a, p. xxxiv)

In recent decades, international legal scholars are among those who have noted a widespread proliferation of soft law, along these lines, across many fields of legal work (d'Aspremont, 2008; Klabbers, 2006, 2001; Kirton and Trebilcock, 2004; Dupuy, 1991). In light of this, some literature on international and regional law has exhibited a reflex bias towards hard law, including in relation to transboundary water resources. It is frequently contended, for instance, that one should aspire to govern river basins internationally, through legal norms and institutions of the utmost firmness, to overcome the perceived weakness and pliability of soft law (for example, Toope, 2007, p. 108).

Elsewhere, international lawyers have argued for the distinctive merits of soft law over hard law (Chinkin, 1989; Bilder, 2000; Meyer, 2009). In the field of transboundary water management specifically, it has been argued that 'informal, non-binding norms may come to shape practice quite effectively' – more effectively, indeed, than formally entrenched and highly specific 'hard' norms (Toope, 2007, p. 119). Soft law may be seen as less threatening and more likely to be adopted than hard law, while achieving comparable effects on behaviour, as evidenced by its growing prevalence in the environmental realm, for instance (Anton and Shelton, 2011, pp. 64–65).

This debate concerning the relative advantages of 'softer' and 'harder' forms of law seems to us 'stuck in untenable positions' (Trubek and Trubek, 2005, p. 355). Related scholarship is prone to underestimate both the 'hardness' of soft law (the impact of pressures to conform and discursive transformations effected by soft law) and the 'softness' of hard law (the pervasiveness of prerogative power in its midst, for instance) (Trubek and Trubek, 2005, pp. 355–361). The obtuseness of the terms 'hard' and 'soft' obscures the specific normative characteristics and effects of particular legal techniques and materials (Pronto, 2008, pp. 613–614). Debate framed at such a level of generality often distracts, too, from what is at stake in the legal regime or regimes in question.

It is for the foregoing reasons that this book more or less sidesteps debate as to the relative merits of hard or soft law, and chooses for the most part not to foreground or police this distinction. It does not seem a productive line of argument or a useful point of analytical distinction for our purposes. Instead, the book focuses on a wide array of practices in which legal vocabularies, instruments and institutions are invoked in the Mekong River Basin, focusing on their particular and variegated effects rather than any generic typology of their properties.

International, regional, national and subnational laws

The types of law with which this book is concerned vary not just in quality, source and form, but also in scale. While the book's focus is on legal norms and agents operating basin-wide, it touches upon norms specific to subnational communities (those enacted or established by custom at the village level, for instance), national laws and regional laws, as well as international laws. In the final category of the 'international', our inquiry embraces laws that might be dubbed 'transnational', as well as law and policy cast as 'global' in scope. Each of these terms has a distinct lineage and set of implications.

Transnational law was a concept most famously elaborated by the U.S. international lawyer Philip Jessup in his Storrs Lectures at Yale Law School in 1955. In Jessup's version, the term includes 'all law which regulates actions or events that transcend national frontiers. Both public and private international law are included, as are other rules which do not wholly fit into such standard categories' (Jessup, 1956; Zumbansen, 2008).

Public international law, in this context, refers to the law that primarily binds states in their relations with one another. Private international law refers to the law – mainly domestic law or law specific to one nation-state – that addresses the dealings and disputes of individual natural or legal persons (that is, human or corporate persons) spanning more than one jurisdiction. Jessup's 'transnational law' encompasses both.

Global law, on the other hand, implies law understood to be operating more or less without recourse to the nation-state, as 'various sectors of world society … are developing a global law of their own' (Teubner, 1997, p. 7). Rules and conventions that arise from global commercial practice – principles of *lex mercatoria* – present one possible example of global law (although these may be shown to rely upon the state in a range of ways). Laws can be seen to be operating in both transnational and global modes in and around the Mekong River Basin, as subsequent chapters will show.

Selected components and features of these multiple legal orders will be discussed in the chapters that follow. As a reference for readers following this discussion, we have included, in appendices to this book, selective lists of international, regional and national legal instruments and guidelines with particular significance for hydropower development. Not all of the instruments listed in the appendices will be the focus of explicit attention in this book, nor are these tables exhaustive in capturing all relevant sources of legal argument and obligation. Moreover, as this book will go on to show, the instruments set out in these tables encapsulate only a limited set of the effects that law, legal vocabularies and legal agents are having in conflict over hydropower development in the Mekong River Basin. Nonetheless, we include them to convey a snapshot of the multiple legal orders with which this book is concerned, to be intercut, recut, overlaid and sometimes fused in the course of the book's pages.

Laws and policies of national hydropower decision-making

Hydropower dam decision-making in the Mekong countries is structured and informed by an array of formal legal frameworks. By way of systemic background, Thailand established a constitutional rights-based democracy in the 1990s, but continues to be punctuated by military coups, the latest in 2014. Cambodia has formally been a democratic constitutional monarchy since 1993, but in practice is strongly steered by its authoritarian prime minister, Hun Sen. Vietnam and Laos formally guarantee constitutional rights and participation in government, but remain one-party communist states, despite significant economic liberalization.

A selective list of national laws relevant to hydropower decision-making and regulation is provided in an appendix to this book. Despite their specificity, national hydropower laws share certain broad similarities across Mekong countries for a combination of reasons. Factors contributing to this confluence include: the standard-setting influence of donors, financiers and development bank safeguard policies; the Mekong Agreement and its institutions; the interactions and portable, shared vocabularies of transnational scientific experts who work on water, hydropower, environment, land and resettlement; the background influence of international water norms and technical concepts; legal modernization or law reform initiatives that engage foreign expertise and borrow or transplant legal ideas; and transnational civil society advocacy. By these means and others, law in the Mekong surrounding hydropower development reflects the 'hybridization of law' (Rajagopal, 2005, p. 348).

In all four countries, decision-making about large hydropower dams is highly centralized at a national level and coordinated through energy ministries. In some countries, hydropower decisions are even channelled and chaperoned through the highest political levels of government – as in Cambodia, where proposals for hydropower development are both initially and finally approved by the prime minister or senior ministers (Suhardiman, de Silva and Carew-Reid, 2011, p. 82).

Such configurations reflect the high priority accorded to energy production in strategies of economic growth and development in the region (Suhardiman *et al.*, 2011). The key authorities include Laos's Ministry of Energy and Mines (under the Law on Electricity 1997); Cambodia's Ministry of Industry, Mines and Energy (under the Electricity Law 2001); Vietnam's Ministry of Industry and Trade (under the Electricity Law 2005); and Thailand's Ministry of Energy (under the Energy Industry Act 2007).

Hydropower decision-making is nevertheless practically polycentric in each country. A complex amalgam of state actors is involved in the sectors affected by hydropower, including (in addition to energy) finance, investment, foreign affairs, planning, industry, trade, commerce, science, technology, development, agriculture, environment, water, land, forestry, fisheries, dispute resolution and resettlement. In most countries, each of the aforementioned sectors is generally regulated by sector-specific legislation and managed by a corresponding ministry, department, agency or other authority. Occasionally, however, as in Vietnam's Electricity Law 2005, national legislation gives holistic attention to matters of environmental protection, resettlement, compensation and land at stake in hydropower development.

The roles allocated to each actor in relation to hydropower vary between countries and across the different phases of a project: from advice, negotiation and consultation through to the licensing, approval, regulation or monitoring of certain project-related impacts (such as impacts on fisheries, forests, land or people) and, in some instances, the maintenance of relations with surrounding nations. Different elements of decision-making authority can therefore be dispersed across different state entities. Typically, the energy (and often the finance) authority in the nation-state in which a hydropower project is developed will retain overall control of that project, whether formally or in practice. In Laos, for instance, the National Land Management Agency is legally responsible for issuing land concessions for hydropower development projects. In practice, however, these concessions are negotiated by the Ministry of Energy and Mines (Suhardiman *et al.*, 2011, p. 40).

Of the above-mentioned sectors, finance and environment ministries tend to be the most important actors after energy ministries. Financial authorities are closely involved because larger projects are often funded by a combination of public and private finance. Projects may also be guaranteed by the state, either directly or by virtue of the state having made certain enforceable representations to the parties involved, creating risk of financial liability. Concession agreements and associated financial instruments have long-term implications for state revenue, expenditure, financial risk, credit ratings and development policy.

As discussed in Chapter 5, environment ministries have a special role in oversight of the environmental impact assessments (EIAs) that are legally mandated in each country (as well as associated tools such as social, cumulative or strategic assessments). The sequencing of EIA in the approval process can, however, vary. In Laos, for example, an EIA is conducted in parallel with the separate issuing of a concession agreement, such that the late sequencing of the EIA in the decision-making process renders it more difficult to disrupt the project's momentum or to question the earlier economic cost–benefit analysis on which the project has proceeded, thus making the EIA process somewhat redundant (Suhardiman *et al.*, 2011, pp. 41, 133). The Water Resources and Environment Administration is also institutionally weaker than the Ministry of Energy and Mines in Laos, making it difficult for it to utilize the EIA to challenge the latter's views. This is also a problem in Cambodia (Suhardiman *et al.*, 2011, p. 90).

In all countries, subnational authorities tend to play a subsidiary role in decision-making processes concerning large hydropower projects. Such authorities may be (but are not necessarily) consulted by national authorities during the early consideration of projects, and are often engaged at a later stage when EIAs are conducted. Usually their consent is not, however, legally required for a large project to be approved. Once a project is underway, local authorities are then required to cooperate with national authorities, and are often assigned particular tasks during the implementation of the project (as in relation to resettlement of people displaced by the project).

In addition, national laws often permit smaller hydropower projects to be decided at a subnational level, often subject to less demanding criteria and procedures. Such decentralization and pluralism in hydropower can be considerable. In Laos, for instance, there is a graduated scale of decision-making authority, with the largest projects (over 100 megawatts) requiring National Assembly approval, large projects (five to 100 megawatts) being approved by the national government, medium projects (100 kilowatts to five megawatts) being approved at the provincial level, and small projects (less than 100 kilowatts) being approved at the district or municipality level. This is notwithstanding the fact that the cumulative impact of a number of small projects may be considerable, as highlighted in Chapter 5. A graduated scale also exists in Vietnam. So-called pico-hydropower (composed of hydroelectric turbines installed at the household or village level that generate up to one kilowatt of electrical power, or up to five kilowatts according to some definitions) may operate without any formal regulatory approval at all (Smits and Bush, 2010).

National laws also typically provide for regulatory supervision of hydropower projects on an ongoing basis. The level of regulatory independence varies. In Thailand, the Energy Regulatory Commission (ERC) was established by the Energy Industry Act 2007 to separate regulatory functions from the energy policy development roles performed by the Ministry of Energy and the National Energy Policy Council. Other statutory regulators include the Electricity Authority of Cambodia and Vietnam's Electricity Regulation Authority.

State electricity utilities involved in hydropower are also typically established and regulated by national laws. These include the Electricity Generating Authority of Thailand, Electricity of Vietnam, Électricité du Laos and Électricité du Cambodge. Private hydropower investors, builders and operators are typically subject to a combination of contractual legal arrangements governing a specific project, national laws governing the various sectors affected by a project (mentioned earlier), and national corporate, taxation and investment laws. Private investors that are public companies or have otherwise raised funds via a stock exchange will also be subject to disclosure and reporting requirements, in relation to their hydropower investments, under applicable securities laws and stock exchange listing rules. In many instances, as in the Nam Theun 2, Xayaburi and Don Sahong projects discussed in this book, private investors will obtain concessions from governments to exempt them from certain ordinary laws and limit their exposure to subsequent changes in national law (Johns, 2015).

In some of the Mekong countries, there are also specific remedial procedures provided for hydropower-affected communities. For example, the Lao National Hydropower Policy 2005 provides for the establishment of a 'grievance [or] dispute mechanism accessible to project-affected people'. It also requires provisions in hydropower concession agreements for 'immediate corrective action and financial penalties' for developer non-compliance. In other countries, as in Thailand, constitutional or statutory remedies may be available to challenge hydropower decision-making on administrative grounds, for non-disclosure of information or for lack of public participation, for example. The way in which civil society utilizes some of these remedies is discussed in Chapter 7.

'Governance' in environmental management and legal scholarship

Even before one enters into the various modes of 'doing' with which subsequent chapters are concerned, it will be apparent from the foregoing account that law is assigned a wide range of purposes and varying levels of potency in the Mekong River Basin. Law has been – and seems still to be – a central medium of argument as to how the four nations examined have or have not changed, and how they might yet change, individually and in combination. Nevertheless, increasingly apparent in such argument today is a blurring of 'law' into 'governance' – a tendency to which this chapter will now turn.

If 'law' as such has, as suggested above, received relatively cursory attention in scholarship on transboundary water issues in the Mekong River Basin, 'governance' has, in contrast, been a matter of increasing concern, as Chapter 4 will explore further (for example, Rajesh *et al.*, 2013). This is notwithstanding persistent instability regarding the precise scope and meaning of 'governance' (Robichau, 2011). 'Governance', accordingly, demands some disarticulation from 'law' and merits location in the legal and social science literatures in which it has gained purchase. Here, we present three different versions of the term.

First, in environmental management literature, including in relation to river basins, governance has signified a 'dynamic political process' involving 'managing and networking of issues, interests and actors to produce actions that have to be transparent in process and effective in achieving the stated goals of regimes' (Myint, 2012, p. 11). In contrast to 'government', the multisited, multiscalar operations of 'governance' are said to address 'societal problems … appearing to be too interlinked, too complex, but also too overwhelming for any single nation-state to address them alone' (Finger *et al.*, 2006, p. 1). Whether as a matter of extant practice or yet-unrealized aspiration, governance has been taken to imply management, coordination and regularization in the absence of overarching authority. In other words, governance serves as a master descriptor for the efforts of environmental management scholarship and practice to regularize complex phenomena.

Second, in contrast to the descriptive valence it is given in environmental management literature, governance in legal thought has tended to carry an explicit normative inflection. In legal scholarship, the term 'governance' entails viewing 'different sectors – state, market, and civil society – as part of one comprehensive, interlocking system' (Lobel, 2004, p. 375). In addition, some legal scholars generating so-called 'new governance' scholarship deploy the term to put forward an affirmative argument for '[n]ew participatory arrangements … at all levels of government and non-government action' in order to 'democratize political decision-making through the bottom-up production of law' (Lobel, 2004, p. 375; Cohen, 2010, p. 358).

New governance literature promotes institutional arrangements and forms of governance that are designed to foster 'participatory, deliberative, locally-informed, and adaptive problem-solving' (de Búrca *et al.*, 2014, p. 7). In this respect, it has frequently been identified with neoliberal economic and political thought and a turn against the centralized policymaking of a welfare state (for example, NeJaime, 2009, p. 343). On the contrary, however, Amy Cohen has argued persuasively that new governance thinking is far more closely aligned with 'left-of-center liberal legalism' than it is with neoliberalism, albeit in a mode that stops short of explicit engagement with law's 'distributive dimensions' (Cohen, 2010, pp. 360, 386). New governance scholars, Cohen has shown, strive to introduce a 'formalizing, reason-seeking … law-seeking' overlay to the 'informal flexible' model of social life popularized by neoliberal thinkers such as Friedrich Hayek (Cohen, 2010, p. 361). Cohen succinctly captures the descriptive and normative goals that legal scholars in the new governance field combine, in a way that merits direct quotation:

> [N]ew governance literature is explicitly concerned with conceptions of the social or public good that might be achieved through private–public associations and networks compelled by a host of new regulatory frameworks and forms … this work typically combines a normative effort to reconceive governance as a series of networked or 'negotiated relationships' that involve both public and private actors … with efforts to identify, describe, and theorize actual, if discrete and disparate, instances of negotiated self-governance within and across state boundaries that are unfolding on the ground.
>
> (Cohen, 2008, p. 512)

This unapologetically normative dimension sets the thinking of 'governance' in legal thought somewhat apart from its treatment in literature on environmental management, including river basin management. As highlighted by de Búrca and Sabel (two influential figures in new governance scholarship, writing in this instance with political scientist Robert Keohane), scholarship on adaptive environmental governance has tended to focus on the field observation or 'testing' of regulatory arrangements and devices, rather than organizing or advocacy for particular versions of them (de Búrca et al., 2014). In contrast, new governance literature in law advocates *for* governance in a particular mode.

It might appear to some that the second of these modes of governance is unlikely to have had much, if any, purchase in the Mekong River Basin. An impulse towards democratic experimentalism in a 'new governance' mode is not one readily identified with the Hun Sen administration in Cambodia (in power since 1998), or with the Lao People's Revolutionary Party or the Communist Party of Vietnam. Somewhat surprisingly, however, 'negotiated relationships' *have* been a growing feature of hydropower development in the Mekong River Basin, as later chapters – Chapter 6, for example – will make clear. The imperative of trying out 'new governance' arrangements has been among the ways that the increasing recourse of governments to public–private partnerships in hydropower development in the Mekong River Basin has been justified and explained. In a rather unlikely way, Laos has positioned itself in the regional vanguard of creativity of this kind. Thanks in part to its high score for 'export sophistication' (a euphemism for the growth in power exports arising from hydropower development), for instance, Laos was ranked ninth among 24 countries surveyed on a Creative Productivity Index in

Asia, produced by the Economic Intelligence Unit for the Asian Development Bank in August 2014 (Economist Intelligence Unit, 2014, pp. 19, 61).

Third (and finally for our purposes), 'governance' is a term the proliferation of which may be read critically by reference to 'governmentality', and has been so read in some legal and social scholarship. Governmentality directs attention towards 'particular mentalities, arts and regimes of government and administration', including the shaping of experiences of subjectivity, as much as the 'calculated direction of human conduct' through legal and other means (Dean, 1999, p. 2). So construed, governance is a marker for the importance of governmental, subject-crafting 'arts' alongside institutions of disciplinary rule in modernity.

In brief, work on governmentality takes as a starting point the insights of Michel Foucault concerning the emergence, in the eighteenth century, of practices of regulation (and self-regulation) associated with 'transition … from a regime dominated by structures of sovereignty to one ruled by techniques of government' (Foucault, 1991, p. 101). Despite the 'from' and 'to' of the immediately preceding sentence, that work elucidated 'a plurality of forms of government' operating in parallel, and emphasized continuity and interdependence among structures and techniques, rather than their chronological succession (Foucault, 2007, pp. 8–10, 93).

Attention to governance, read in the key of governmentality by reference to Foucault's work, enables one to track the operation of laws within specific programmes and technologies, as it is 'incorporated into a continuum of apparatuses (medical, administrative and so on) whose functions are for the most part regulatory' (Foucault, 1979, p. 144). Governance in this sense denotes a variant of what Sheila Jasanoff and Sang-Hyun Kim have termed 'socio-technical imaginaries', operating as much in the realm of the material as in that of the ideal (Jasanoff and Kim, 2013). Or, as Stephen Collier has argued:

> [G]overnmentality designates the genus – diagrams of political rationality, 'govern-mentalities' – of which specific political rationalities, such as advanced liberalism, are species. The concept is most valuable in understanding the conditions of possibility of certain ways of understanding and acting; for drawing insightful distinctions among diagrams of power; for understanding what is general to diverse governmental forms in disparate sites.
>
> (Collier, 2009, p. 99)

In so saying, it is important to emphasize that Foucault's account of governmentality does not offer 'a generalisable theory or system of thought' through which to grasp the workings of the Mekong River Basin (Neal, 2009, p. 543). Rather, it offers critical purchase on the way practices of rule and regularization surrounding hydropower development have operated to 'educat[e] desires and configur[e] habits, aspirations and beliefs' as well as 'set conditions' for life in the Mekong River Basin (Li, 2007, pp. 5, 258). Pivotal among the conditions so set has been loyalty demanded of those in the Mekong River Basin to the notion and necessity of 'development'. In this respect, successive cycles of law and development work have played a critical role. It is to these that we will now turn.

Law and development work

As Chapter 3 shows, the story of hydropower on the Mekong has long been embedded in political and economic ideas about the 'development' of states, their peoples, their natural resources, and of regions. The notion of development is deeply contested in practice and scholarship. Understandings of development range widely (Trainer, 2000). To some, it signifies national economic growth towards attainment of developed country status (Gereffi, 1996). To others, it is a matter of expanding human freedoms and life opportunities, or fostering 'movement upward of the entire social system' (Sen, 1999; Myrdal, 1974, p. 729). Still others contend that development implies 'takeover of the commons' (Goldsmith, 1992). The concept and goals of development with which one begins tend to drive the means by which it is pursued or engaged, whether through 'modernizing' economic policies, through focusing on 'good governance' and the 'rule of law', or by other routes.

Just as the idea of development is contested, so too is the relationship of law to that idea. Governments and other public and private actors often seek to use law instrumentally to advance economic goals, whether those goals are articulated in terms of development or otherwise. Many legal debates and fields may be shown to have development ramifications. In practice and scholarship, however, law is usually discussed in relation to development in a few key ways.

First, what became known as a 'law and development movement', active from the 1960s onwards, involved Western states and donors in the provision of legal technical assistance to developing states in Latin America and Africa (Trubek and Galanter, 1974, pp. 1065–1069). The aim was to reform laws and institutions to create conditions favourable for capitalist economic development. Following a Western, liberal, rule of law model, emphasis was placed on protecting property rights through contract and property law, and on ensuring the stability and predictability of economic transactions through an independent judiciary. Efforts were also made to change or 'modernize' the legal cultures of recipient states through legal education and training.

These transnational aid practices were accompanied by comparative legal scholarship on law and development that sought to theorize and critique them. One review of the law and development literature in 1977 concluded that there was a 'lack of consensus on everything' (Burg, 1977, p. 528). Critical scholars identified difficulties in the unselfconscious, ethnocentric transplanting of legal ideas and practices from one culture to another; exposed the questionable and partial political and economic assumptions underlying them; criticized the emphasis on formal law and state institutions and the neglect or suppression of customary law and informal structures; observed a lack of participation and engagement by local personnel; and highlighted the lack of empirical data supporting assumptions or evidencing success (Trubek and Galanter, 1974).

By the 1970s, this law and development enterprise came to be increasingly viewed as imperialistic social engineering. It was also widely seen as unsuccessful, since law reform had been unable to transform social realities as envisaged. The critiques highlighted above, alongside changes in American foreign policy, led to a relative decline of law and development practices in the late 1970s and 1980s (Trubek, 2001).

As explained in Chapter 3, the Mekong countries considered in this book were relatively untouched by this initial law and development push of the 1960s and 1970s. Asia was not an area of focus for law and development donors, largely because of the obstructive conflicts in Vietnam, Cambodia and Laos, and the relatively authoritarian rule prevailing in Thailand.

Law and development programmes were, nonetheless, revived in the 1990s at the end of the Cold War, as a means of aiding the transition of former European communist states into market democracies: states that became known as 'transitional economies' (Trubek, 2001). In this iteration, their impact did come to be felt in the Mekong River Basin. Law and development featured, for example, among UN efforts to rebuild post-conflict societies, including, in the Mekong region, in Cambodia. A range of agents supported these initiatives, including Western states, the UN, development banks and NGOs. The expressed goals of such interventions broadened beyond the focus of earlier decades on laws supportive of economic reform to include, for example, efforts to promote human rights, democratization, good governance, anti-corruption and access to justice. Chapter 3 discusses the influx of 'experts' that the law and development programmes of the 1990s brought into the Mekong River Basin seeking to realize this broadened agenda.

Despite the turn to the language of transitional justice and democratization apparent in the 1990s, many law and development programmes retained an explicit or implicit economic reform agenda. Good governance and anti-corruption, for instance, remained strongly identified with the promotion of free markets and private enterprise (Thomas, 1999). An insistence upon economic progress understood in neoliberal terms was central to the World Bank's rule of law promotion throughout this period (Krever, 2011; and as explained in Chapter 3).

From the mid-1990s onwards, the accession of the four Lower Mekong countries on which this book focuses to the World Trade Organization (Thailand in 1995, having been party to the General Agreement on Tariffs and Trade before then; Cambodia in 2004; Vietnam in 2007; Laos in 2013) was also associated with external engagement in law reform. Cambodia's 2004 accession, for example, was conditional upon its adoption of an ambitious National Legislative Action Plan anticipating some 40 new laws; development assistance was deployed to assist in this endeavour (Siphana, 2005). Various technical assistance projects also targeted law reform in specific economic sectors as even the most closed states – Laos and Vietnam – liberalized economically over the course of the 1990s, thanks in part to the impact of World Bank and International Monetary Fund (IMF) structural adjustment programmes.

In theory and practice, plurality and confusion nonetheless continued to characterize the way many of these programmes related law to the ends and means of development (Carothers, 1998; Belton, 2005). Some scholars doubt whether the lessons of the 1960s law and development movement have been learned by those reviving such initiatives since the 1990s (Trebilcock and Prado, 2011; Trubek, 2001). Others caution that the 'emergence of the rule of law as a development strategy has become an unfortunate substitute for engagement with the politics and economics of development policy making' (Kennedy, 2004, p. 167).

Law and development in international legal doctrine

So far we have discussed law and development primarily in terms of the aid, loan and technical assistance programmes of various nations and international institutions, the experiences of the intended beneficiaries of these programmes, and scholarship engaging with this work. In international legal doctrine, the relationship between law and development carries its own distinctive set of meanings. Strictly speaking, there is probably no such thing as 'international development law' as such, although the idea has its proponents (see, for example, Mahiou, 2013). There is, nonetheless, plenty of international legal doctrine that affects development or expresses development aspirations, for better or for worse. For example, international laws on trade, investment, finance, taxation, intellectual property, sanctions and the settlement of commercial disputes all have important impacts on, and convey particular understandings of, development. So too does international law in the fields of global public health, law enforcement cooperation, disaster relief and migration betray development goals and assumptions, albeit far more obliquely.

In light of this, it is unsurprising that international legal doctrine has long been a site of intense political struggle over development, from the 'mandate' system of the League of Nations to the trusteeship system of the United Nations after 1945. Both of these legal regimes were designed to facilitate the transition of specified territories from colonial rule to political independence and both relied upon and mobilized a discourse of development. Development featured prominently in those international legal texts considered foundational of the post-World War II global order. The UN Charter (1945) calls for international cooperation for economic and social progress and development (Article 55), while the Universal Declaration of Human Rights (1948) states that everyone is entitled to a social and international economic order in which their human rights can be realized (Article 28). The modalities of attaining these goals were not, however, prescribed in these texts, other than through discussion in the Economic and Social Council and General Assembly.

Especially from the 1960s onwards, as a result of decolonization, newly independent developing states gained a voting majority in the UN General Assembly and sought to reorient international legal doctrine and institutions affecting them. Developing and socialist states, including some in the Mekong region, pushed for a 'New International Economic Order' (NIEO) to transform economic relations with the developed world. As Anghie has observed: 'The end of formal colonialism, while extremely significant, did not result in the end of colonial relations' (Anghie, 2006, pp. 748–749). In part, this was because international law helped to preserve or validate the economic dependency of some states upon others. The NIEO aspired to counter the market's emphasis on efficiency and openness by reference to other values, such as fairness and distributive equity (Gathii, 2008; Miles, 2013).

The push for an NIEO extracted concessions for developing states in various areas of trade, investment, finance and development law and policy (Mahiou, 2013). A new legal principle was recognized in non-binding General Assembly resolution 1803 (XVII) of 1962 on 'permanent sovereignty over natural resources', acknowledging the rights of developing states to control their own resources. Further, albeit contested, normative changes were achieved in the non-binding Charter of Economic Rights and Duties of States (1974). There were some adjustments in law on the nationalization and expropriation of foreign property favourable to capital-importing or developing states (Higgins, 2003). New development institutions were established by states' agreement, such as the UN Conference on Trade and Development (UNCTAD) (1964), the UN Development Programme (UNDP) (1966), the UN Industrial Development Organization (UNIDO) (1966), the International Fund for Agricultural Development (1977) and the Common Fund for Communities (1989) (Mahiou, 2013).

Developing states were, however, unable to fundamentally transform the international economic order or establish an NIEO. The world economic system remains capitalist and revolves, to some degree, around the post-World War II Bretton Woods institutions (including the World Bank and the IMF and the institutional configurations of which they are part, such as the 'Troika' composed of the IMF, the European Commission and the European Central Bank), notwithstanding the persistent criticism by developing states of these bodies and their work. The world trade system, while much liberalized, still allows protectionism in economic areas, such as agriculture, crucial for many developing states; conflict on this issue contributed to the breakdown of the Doha Development Round of World Trade Organization negotiations (Jones, 2010). The advent of financialization has arguably increased the vulnerability of 'emerging market' economies (Epstein, 2005).

Another doctrinal front has nonetheless since opened in the discourse on law and development, centring on human rights. In 1986, the General Assembly adopted a Declaration on the Right to Development (resolution 41/128). This cast development less as a strategic goal for states or international institutions than as an individual and group right 'to participate in, contribute to, and enjoy economic, social, cultural and political development, in which all human rights and fundamental freedoms can be fully realized' (Article 1(1)). The Declaration sought to make people the subject of development and to impose on states, individually and cooperatively, an obligation to realize development goals. The legal status of the right to development remains controversial, with states divided on it, and conceptual debates about its nature as a 'right' (Alston, 1988). Even so, the shift towards a human rights orientation in development may have had certain normative effects on the practice of states, regional bodies and international organizations (Marong, 2010). Its influence may be discerned in the later chapters of this book.

Alongside efforts to affirm a right *to* development, there has been much recent emphasis in practice and scholarship on the importance of ensuring respect for human rights *in* development processes (Robinson, 2005). This has been the case regardless of whether one takes the further step, as Sen does, of identifying the ends and means of development as the fulfilment of rights and freedoms (Sen, 1999).

Those advancing this rights in development agenda claim that states must ensure that development projects in which they are involved – including dams – comply with international human rights standards. Human rights doctrine would suggest, further, that states must regulate not only their own activities, but also those of private actors within their jurisdiction. This extends not only to civil and political but also to economic, social and cultural rights, as well as collective rights of indigenous and minority groups. These claims have, of course, provoked resistance. Many states – including some in the Mekong region – emphasize the primacy, or the temporary precedence, of collective economic and social progress over individual civil and political rights in development processes.

The push to foster respect for human rights in development has potentially broad implications. It implies that states should ensure that their extraterritorial activities, and the global activities of companies or NGOs organized under their laws, conform to human rights standards. Such demands concern state activities that directly impact on rights extraterritorially (such as the deployment of state agents). They also concern actions that may facilitate rights violations by others, as where the state supports the private sector financing of a foreign project with adverse human rights outcomes. States may also bear specific obligations under the International Covenant on Economic, Social and Cultural Rights, for example, to cooperate internationally to realize economic, social and cultural rights in developing states.

Quite a few international organizations have been influenced by this movement to mainstream human rights in development. Indeed, this has been a feature of the so-called post-Washington Consensus, with its focus on egalitarian and democratic development (Stiglitz, 1999). Thus, the World Bank group underwent protracted internal legal debates about squaring its economic development mandate with what some saw as extraneous human rights considerations (Palacio, 2006). The evolution of social and economic safeguard policies and complaints procedures within development banks reflects the institutionalization of human rights, albeit subject to significant limitations. These policies have, in turn, informed private sector initiatives affecting development financing, such as the Equator Principles mentioned above – voluntary social and environmental guidelines designed for the banking sector (Sarfaty, 2005).

To some, this institutionalization of human rights amid law and development has had a tempering or diluting effect, filtering new considerations and questions through old certainties. To others, the spread of international human rights doctrine has served to regularize and defuse political claims and conflicts surrounding development, such as those once articulated in and around calls for an NIEO. In part under the auspices of human rights, demands for wholesale restructuring of the economic order towards greater distributive justice seem, at times, to have slipped quietly off the international legal table (Kennedy, 2004, pp. 8–10).

Without coming down, in a wholesale way, on either of these rather schematically stated 'sides', it will be apparent over the course of this book that the burgeoning emphasis on human rights to and in development has helped to sustain the economic reform agenda that has long underpinned the law and development movement, as much as calling it into question. Arguments for uniform property rights formalization, for instance, on which law and development advocates have long dwelled, can be and have been discursively reformulated quite easily in human rights terms (Jacobs, 2013). Similarly, discursive confluence emerges, as Chapter 6 describes, between the emphasis of states and developers on transparency in connection with economic development, and the calls for greater transparency made by those protesting development projects in the Mekong River Basin.

Moreover, synergies of this sort rarely play out according to either a human rights-focused or an economic modernization script. 'Bright line' property rights may sometimes aid those fighting land-grabbing for development in the Mekong River Basin, for example, but in many instances, and for a wide range of reasons, that will not be the case (see, generally, Fitzpatrick and McWilliam, 2013). The same unforeseeability afflicts transparency claims, as Chapter 6 shows. In summary, human rights and development are as frequently fellow travellers – and amicable, mutually supportive ones at that – as they are adversaries, yet their relationship can be unpredictable.

The post-Washington Consensus notwithstanding, arguments surrounding the development imperative and its relationship to law thus remain unresolved. This irresolution will be apparent throughout this book. States, developers and those identified with civil society continue to promote and to contest hydropower development in part by recourse to claims about law and development. Also apparent in this book is the borrowing of development-related legal ideas and vocabularies from one jurisdiction to another, with divergent effects, as in the area of EIA, on which Chapter 5 focuses. Conflicting ideas about 'sovereignty' over natural resources development – and thus politico-legal control over such resources and mechanisms appropriate (or not) for their monetization – are also central throughout this book. The human rights and development theme highlighted above is similarly recurrent; the safeguard policies of development banks and donors, corporate social responsibility initiatives, and the work of Thailand's National Human Rights Commission are just some settings in this book in which human rights arguments about development feature, and in which the ambivalence of what 'human rights protection' could and should entail becomes apparent.

Conclusion

The adoption of a socio-legal approach to the Mekong River Basin is plainly, then, about much more than introducing social data into a legal calculus. Although it might occasionally slip into one or other of these registers, this book's goal is neither to measure social realities against legal yardsticks nor to reveal the perennial shortcomings of law in answering worldly demands. Rather, the socio-legal approach put forward here is adversarial. It questions some commonly held assumptions about the actual and potential role of law in the Mekong River Basin. In particular, it takes issue with two prevailing notions.

First, this book refutes claims that the Mekong River Basin awaits the messianic coming of law. Against prevailing ideas that life in the Mekong River Basin requires greater juridification – whether to displace unrestrained sovereign politics or to normalize activities in the basin against international benchmarks – this book presents an account of the basin as legally saturated, with arguments concerning hydropower in particular shown to be, already, highly juridified in various ways. Indeed, it depicts a wide range of actors in the Mekong River Basin articulating often sophisticated and highly nuanced accounts of law, its limits and its possibilities. If there are decisive answers that might be found to the conundrums confronting many in the Mekong River Basin in relation to hydropower (as they grapple with aspirations for economic growth and political change, while seeking to maintain livelihoods), they do not reside among some catalogue of legal norms already on hand, only yet to be fully translated or implemented.

Second, this book questions the reliability and self-evidence of development templates, especially with regard to their legal dimensions. As this book shows, decision-makers concerned with development in the Mekong River Basin – whether within governments, the Mekong River Commission or affected communities, or among donors and NGOs – have long relied upon models in seeking to deliver on reform and development promises. Financial models, scientific models and models of public participation, environmental assessment, poverty alleviation and resettlement: daily these shape 'socio-technical imaginaries' in the Mekong River Basin (Jasanoff and Kim, 2013). Such models act as vehicles for the conveyance and entrenchment of normative understandings, with tremendous import. The argument of this book is that these representational schemas should be recognized for their political significance and their negotiability, and navigated tactically on that basis, rather than being dispensed with, on the one hand, or extended unthinkingly, on the other. We seek to show that socio-legal inquiry, mobilizing a plural understanding of law, can aid this endeavour.

Framed in the affirmative, this book's socio-legal approach also enlivens a sense of law that has been in abeyance in scholarship concerning the Mekong River Basin, to the extent that it has been discernible at all. That sense of law is as something with which a wide range of people is making and doing a wide range of things – as, in some respects, an unruly and ungovernable material or form of knowledge. At the same time, this book holds in view a question as to whether all this 'making and doing' can and should be understood instrumentally – as a matter of handling and deploying law wilfully as a tool, to accomplish certain tasks or purposes. Much in this book confounds the expectation that law can be used in a directed and directive way with any confidence as to outcomes. Many intended consequences of the 'use' of law emerge from the law stories in this book, but so, too, do many unintended consequences. It is, ultimately, through the interconnected stories of legal claims and counterclaims, instruments and institutions, allegiances and disaffections in the Mekong River Basin that we seek to enliven this sense of law as unwieldy and yet still filled with potential. So, it is to these stories to which we will now turn, beginning with an account, in Chapter 3, of the seemingly endless cavalcade of experts who have made the Mekong River Basin their field of expertise and locus of authority.

3 Making the Mekong River Basin

Donors, developers and experts

The significant role that international funding and expert knowledge have played in shaping development and governance in the Mekong River Basin has been much discussed (Bakker, 1999; Hirsch *et al.*, 2006; Molle *et al.*, 2009; Sneddon and Fox, 2011; Sneddon, 2012; Öjendal *et al.*, 2011). Less well explored is the role of law and legal actors in creating and circulating norms that shape how development and governance are understood, practised and contested. In particular, the impacts of transnational laws and institutions and their normative effects on states, domestic and international organizations, regional bodies, and diverse public and private actors have not featured in stories about changing conceptions of the river basin (cf. Bakker, 1999; Sneddon and Fox, 2012). As noted in Chapter 2, this lack of attention to transnational legal influences has sometimes led to the perception that development in the Mekong River Basin is lightly or inadequately regulated by international law and more heavily dependent on 'informal negotiation' and 'ad hoc' interventions (Interview 2, 2011).

As this chapter demonstrates, various law reform agendas and 'rule of law' interventions have in fact been crucial in shaping the development enterprise in the Mekong River Basin. Foreign governments, multilateral agencies, technical experts, corporate interests and NGOs have all worked with (or exerted pressure on) national governments and regional bodies to create legal frameworks and institutions mostly conducive to their preferred conceptions of development. Legal reforms have addressed many areas of private and public law at the level of bilateral, regional and multilateral legal relations and institutions. Even customary law has been a focus of the development enterprise.

In this chapter, we trace key periods of legal reform and set these within broader, shifting geopolitical contexts, including colonialism, socialism and neoliberalism in their Mekong-specific permutations. We track the contributions that funding bodies and communities of experts have made to changing conceptions of the Mekong River Basin and, in particular, to the legal norms and institutions that have governed development on the Mekong River since the late nineteenth century.

We begin by outlining some early experiments in transplanting Western legal models into French colonial Indochina, and the subsequent pushback against such experiments by communist governments (which also experimentally borrowed legal ideas from elsewhere). Despite the unreceptiveness of Mekong countries to

Western-inspired law and development models for much of the twentieth century, US engineers from the Tennessee Valley Authority managed to export to the Mekong River Basin a modernist project of 'hardware' development. This took the form of concrete structures that could impound, divert and store water for energy – and associated planning and coordination practices modelled on the Tennessee Valley Authority's programme of 'multipurpose' development. Although the economic potential of the Mekong River featured strongly in colonial and postcolonial state discourses of development, commercial exploitation of the river was nevertheless limited due to the broader geopolitical forces that came to bear during the region's experience of prolonged conflict and war.

Following the reaping by nation-states of a 'peace dividend' or economic boost in the early 1990s after the end of the Cold War, global attention was again directed to legal reform in the Mekong region. Led by multilateral and bilateral donors and lending institutions, these efforts have been influenced by neoliberal ideas concerning the importance of law for the operation of private markets, and by development policies oriented towards market liberalization and foreign investment. In the Mekong River Basin, international funders and experts have played a critical role in building a complex and plural legal architecture to support cross-border hydropower schemes, with a view to developing a regional power grid since the mid-1990s. Reforms have been oriented towards deregulating national electricity sectors to build competitive regional energy markets, and promoting private sector involvement in these sectors.

In recent decades, neoliberal conceptions of development have, however, come under pressure from new ideas about 'human development', 'public participation' in decision-making and 'good governance', including in the management of water and other natural resources (see Chapters 4 and 5). This, together with changing practices of donors, has exerted pressure worldwide to change the dynamics of the role of law in development, with regards to both its means (or processes) and ends (or objectives), and whom it serves (whether states, private entities, communities or individuals). Among donors, law is no longer seen only as a means to create and protect markets and foster economic growth, although this continues to be a primary concern. Many in the development sphere have come to believe that law must also promote social development goals, including democracy, human rights and social and environmental justice; indeed, development is now often seen as the 'freedom' to choose our own development futures and expand our life opportunities, both socio-economic and civil and political (Sen, 1999). Reconfigured in this way, law reform is not just a means to achieve human development, but also an end in itself (Trubek, 2006).

As the hydropower industry in the Mekong River Basin has grown to become a commercial operation led by private sector actors, international donors have similarly shifted their focus to strengthening legal frameworks and surrounding infrastructure for achieving ecologically sustainable and socially equitable hydropower development. The underlying logic of many involved in the development enterprise is that hydropower development can achieve economic growth and social development objectives *provided that* Mekong countries build strong, autonomous, authoritative and legitimate legal systems.

As this book makes clear, however, laws and institutions enabling and regulating hydropower development in the Mekong River Basin still often fail to serve and support the interests of communities impacted by hydropower development and thereby fail to realize the promise of human development. This is so despite the substantial incorporation of international norms on human rights and the environment into national legislation in much of the region (see Chapters 5 and 6) and the promotion of such norms by an active and transnationalized civil society (see Chapter 7). Rather than protecting rights, laws and policies built partly with donor support have often served to strengthen the role of central governments in decision-making, fostering a kind of (neo)liberal authoritarianism in some Mekong country contexts, as we discuss below.

Law and colonialism in French Indochina

In the period of French colonial expansion into Southeast Asia between the late nineteenth and mid-twentieth centuries, the Mekong River offered commercial opportunities and invited the prospect of French advance into new territories. To the nineteenth-century French explorers who travelled up the Mekong seeking to establish a navigable trade route into China, the river acquired an almost mythical quality as a conduit that would lead to unimaginable riches and bring glory to the French Empire (Osborne, [1975] 1997). However, French hopes for the Mekong to serve as a commercial highway northward to China were never fulfilled due to the numerous formidable rapids on the river that impeded navigation.

It was, rather, the Mekong Delta in South Vietnam that became the crown jewel of French Indochina, echoing its apparent significance to earlier rulers, such as those of the trading, canal-building kingdom of Funan between the first and sixth centuries (Bishop *et al.*, 2004). This is where the first modern hydraulic works began (Nguyen, 1999). Applying river-dredging technology from the West, the French colonial administration built a network of canals and dykes for drainage and irrigation, converting previously uncultivable swamplands into rice fields and areas suitable for human settlement. Along with rubber, commercial rice exports became a key source of revenue for the French colony up until commodity prices crashed during the Great Depression of the 1930s (Hardy, 1998).

The canals also served as important waterways for the growing number of vessels that transported agricultural products from different parts of Indochina to the port of Saigon. Indeed, the Mekong River and tributaries were key arteries linking France's colonies in Vietnam to its protectorates in Laos and Cambodia, giving unity to the French Empire in the region (Nguyen, 1999). Securing access to the Mekong River for navigation and commerce was a central feature of a number of international agreements between French Indochina and Siam (Thailand). The Treaty of Friendship, Commerce and Navigation (14 February 1925) and the Franco–Siam Convention Concerning Relations Between the Two Countries (25 August 1926) established a Permanent Franco-Siamese High Commission for the

Mekong tasked with defining borders (for which the Mekong River was to form a frontier) and establishing rules and regulations for navigation. As France began losing its grip on power in mainland Southeast Asia, regulating navigation on the Mekong River and the maritime harbour of Saigon was the subject of two more treaties between France, Thailand and the newly independent 'Associated States' of Vietnam, Laos and Cambodia, concluded in 1950 and 1954 respectively.

Besides introducing norms governing navigation on transboundary watercourses, as well as water management technologies for commercial-scale agriculture, the French created many other laws and legal institutions to regulate the social, political and economic lives of their colonial subjects. Much has been written about the assumptions and contradictions of the colonial civilizing mission, particularly with regard to the role of law (Fitzpatrick, 1992; Anghie, 2005). The introduction of modern law was fundamental to the colonial project of remaking 'primitive' cultures along the lines of political and economic development in France, including bestowing freedom and democracy upon those cultures. Yet these same technologies of governance were also central to the process of colonial domination and enabled the forcible acquisition and underwriting of an empire. As subjects of the French colony, most Vietnamese were denied basic civil liberties, including freedom of assembly and association, rights to travel without permits, equal opportunities for education and employment, and the prospect of writing without censorship (Long, 1973). People subject to French rule were taxed at exorbitant rates and subjected to *corvée*, or unpaid labour (Long, 1973; Hardy, 1998). People were commonly imprisoned without a fair trial, particularly once rebellion against the colonial regime took root (Zinoman, 2001).

While law often served as an instrument of control for colonial authorities and local elites, it would be misleading to portray French colonial authorities as omnipotent administrators able to unleash the full power of the law onto populations with predictable and desired effects. Colonial policies did not simply emanate from Paris for immediate transplant and uptake into the colonial peripheries and the countryside. Rather, French laws and legal institutions interacted with pre-existing legal norms and power structures and, in the process, were transformed in their locally specific effects. Not unlike the plight of legal reforms that would follow in subsequent decades, accounts of legal 'transplants' in French Indochina are full of tales of resistance, distortions, irritants and downright failings (Hooker, 1975; Popkin, 1976; Gillespie, 2004, 2005; Rose, 1998).

Gillespie (1994, p. 329), for example, notes that French rights-based law 'did not find a receptive social soil' in Vietnamese neo-Confucian legal culture. The French colonial administration quickly realized that a continuation of Vietnamese customary law would be a necessary condition to rule and passed a decree on 25 July 1864 providing for its continued application (Hooker, 1975, p. 229). A parallel legal system was thus created, wherein a civil law system governed French citizens and a small group of elite Vietnamese, while the Nguyen Code (a Confucian code, adapted from the Qing Dynasty that predated the incursion of French law) and customary law governed the rest of the population (Gillespie, 1994, p. 334; Rose, 1998, p. 96).

Those who did acquire knowledge of the French legal system – Vietnamese nobility, bureaucrats, judicial officers, lawyers and merchants – were often able to use it to their own advantage (Rose, 1998). It was not French concessionaires who benefitted most from the rich agricultural land that was made available through massive drainage programmes in the Mekong Delta: rather, it was a growing class of land-rich Vietnamese well-versed in colonial legal and administrative systems, some of whom were able to manipulate rules to appropriate land from peasant families (Rose, 1998; Bergling, 1999; Osborne, 2000).

It was the growing discontent of the landless rural peasantry that communist insurgents would later tap into in their fight against French domination. Tax reforms introduced by the colonial administration, to ease the debt burden on poor farmers, proved futile because the village elites did not support reform and instead continued to collect tax as they had always done (Popkin, 1976). Tax reform resulted in distortions that increased rather than diminished the tax burden on poor farmers, further inflaming the conditions for revolution. These were some of the early lessons in challenges surrounding any 'export' or 'import' of legal models (for later studies on this theme, see Dezalay and Garth, 2002).

Socialist-style law and development in the Mekong

Marxism–Leninism first entered French Indochina in the 1930s via the Soviet-sponsored Indochinese Communist Party in Hanoi. Under the leadership of Ho Chi Minh, Vietnamese nationalist–communist forces grew to become an army that defeated first the French and later the United States in the Indochina Wars (Lawrence and Logevall, 2007; Brocheux and Hémery, 2009).

Following the 1954 Geneva Accords, Vietnam was divided into two pro- and anti-communist zones, which produced an eclectic legal landscape. The northern Democratic Republic of Vietnam adopted a socialist legal system in accordance with a Soviet-style command economy. The southern Republic of Vietnam retained the French legal system, with some US influences (Rose, 1998, p. 97). Cambodia and Laos were also granted independence and became embroiled in their own Cold War proxy wars. Despite aggressive military action by the United States and its allies to defeat communist forces in Vietnam (including unsanctioned aerial bombing and spraying in Laos and Cambodia on the fringes of the Vietnam War), all three countries that formerly made up French Indochina eventually fell under communist control in 1975.

These communist regimes brought land and industry under state and collective ownership and adopted centralized party rule systems modelled on those in the Soviet Union and China. Those with links to the old regimes, including lawyers educated in France, were dispatched to 're-education' camps and in Cambodia many were executed. Thousands of people went into exile. In Vietnam and Laos, a new generation of public administrators was sent to the Soviet Union to receive training in socialist ideology, including in socialist-political legal thought. The experts charged with legal responsibilities were selected based mainly on their 'excellent political and revolutionary credentials', rather than on the basis of their formal legal training (Nicholson and Low, 2013, p. 9). Curiously, the same unreflective legal borrowing and import–export took place in Vietnam in the 1960s and 1970s under communism as had occurred under French colonialism in the decades prior (Gillespie, 2005). Unlike the French liberal rights-based model of the colonial era, which classified individuals in relation to property and cast law as a system of dispute resolution between private individuals, the Soviet legal template emphasized class consciousness and nationalism, and oriented law towards state interests and those of the Party (as representative of the will of people).

Initially, the new communist regimes displayed contempt for or disinterest in law, with few attempts to formally repeal colonial laws or build a new legal system. In post-war Vietnam, an explicitly 'antilegalist attitude' was discernible among the country's new leadership – a conflation of 'anti-colonial Confucian-influenced ... antipathy for legalism' (Gillespie, 1994, p. 332) with a Marxist belief in notions of 'justice' and 'rule of law' as expressions of false consciousness. It was not until the Third Party Congress in 1960 that Vietnam adopted the Soviet 'socialist legality' doctrine (Gillespie, 2005). This concept was defined in Vietnamese writings during the 1960s as 'a tool of proletarian dictatorship ... to defeat enemies and to protect the revolution and collective democratic rights to organise, manage and develop a command economy' (Gillespie, 2005, p. 47). Since the Communist Party determined the content of law, law was not above the state but emanated from it. Further, '[t]he conflation of Party policy and law enabled the Party and state to use law as a "management tool" ... to adjust or balance ... social relationships', in ways that resonate obliquely with later iterations of global law and policy on which much of this book focuses (Gillespie, 2005, p. 47). In other words, under socialist legality, law could facilitate but scarcely constrain power.

The Pathet Lao who took control in Laos in 1975 drew much of their inspiration and direction from Vietnam, including its political and legal institutions. However, because the unified Lao state and Lao nationalism emerged from, and are rooted in, elite-led opposition to French colonialism, the anti-colonial movement in Laos was driven by educated classes rather than achieved through 'class struggle' (Lockhart, 2003, p. 152; Ivarsson, 2008). Partly for this reason, Marxist ideology has sat somewhat more uncomfortably as a source of party legitimacy in Laos than in Vietnam, despite various attempts to reconstruct a national historiography that better fits the ideal of a revolutionary movement with deep roots in popular resistance to foreign domination (Stuart-Fox, 2003; Ivarsson, 2008). Perhaps also for this reason, less emphasis was placed in Laos on building an ideologically coherent socialist legal state apparatus. It took 16 years for the Lao PDR to adopt a constitution after the communists came to power and there was very little other legislative activity over the same period (Bogdan, 1991; Huxley, 1991). By the early 1990s, ideological justifications for the state's reconstructions of history, society, economy and law were 'already breaking down under the impact of international political and economic forces that were opening Laos to modernizing currents it was ill prepared to negotiate' (Stuart-Fox, 2003, p. 89). Laos, it seems, was ripening for another round of law and development reforms.

Meanwhile, in Cambodia, 'socialist law' took a sinister and tragic turn towards 'lawlessness', enabling the Khmer Rouge to commit genocide and crimes against humanity. A UN intervention in Cambodia in the early 1990s, discussed further below, was eventually made possible by the end of the Cold War, the unquestionable political illegitimacy of the Pol Pot regime from 1975 to 1979 and the ongoing conflict after Vietnam's withdrawal in 1989 (following its invasion of 1979). The UN objective included not only to bring peace and an end to conflict, but also to install a new legal and institutional architecture to rebuild the country.

Compared to colonialism, where law often justified and legitimated conquest and control, law played a relatively modest part in sustaining the grand myths and narratives of communism (Krygier, 2014). Certainly, there was the same legal borrowing and replication of development templates with little regard for local conditions of a kind that we have seen elsewhere. Law under the communist regimes from the 1950s to the 1980s was, nevertheless, at best seen as a tool for ensuring social compliance or promoting Communist Party policy. More often, legal norms and institutions were treated with suspicion or simply ignored. It should also be noted that throughout this period (and to the present day), customary law and practice played a role in determining the allocation of rights to use and manage common resources, such as forests and wetlands, and in solving disputes at the local level (Ahmed and Hirsch, 2000), but under communist regimes they were not considered to have any legal register.

In this context, and in the midst of an ideological war, there was simply no room for legal ideas and programmes associated with the first law and development movement, emerging out of the United States in the 1960s and 1970s, to gain any traction in the Mekong region. Even Thailand, which maintained its independence from France and became a staunch US ally during the Cold War, did not provide receptive ground for US-inspired legal reforms. Despite receiving large sums of international development assistance during this period, Thailand was ruled by a strong executive military authority that did not have much regard for institutional and legal reform in the development process (Phongpaichit and Baker, 2009).

While US legal models did not penetrate easily into the national legal cultures of Mekong countries for most of the twentieth century, US engineers from the Tennessee Valley Authority were nonetheless able to export to this region a vision of modernity, economic reform and associated legal infrastructure based on large-scale river basin development. As examined in the following section, it was this common vision for development – in which large dams were axiomatic – upon which Mekong governments and Western donors were able to converge between the 1950s and the 1970s, creating an international legal framework for cooperation and development that carries on to this day. Examination of this framework – an important precursor to contemporary arrangements – requires us to step back in the chronology of our narrative to the mid-1950s.

The making of an international water resources management regime in the Mekong

In the brief and uncertain period of peace following the signing of the 1954 Geneva Accords, US policymakers were beginning to formulate plans involving the

Mekong River to contain the spread of communism. These plans were informed by studies conducted by the US Bureau of Reclamation (1956) and the UN Economic Commission for Asia and the Far East (ECAFE) (1957), which identified potential sites for a cascade of 'multipurpose' dams to harness the Mekong's waters for electricity, irrigation and flood control. The studies produced a vision of Mekong River Basin planning and development on a grandiose scale, based on the Tennessee Valley Authority (TVA) model.

Such was the persuasiveness of the TVA model that it garnered the support of Mekong governments, then on opposite sides of the ideological divide, which rallied behind a common vision for development. The global embrace of the 'Mekong Project', as it was then called, is symptomatic of a time when there was great optimism about the capacity of governments to realize complex development objectives through technical planning and engineering prowess (cf. Schaaf and Fifield, 1963). During the Cold War, the TVA became a blueprint for aid programmes to reduce poverty in regions strategic to the US fight against communism, with the Bureau of Reclamation operating as 'geopolitical agent' (Sneddon and Fox, 2011, p. 450).

The Mekong Committee for Coordination of Investigations of the Lower Mekong River Basin (Mekong Committee) was established in 1957 under the auspices of ECAFE. Its membership included Cambodia, Laos, Thailand and South Vietnam. North Vietnam, Myanmar and China (at the time not a member of the United Nations) did not join. The committee had a principal mandate in water resource development planning, with core funding from the United States, Japan and European countries. The Mekong Secretariat created alongside this committee was expected to mobilize additional technical and financial resources to support water resource infrastructure development. The statute forming the Mekong Committee, drafted with the help of the UN Office of Legal Affairs, became in effect the first 'constitutional' document of the Mekong River Basin regime (Browder and Ortolano, 2000, p. 505) of which the Mekong River Commission (MRC) is a later iteration.

Throughout the 1960s, the Mekong Committee engaged in a massive programme of investigation and planning surrounding prospects for water resource development. This programme followed the recommendations of Lieutenant-General Raymond A. Wheeler, a retired senior officer from the US Army Corps of Engineers. Plans for projects were developed, refined and revised in studies and reports coordinated by the Mekong Secretariat, a body composed largely of foreign experts from the UN and donor countries. Even so, by the late 1960s, it was clear that Lower Mekong development was not proceeding as fast as had been hoped. Enthusiasm from the international community started to wane as the challenges of constructing complex mega-projects on transboundary river basins in the midst of war and conflict began to surface (Jacobs, 2002). By 1970, US$60 million had been spent on planning and investigation work alone, but no construction had started on the Mekong River itself (MRC, 2013). The Nam Ngum dam in Laos, completed in 1971, was the only internationally significant dam developed under the framework of the Mekong Committee. As the first cross-border power project supplying electricity to Thailand, it became a model of how countries in the Mekong region could overcome political tensions through economic integration and infrastructural linkage.

A 1970 Indicative Basin Plan released by the Mekong Secretariat tried to shift the emphasis from planning to implementation, breathing new life into the projects then on the drawing board. This plan presented a menu of water resource development projects to the donors, including a cascade of seven large dams on the Mekong mainstream capable of storing a third of the Mekong's annual flow (Mekong Secretariat, 1970). International donors, it seemed, were ready to invest in the 'Mekong Cascade'. In the words of a consultant working for the Mekong Secretariat at the time, 'there was a feeling that USAID was going to dump a lot of money. Australia was dumping money; everyone was dumping money in the Mekong' (Interview 25, 2011).

Yet a remaining concern, mainly among donors, was that without a set of legal principles outlining the rights and responsibilities of riparian countries, collaboration between the jurisdictions needed for the implementation of transboundary projects would be difficult. With encouragement from donors, and after intensive negotiations (Browder and Ortolano, 2000), the four riparian countries signed the 1975 Joint Declaration of Principles for Utilization of the Waters of the Mekong River Basin. The 1975 Joint Declaration gave meaning to the principle of 'reasonable and equitable use' of the 1966 Helsinki Rules on the Uses of the Waters of International Rivers by prohibiting 'unilateral appropriation' of mainstream waters without 'prior approval' (Article 10), particularly for extra-basin diversions. In other words, the consent of all member states was required when any mainstream diversion was proposed. While the 1975 Joint Declaration was more a statement of intent than a binding treaty (Browder and Ortolano, 2000, p. 509), its suggested 'veto power' became a major sticking point when its successor instrument was negotiated in the early 1990s: the legally binding 1995 Mekong Agreement (Makim, 2002). As explored in Chapter 4, the softer Procedures for Notification, Prior Consultation and Agreement, adopted in 2003 under the 1995 Mekong Agreement, have been mired in controversy in recent years, partly due to ongoing legal ambiguity and unresolved disagreement on this matter.

Importantly, the impetus for incorporating legal norms from the Helsinki Rules into the Mekong legal regime came as much from demand by Mekong governments for law as from the demands of donors. A former legal adviser to the Mekong Committee explains:

> The role of law back then was perceived as very important, if nothing more than as a guide. The [member countries] thought, 'this is what countries do, this is what happens in other parts of the world', so they wanted to have it [the 1975 Joint Declaration]. Of course it didn't hurt any that President Johnson in the US felt that the Mekong could dwarf the TVA if they ever got their act together … the four countries basically just agreed with whatever the Secretariat proposed because it all had to do with donor funding.
>
> (Interview 25, 2011)

Just when all the parts seemed to finally be in place to transform the vision of large-scale Mekong dam development into a reality, a series of communist victories in the region suddenly shifted the geopolitical landscape, and the Mekong Committee collapsed in 1975. Internal turmoil in Cambodia meant that it was no longer able to participate and in 1978 the committee became a three-member 'Interim Committee' composed of Laos, Thailand and Vietnam. Plans for the Mekong Cascade were shelved.

By the time Cambodia was readmitted to the committee in 1991, the appetite for dams on the Mekong mainstream had diminished. A shift in the region's ecopolitics made the environmental impact and scale of resettlement required by the Mekong Cascade, as per the 1970 Indicative Basin Plan, much less politically acceptable. In response to these concerns, modifications to the Mekong Cascade plans appeared in the 1987 Revised Indicative Basin Plan and again in a 1994 study by the Mekong Secretariat. The 1994 study reconfigured the original Mekong master plan into a cascade of 11 'run-of-river' dams, reducing the resettlement numbers substantially.

Despite these adaptations, the changed geopolitical context made tributary developments a priority over basin-wide cooperation for the nations of the Lower Mekong, giving rise to tensions between national and regional development agendas (MRC, 2013). In particular, Thailand was concerned that the legal framework under the Mekong Committee, particularly provisions giving 'veto power' to neighbouring countries under the 1975 Joint Declaration, could stand in the way of its then current plans to extract water from the Mekong River in the dry season through its Khong-Chi-Mun irrigation project, discussed in Chapter 1. Vietnam, on the other hand, concerned about the impacts that reduced flows would have on the Mekong Delta, wanted to maintain firm rules for water allocation (Makim, 2002). Persistent disagreements about the constitutional structure and allocation of power within the 1975 Mekong regime made it clear that a new agreement was needed.

In anticipation of bringing such a new agreement into being, US legal advisers provided legal training to the four member countries, particularly Cambodia and Laos, which had 'no lawyers' or had 'lawyers [with] no idea about water' (Interview 25, 2011). The four parties also retained external (US-trained) legal counsel to advise them individually and collectively throughout the negotiation process (Interview 25, 2011). Finally, after four years of discussion in which negotiations for an agreement came very close to being abandoned altogether, Cambodia, Laos, Thailand and Vietnam signed the 1995 Mekong Agreement, establishing the MRC. We elaborate on the MRC as the key international governance framework for the Mekong in Chapter 4, emphasizing the socio-legal workings of the commission.

Post-Cold War 'neoliberal' law and development in the Mekong

Despite the regional watershed that the 1995 Mekong Agreement represented, larger forces were at work sculpting the politico-legal terrain of the Mekong River Basin. The collapse of the former Soviet Union inspired a shift towards market-oriented economic policies throughout the former Soviet bloc. This converged with mounting pressure from the international community to peacefully resolve the conflict in Cambodia (also associated with the growth of the international human rights movement) to resurrect the project of donors and experts delivering law and development assistance in the developing world, and in the Mekong River Basin specifically (Trubek, 2006).

The phenomenal increase in foreign legal assistance in the Mekong River Basin from the late 1980s onwards was associated with renewed global interest in the role of law and legal institutions in promoting development, particularly the development of markets. As outlined in Chapter 2, the first wave of the law and development movement, spearheaded by the United States in the 1960s and 1970s, was largely premised on the idea that a strong, effective, democratic state would be needed to initiate and manage the economy. In contrast, the second wave of law and development reforms in the 1990s was heavily influenced by 'neoliberal concepts' of economic relations and governance. The latter reforms emphasized constraining or downsizing the state, as well as deploying state regulatory authority to empower private actors in the economic sphere and in the provision of social welfare (Trubek and Santos, 2006).

With the term 'neoliberal concepts', we encapsulate a mode of thought with 'many authors, many birthplaces' (Peck, 2010, p. 39). Broadly characteristic of that mode is a commitment to active state promotion of competitive capitalism on a global scale. Neoliberal ideas of development are premised on a rejection of both interventionist Keynesian policy (and associated expansion in government-administered economic and social relief) *and* classical economic liberalism's call for passive state deference to the wax and wane of exchange. The 'neo' of neoliberalism relates, thus, to a dismissal of what Hayek (1944, p. 13) termed the 'wooden insistence of some liberals on rules of thumb, above all the principles of *laissez-faire*'.

Consistent with Peck's account quoted above, recent scholarship has emphasized the importance of avoiding overly simplified depictions of neoliberalism as global doctrine, highlighting the multiple and, in some instances, divergent ways in which neoliberalism is understood. Bakker (2010, p. 716), for example, notes that neoliberalism has been conceptualized as a political doctrine, economic project, regulatory practice or mode of governmentalization; she calls for greater specification of the processes at work in 'actually-existing neoliberalisms' (Bakker, 2010, p. 720). Along this vein, scholars of Southeast Asia have for some time now been describing a peculiar mix of neoliberal and non-liberal (or illiberal) modes of governance that have emerged in some countries.

Writing in the law and development tradition, scholars have noted that the economic success of Asian developmental states with statist legalist systems deviates significantly from the expectations of many models of law and development (Ginsburg, 2000; McAlinn and Pejovic, 2012). Some scholars rely on essentialist conceptions of an 'Asian culture' to account for these 'idiosyncrasies' (Jayasuriya, 1999). Others have described 'new forms of political and social oligarchy ... consolidated in systems of authoritarian and discretionary governance' that have arisen with the advance of neoliberal markets in places such as Cambodia and

Laos (Robison, 2009, p. 20; Springer, 2011; Barney, 2012; Middleton *et al.*, 2013). We too wish to dispense with notions of neoliberalism as an ideal type in favour of recognizing locally negotiated articulations of this 'rascal concept', notoriously 'promiscuous in application' (Peck, 2013, p. 133).

These locally negotiated articulations were often seeded in, or responsive to, loan and aid conditions. Initially, policy prescriptions circulating in the Mekong in the late 1980s and 1990s focused on economic reform of the like promoted by the so-called 'Washington Consensus' (Williamson, 2004).The International Monetary Fund's (IMF) structural adjustment programmes demanding economic liberaliza-tion, fiscal austerity, free trade and privatization were seen as cures for economies transitioning from socialism; the aforementioned demands were pushed through under the rubric of IMF loan covenants and adopted by some bilateral donors in their aid conditions. Among the new laws adopted by Lower Mekong coun-tries during this period were legislative instruments on foreign investment, con-tract law, property law and laws on extra-contractual liability, civil procedure and insurance. Also adopted during the 1980s and 1990s were legal and institutional reforms specific to the electricity and water sectors designed to facilitate hydro-power dam construction and cross-border electricity trade in the Mekong River Basin, as examined below and in the following chapter.

Despite its early embrace, growing disillusionment with the inability of structural adjustment to deliver the growth that was promised, while it arguably contributed to growing inequality, shifted the attention of international donors to the role of institutions, particularly legal institutions, in accelerating economic and social devel-opment (North, 1991). A 1994 World Bank report on governance contended that 'sustainable development' requires 'a predictable and transparent framework of rules and institutions for the conduct of private and public business', including guaran-tees of property rights, assurances of the enforceability of contracts, and protection against government interference or arbitrary use of power (World Bank, 1994, p. vii).

Institutional and legal frameworks that offered stability, transparency and account-ability to international investors had, by the early to mid-1990s, become ideal standards of 'good governance' against which future reforms would be assessed and measured (Krever, 2013; see also Chapter 6).While these still tended towards facilitat-ing free market transition, programmes promoting 'good governance' and the rule of law also came to be seen as a way of curbing corruption, displacing patronage-ridden government institutions, and promoting democracy and human rights.

An important aspect of this turn to 'governance' (as opposed to 'government') was the challenge that it posed to the primary authority and legitimacy of the state. This has allowed non-state actors to emerge as key players shaping development in the Mekong River Basin, in part by generating and entrenching relevant norms. The development enterprise thus came to engage and depend on not only states and international donors, but also a diverse mix of civil society and private sector actors. This diversification of interests was reflected in the broadening of scope of law reform agendas throughout the 1990s and 2000s to include the promotion of human rights, democratization, social justice and environmental protection, mov-ing well beyond the economic growth imperative that had earlier prevailed. In the

Mekong River Basin, for example, civil society agitation over the negative impacts of dam construction prompted donor agencies and institutions to adopt social and environmental standards. In turn, donors worked with states (or parts of states) to incorporate legal protections for affected communities and the environment, such as laws on environmental impact assessment, resettlement and compensation, and information disclosure; (see Chapters 5, 6 and 7).

By promoting good governance and a strengthened rule of law (including by encouraging independent courts, judicial training and legal aid), legal programmes funded under bilateral and multilateral aided arrangements during the late 1990s and 2000s promised to protect both businesses and citizens from discretionary abuses of power. In other words, the turn to law in the era of 'chastened neo-liberalism' (Kennedy, 2006a) has promised 'advances [in] both principles and profits' (Carothers, 1998, p. 95).

In the Mekong region, the neoliberal turn in law and development and its 'chastened' aftermath have nonetheless been temporally and spatially differentiated. For this reason, the market integration trajectories – and associated law and governance reform paths – followed by each of the four countries on which this book focuses merit separate attention (with the exception of Laos and Vietnam, considered together), just as they received in Chapter 1.

Laos and Vietnam

In the 1980s, Vietnam and Laos faced severe economic decline and food shortages, particularly after the Soviet Union began scaling back its support to the communist world. In 1986, Vietnam and Laos adopted policies aimed at transitioning towards socialist-oriented market economies. Law reform was a key component of these market liberalization policies.

The Communist Party leaderships in Vietnam and Laos made explicit commitments to run their countries according to law, and shifted away from prioritizing bureaucratic management (Rose, 1998; McGillivray *et al.*, 2012). In 1991, for example, Vietnam adopted a policy to construct a law-based state (*nha nuoc phap quyen*), which resulted in the constitutional reforms of 1992 (Gillespie, 2005; Nicholson and Low, 2013). The Vietnamese Constitution was amended to provide legal recognition of private ownership of property and the private sector economy. It also included a clause requiring respect for human rights. More fundamentally, Article 2 was amended by requiring the people and the party to be bound by law. Similarly, Laos adopted a Constitution in 1991 that stated that party and state organizations 'must function within the bounds of the Constitution and Laws' (Article 10). Nevertheless, the supervening authority of the party over the executive and the judiciary remained constitutionally entrenched in both countries.

In the early 1990s, Vietnam and Laos also opened their doors to international legal assistance, initially only focusing in the areas of trade and investment (Rose, 1998; McGillivray *et al.*, 2012). The scope of legal assistance soon expanded to include the fields of civil law, labour law, environmental law, property law and family law. By the late 1990s, legal assistance programmes also included the more sensitive areas of administrative reform, constitutional law, criminal law and human rights (Rose, 1998; McGillivray *et al.*, 2012). Activities funded during this period focused on legislative drafting, the training of lawyers and judges, institutional capacity-building, and dissemination of laws and legal information. Legal assistance included bilateral (notably Swedish, Danish, Canadian, French, Japanese, Australian and, later, US) support and multilateral funds (especially from the UN Development Programme, the Asian Development Bank and the World Bank).

Contrary to expectations about the political spillover effects of foreign legal assistance, the transition to market-based legal systems with democratic flourishes has not fundamentally disrupted central Communist Party leadership in Laos or Vietnam, or served as a catalyst for further reform. On the contrary, there are signs that various donor-supported legal and judicial reforms have served to entrench the power of these central governments.

Nicholson and Low (2013, p. 11), for example, note that a judicial reform programme in Vietnam begun in 2005 has actually been 'a reform of state power to entrench political stability and preclude any one arm of the justice sector acquiring power', thus ensuring that the leadership of the Party was not compromised. Referring to law reform more generally, Gillespie has observed that 'Vietnamese policy makers are attracted to neoliberal modes of law and development advocated by multilateral agencies precisely because they hold out the possibility of centralized economic and legal mechanisms' (Gillespie, quoted in Ginsburg, 2000, p. 846).

Donors to these countries appear to be concerned about the concentration of state power, but most are reluctant to push governments too hard on issues of human rights or other sensitive issues that might suggest the imposition of a political agenda. Most donors perceive confrontational approaches to be 'counterproductive' (Interview 2, 2011) in dealing with governments in the Mekong, and can leave one 'high and dry' (Interview 50, 2012).

The ruling parties in Laos and Vietnam have continued to show reluctance to accept legal limits on their power in the management of more sensitive political matters, or to allow a more autonomous role for legal and non-governmental institutions. Central government control has allowed governments to disregard existing laws or to use discretionary power to interpret laws relating to the protection of rights of communities impacted by hydropower and other developments. Yet, these Mekong governments have shown a lot more willingness to accept legal conditions set by donors and others to facilitate market access and participation, such as contract terms attached to the financing of public–private partnerships (PPPs), as discussed in the next section and in Chapter 6. Particularly emblematic of the latter attitude has been Vietnam and Laos joining the World Trade Organization (WTO) in 2007 and 2013, respectively.

Cambodia

Cambodia's contemporary legal system emerged in the context of the UN Transitional Authority in Cambodia (UNTAC) in 1992–1993. After decades of violence and international isolation under the Khmer Rouge regime, and instability following the Vietnamese withdrawal in 1989, the Cambodian People's Party had little choice but to accept the terms of the 1991 Paris Peace Agreements and subject itself to the supervision of the UNTAC force led by major Western countries. A new Constitution, drafted in 1993, embraced not only a market economy (Article 56), but also electoral democracy (in the form of a constitutional monarchy) (Articles 1, 51), an independent judiciary (Article 128) and respect for human rights (Article 31). In practice, however, the party maintained control over the state administration, stalling many of the reforms advocated by UNTAC, particularly judicial and court reforms (Nicholson and Low, 2013).

Nevertheless, the rewriting of Cambodia's legal system by UNTAC opened up political space for international development agencies, lending institutions and NGOs, which flocked to the country looking for opportunities for aid and investment. Since UNTAC, bilateral and multilateral donors and lending agencies have funded a wide range of legal and judicial reform programmes. These include, for example, programmes supporting the drafting and enactment of the Criminal and Civil Codes and related procedural codes, together with a range of legislation related to economic reform, private investment, environmental protection, natural resource management and social issues. Donors have also funded judicial training and capacity-building, established the Bar Association of Cambodia and encouraged legal aid, legal education and alternative dispute resolution (Nicholson and Low, 2013, p. 17). Unlike in Laos and Vietnam, NGOs have played an active role in the provision of legal and judicial reforms in Cambodia (using donor funds) and have been critical of the government's performance with regard to justice, human rights and the environment, as seen in Chapter 7.

Despite large amounts of aid having been spent on foreign legal assistance, today Cambodia is still far from having judicial independence. The government's commitment to the standards of democracy and respect for rule of law inscribed in the 1993 Constitution is seriously doubted (Ear, 2007; Un, 2006). As one member of the international donor community noted gloomily:

> There has been no experience here of a developed functioning legal system, ever. The lack of independent institutions, the lack of tolerance for opposition, it has been like that for so long, during every phase of this country. So it really is a cultural mind shift that [is needed], which you can't be sure will happen with the new generation.
>
> (Interview 4, 2011)

Donors and experts we interviewed acknowledged that the many reforms proposed for the legal system in Cambodia have failed to gain government support, particularly reforms that would touch on key interests of the political elite (such as creating an independent judiciary or pursuing land reform). The design of more 'demand-driven' reforms and the exertion of pressure on the legal system from the citizenry are sometimes suggested as possible ways forward (Interview 2, 2011; Interview 4, 2011; Interview 5, 2011). Nevertheless, donors continue to support whatever pockets of legal reform may take hold within an otherwise patronage-based ruling system. As in Vietnam and Laos, there seems to be willingness to adopt law reform when it comes to investment and free trade measures, such as those required for Cambodia's WTO accession in 2004. Thanks, perhaps, to these reforms, the country has enjoyed a fast growth rate, particularly in the decade 2000–2010, but this growth has been accompanied by increasing inequality (Hill and Menon, 2013; Ear, 2007).

Thailand

In Laos, Vietnam and Cambodia, post-Cold War efforts to integrate economically into the global market were bound up with the movement of these states from socialist command economies towards market economies. In Thailand, the changing dynamics of the role of law in development articulated more explicitly with globalization and interspersed by the anti-globalization discourses prominent at the time.

Prior to the 1980s, interest groups had generally coalesced in support of a state-led import substitution model of industrialization and economic develop-ment, which had been advised by the World Bank in the late 1950s. Starting in the early 1980s and running through until the 1990s, however, this consensus began to fracture and give way to a new set of interests emphasizing private sector partici-pation (Kitthananan, 2008). Both domestic and international pressures resulted in a number of law reforms being adopted in the economic sphere, emphasizing privat-ization and greater integration into the global economy. Linked to and facilitating economic reform was a process of transition (wavering as it has been and now with an uncertain future) from military dictatorship towards democracy.

International pressure for law reform in Thailand came mainly from World Bank and IMF loans under structural adjustment programmes that were made condi-tional on reforms to 'fiscal policy, tariffs, investment incentives and institutional arrangements designed to reorient the Thai economy towards an export-oriented strategy' (Kitthananan, 2008, p. 92). A key tenet of such neoliberal prescriptions was financial liberalization, which opened up Thailand to global financial markets. No one predicted that this would lead to structural and policy distortions and trigger the 1997 Asian financial crisis (Corsetti *et al.*, 1998). Demand for economic reform was also domestically driven by growing business interests in the Thai government, which set the stage for 'a series of governments friendly to foreign investment' (Gracean and Gracean, 2004, p. 519).

The 1997 financial crisis intensified resistance to a globalized ideal of free trade and market-driven development. Civil society in Thailand emerged in the context of post-Rio Summit (that is, post-1992) global environmentalism, the latter expressing disaffection with what was seen to be an unsustainable and unequal development trajectory. As illustrated in Chapter 7, civil society opposition to dam construc-tion has been a key issue shaping demands for greater justice and accountability in Thailand, and an important normative force behind growing expectations for law to play a role in shaping more sustainable and equitable development outcomes throughout the Mekong River Basin. This resulted in changes to the behaviour of donors with regard to the way they would 'do' development (for which their environmental and social safeguard policies typically provide guidelines). Perhaps ironically, World Bank and ADB responses to growing environmentalism in Thailand also led to an abrogation of rights to local resources through the introduction of global regulatory regimes for the environment that lock resource-based populations into the logics of capitalist development (Goldman, 2001). The latter has involved the uptake of environmental valuation methodologies and market-based 'solutions' to natural resource management and conservation (see also Chapters 5 and 6).

At the national level, Thailand's 1997 'People's Constitution', emphasizing democracy, the rule of law and citizen participation, is a stark example of the normative effects of global shifts towards the incorporation of human rights in development. Thailand's apparent embrace of law and constitutionalism, however, ended with the military coup in 2006. In his synthesis of more than a century of Thai legal history, Harding (2012, p. 111) concludes that what appears to be a legal system that emphasizes liberal constitutionalism 'turns out to disguise a model in which it is political power that determines both the content and the practice of constitutionalism'.

Many of the law reforms associated with post-Cold War market opening and integration in Laos, Vietnam, Cambodia and Thailand have also taken on a regional dimension, whether through ASEAN, the MRC or the ADB's Greater Mekong Subregion (GMS) Economic Cooperation Programme. In the following and final section, we examine how the logic of linking resources to markets through large-scale cross-border infrastructure, driven mainly by the ADB and other donors, has transformed the Mekong River into an economic corridor centred on the construction of hydropower electricity for export, at least according to influential understandings circulating in the basin.

Law, hydropower and 'actually-existing-neoliberalism' in the Mekong

In the previous section, we examined how legal assistance programmes promoted in the Mekong region since the early 1990s have expanded in scope in line with changing global concepts of development. Law reform agendas over the past two decades have included such diverse objectives as market liberalization, private sector investment, fiscal austerity, judicial independence, court reform, democratization, protection of the environment and social justice. This broadening of the law reform agendas of governments, donors and NGOs reflects the interplay of power between various actors with different interests and divergent expectations of what law can and should deliver, and in whose interests.

As this chapter has illustrated, however, law reform programmes influential in shaping development outcomes in the Mekong River Basin have often had mixed and unpredictable results. This section further explores how donor and expert support for law and development in the hydropower sector has not necessarily led to expected or desired outcomes – even while the donors and experts concerned may continue to celebrate their hand in transforming the Mekong River Basin into an engine of economic growth, prosperity and social well-being (ADB, 2012a,

p. 2). Efforts to liberalize the region's electricity sector and increase private sector participation in the construction and operation of dams have also contributed, at least in part, to the rise or entrenchment of monopolistic state electricity utilities, the concentration of power among government elites, and adverse impacts on local communities pursuing traditional livelihoods and on ethnic minorities and indigenous peoples. Laws, policies and regulations supporting hydropower development built largely on donor edifices have often resulted in the role of central governments in decision-making being strengthened, rather than their responsiveness to rights-assertion, criticism or claims being enhanced. As noted earlier, it is possible to characterize the resulting dynamic as one of (neo)liberal authoritarianism, at least in some Mekong country contexts.

The Greater Mekong Subregion and regional energy trade

Regional integration has been the principal tenet of the post–Cold War global economic integrative agenda spearheaded by the ADB through its GMS Economic Cooperation Programme. The era of relative peace in the Mekong region from the early 1990s onwards sparked renewed interest in making use of the abundant hydropower potential (particularly in Laos, Myanmar and China's Yunnan and Guangxi provinces) as a means of meeting growing electricity demand at low cost and allowing diversification of the power generation mix. Launched in 1992, the GMS programme liberalized trade and investment and built cross-border infrastructure to support the flourishing of regional markets. The programme emphasized regional physical connectivity – through the construction of cross-border infrastructure such as roads, railways, telecommunications, dams and high-voltage electricity transmission lines – as well as the institutional and legal infrastructure to enable economic integration (ADB, 2012a).

As the most comprehensive plan for regional infrastructure integration developed for the Mekong region at the time, the GMS programme replaced the old Mekong Committee as the principal framework for channelling economic development assistance region-wide (Middleton *et al.*, 2009, p. 28). By the end of 2013, the GMS programme had mobilized more than US$16.6 billion in investment projects and US$330.8 million in technical assistance, a large slice of which supported projects in transport and energy (ADB, 2014).

Regional power trading has been a main feature of the GMS programme. This has included efforts in 'planning and implementing key projects for power generation and cross-border interconnections and for improving individual country transmission systems toward the establishment of regional power trade arrangements' (ADB, 2012a, p. 7). Proponents claim that sharing the region's diverse energy resources and optimizing power supply to meet varying demands across the region would bring economic and environmental benefits to all countries (see ADB, 2012b, pp. 1–2). Poor but resource-rich countries (Laos, Cambodia and Myanmar) can generate much-needed revenue, proponents argue, by selling hydroelectricity to energy-hungry economies (Thailand, Vietnam and China), as they have already started to do. An optimization exercise undertaken as part of an ADB technical assistance concluded that grid interconnectedness in the GMS saves about 19 per cent of total energy costs, or nearly US$200 billion (ADB, 2009). The ADB also claims that the integration of power systems results in slower growth of carbon emissions compared to business as usual, partly because of the substitution of fossil fuel generation with hydropower (ADB, 2009, 2012b).

Civil society groups have, however, criticized the technology of long-distance transmission lines as a form of 'resource colonialism' (Gracean, quoted in *Watershed*, 2008, p. 24), pointing out that benefits flow to people in urban cities in Thailand and Vietnam, while costs associated with dam impacts accrue to local, often ethnic, minority populations in highland regions in Laos and Cambodia, where access to justice is limited (see Chapter 7). Partly in response to growing criticism of dam projects, donors have placed greater emphasis on supporting improved water governance and sustainable hydropower (explored in Chapters 4 and 5).

Initially, the objectives of the GMS were partly pursued through dam financing. Up until the mid-2000s, the ADB, the World Bank and bilateral donors were the main financiers of dams in the Mekong. As well as funding the physical infrastructure for electricity transmission, these donors played a pivotal role in fostering the creation of what they perceived to be the necessary legal and institutional framework to create a regional energy export market. This included the formation, in 1995, of the Electricity Power Forum (EPF), made up of two representatives from each GMS country. In 1998, the EPF established the Expert Group on Power Interconnection and Trade to oversee the preparation of a master plan for regional power integration and to determine the institutional and legal arrangements to develop and manage the interconnected power grid (ADB, 2012b).

The ADB-led Regional Indicative Master Plan on Power Interconnection, completed in 2002 and updated in 2010, laid the initial groundwork for regional power trade by identifying priority interconnection projects up to 2020 (ADB, 2012b). On 3 November 2002, the Inter-governmental Agreement on Regional Power Trade in the GMS was signed by regional leaders, providing the legal framework to implement power trade among GMS members. The agreement also called for the establishment of a Regional Power Trade Coordination Committee, which has since developed two working groups that provided strategic direction and management of the interim stage of GMS power trade (ADB, 2012b). Two subsequent Memoranda of Understanding of 2005 and 2008 set out the guidelines and conditions to be followed in bilateral trading, and a road map for reaching more advanced stages of power market integration (ADB, 2012b). More recently, the ADB has worked closely with Mekong governments to establish a regional coordination centre for power trade to facilitate the exchange of information on the energy sector plans of member countries.

Despite these developments, policy and institutional progress towards a fully competitive regional power market has been slow. A 2011 review of the GMS regional power trade commissioned by the Swedish International Development Cooperation Agency (Sida) concluded that 'despite many good intentions and some clear advances in regional networking on energy-related issues, a regional power market is not really closer to being achieved now than it was when the IGA [intergovernmental agreement] was signed in 2002' (Antikainen *et al.*, 2011, p. 15). The report notes the lack of a clear framework for an integrated regional power trading market, and that electricity imports and exports continue to be approved on a case-by-case basis. Moreover, most national transmission grids still cannot physically support power sales by multiple suppliers to multiple off-takers, as would be required for regional power trading.

The more fundamental barrier to the realization of the GMS vision, however, is not technical but political. With the exception of Cambodia, 'which has a more liberal legal structure for private sector participation in power trading' (Doran and Christensen, 2014, p. 68), state-owned power utilities in the Mekong countries – the Electricity Generating Authority of Thailand (EGAT), Électricité du Laos and Electricity of Vietnam – have vested interests in maintaining their position as the only permitted purchasers of power in their respective nations. As explained by lawyers involved in negotiating cross-border power trade in the Mekong:

> Even if the physical infrastructure were to be constructed, regional power trading would predominantly exclude the private sector, as state-owned utilities are the common practice throughout … Southeast Asia. [Regional power trading] would give these state-owned utilities a monopolistic role as both power supplier and power purchaser of all imports into their country resulting from such regional trading. As the sole distributor and retail power supplier within their power sectors, the key issue for regional power trading is the termination of the monopolistic electricity distribution.
>
> (Doran and Christensen, 2014, pp. 67–68)

The 2011 Sida report also concludes that 'not all the GMS countries, including their national utilities, have the same level of commitments, or the same perception of potential gains from the establishment of a regional power market' (Antikainen *et al.*, 2011, p. 15). To date, all cross-border electricity trade in the Mekong has been done on a bilateral basis, mainly between Laos and Thailand. This currently takes the form of independent hydropower producers (consisting of joint ventures between private companies and sometimes involving government) selling power directly to purchasers (mainly state-owned utilities, such as EGAT) under take-or-pay power purchase agreements. As noted by one senior Lao government official, this can 'lock' countries into sub-optimal long-term power purchase agreements:

> The export of power is project specific and determined by the concession agreement of individual projects … the Nam Theun 2 and Theun-Hinboun dams have to follow EGAT's agreed code, the Xekaman dams have to follow Vietnam's agreed code, and so on … If we had known that the Sepon Copper Mine [located in Savannakhet province in Southern Laos] was going to be developed, we wouldn't have sold all the power to EGAT. We would have supplied 60 megawatts from Nam Theun 2 directly to the copper mine. But

now we have to sell electricity to EGAT at 4.6 cents and then we have to buy it back from Thailand … We want to buy more [electricity] but they said 'no'. The agreement is that the water [from Nam Theun 2] belongs to EGAT.

(Interview 52, 2012)

Despite various setbacks in introducing a fully competitive regional power market, and despite their having been superseded by other financiers in funding dam projects, the ADB, the World Bank, Sida and the French aid agency for development (AFD) continue to fund electricity transmission lines and provide political, legal and technical support for the development of a regional power grid.

Electricity sector restructuring

To facilitate the cross-border sale of private sector-generated electricity, multilateral and bilateral agencies have encouraged privatization and deregulation of the electricity sector, particularly in Thailand and Vietnam (Gracean and Gracean, 2004; Nguyen, 2012). As outlined in the previous section, these reforms reflected something of a new development orthodoxy – privatization, commodification of the environment, decentralization – universalized through notions of 'best practice' (Hirsch, 2006b). However, as alluded to above, these reforms have largely been resisted and have resulted only in the partial 'privatization' of state utilities that remain effectively under state influence.

In Thailand, for example, the electricity sector has shifted from state-owned to a partially liberalized model. As in previous attempts at 'transplanting' legal models that encountered resistance and reorientation, attempts to privatize EGAT and other state-owned utilities in the early 1980s, in 1997 and again in 2001 were met with strong opposition from labour unions and academics and ultimately abandoned (Gracean and Gracean, 2004). In 2006, Thailand's Supreme Administrative Court blocked the Thaksin Shinawatra government from selling shares in EGAT, citing 'conflict of interest', 'improper process' and 'anti-competitiveness' in the preparation for privatization (Tangwisutijit, 2006).

This partial privatization never fully realized the objectives of IMF and World Bank-supported reform – namely, to create a competitive electricity market in Thailand. Nevertheless, EGAT did privatize some of its generation assets after an amendment to the EGAT Act in 1992 allowed the participation of private power producers. Electricity Generating Public Company (EGCO) and Ratchaburi Electricity Generating Holding Public Company were established as independent power producers (IPPs) in 1992 and 2000 respectively, although both are subsidiaries of EGAT in which EGAT still maintains significant shareholdings. As a result, EGAT has become a powerful power purchaser capable of raising financing on the strength of its creditworthiness and extending its grid into neighbouring Laos through its daughter companies, EGCO and Ratchaburi.

Vietnam has also struggled to follow through with its plans to increase competitiveness in the energy sector through the privatization of the state-owned utility

Electricity of Vietnam (EVN), despite legislation passed to allow for that prospect in 2004. In Vietnam, about 40 per cent of the country's installed capacity is produced by IPPs or under build–own–operate–transfer (BOOT) agreements (discussed below), with the remainder generated by the state monopoly, EVN (Nguyen, 2012). With growing energy demand of around 14 per cent per year, Vietnam is under increasing pressure to develop power sources and encourage private investment in the sector (Nguyen, 2012). However, it remains unclear how far Vietnam will go with its plan to create a competitive and more transparent power market; one in which the state plays a less dominant role.

Accompanying these half-realized reforms has been the passing of electricity laws by Mekong governments aimed at building competitive power markets and allowing private investors to enter the power market. Lao's 1997 Electricity Law (revised in 2008 and 2012), for example, provides for public–private arrangements in hydropower construction and operation, and enables the entry of IPPs into the power market (Article 5). Together with other law reform measures, these laws paved the way for the creation of Public-Private Partnerships (PPP) for hydropower development.

The ADB-funded Theun Hinboun hydropower dam in Laos, which began operating in 1998, was the first project to be developed as a PPP under a BOOT contractual arrangement. In BOOT arrangements (a particular form of PPP), private sector developers (usually with the support of third-party financiers) enter into a partnership with the state to create and co-own a special purpose vehicle (SPV) to build, own and operate a hydropower project for a specified and typically lengthy period of time, after which they transfer the infrastructure and related assets back to the state. The duration of the arrangement is designed to allow for the repayment of all project-related debt and the realization by the private sector developers of a minimum return on their equity investment out of project revenues (in which the state shares too, in proportion to its SPV ownership interest). Presently, all the hydropower projects in Laos and Cambodia under construction or at advanced stages of planning are being developed under a BOOT arrangement or some other form of PPP. The proliferation of PPPs in this form in the Lower Mekong has given law reform efforts a new tenor and direction. In place of the grand coherent vision of the GMS, law reform in the Mekong development context has become oriented towards creating project-specific public–private regulatory matrices, the dynamics of which we explore in Chapter 6. Legislative amendments and institutional reforms adopted in this context are directed less towards generalized improvement or harmonization and more towards creating a legal environment in which particular high profile projects can go forward.

Private sector financing

By encouraging national legislatures and executive powers to implement legislative and contractual frameworks that reduce foreign lender and developer risk to an acceptable level, international financial institutions – specifically, the World Bank and the ADB – paved the way for a new phase of mostly private sector-financed hydropower development in the Mekong. This produced a corresponding shift away from public financing of hydropower dams through traditional multilateral and bilateral donors.

Crucial to this shift was the process of law reform that the World Bank and the ADB earlier initiated and oversaw. The Nam Theun 2 dam, in particular, was a catalyst for a number of reforms in Lao's financial regulations required to meet international best practice to the satisfaction of the World Bank and other financiers. This included, for example, improving public financial management practices to increase transparency, clearing up enforcement uncertainties and developing a standardized approach to contracts and regulatory approval processes, as discussed in Chapter 6.

Involvement in complex project financing – Nam Theun 2 and Xayaburi, for instance – has increased the skills, financial savvy and legal competency of Lao government officials (Interview 46, 2012; Interview 51, 2012; see also Mitchell and Souche, 2014). The Lao government's commitment to capacity-building and relatively bold experimentation in financial regulation and practice stands in stark contrast to the poor record associated with the Lao government's regulation of the environmental and social aspects of hydropower development (explored in Chapters 4 and 5; see also Suhardiman and Giordano, 2014).

The transformative politico-legal effects of these projects notwithstanding, they still amount to something less – or something more piecemeal – than what was envisaged and promoted through the GMS. The efforts outlined above have not, therefore, produced the results that international development agencies and other advocates of free market reform envisioned.

In Thailand and Vietnam, for example, resistance to neoliberal reforms in the electricity sector has resulted in a peculiar form of 'privatization without liberalization' (Gracean and Gracean, 2004, p. 527) that seems out of step with the original goals of reformers. Rather than building competitive markets and associated regulatory harmonization, the region's state-owned power utilities, particularly EGAT, have become powerful state 'oligopolies' (Wisuttisak, 2012) capable of foot-dragging and even trumping donor efforts to develop an integrated power grid in the GMS (Antikainen *et al.*, 2011).

In Laos, too, sophisticated financial and legal models for electricity trade and hydropower projects based on 'best practice' approaches coexist with non-market modes of governance and rationalities (Johns, 2015; Barney, 2012). Participation, transparency and accountability remain elusive or ambivalent concepts in Laos, as Chapter 6 will discuss further. This is the case despite billions of dollars spent on regulatory diffusion and the promotion of 'good governance' norms through institutional strengthening, capacity-building and projects assumed to be 'game changers', such as the Nam Theun 2 dam.

The changing face of donors and experts in the Mekong

With the availability of new sources of private capital, regional governments have circumvented multilateral bank loans and the conditionalities they impose with respect to environmental and social safeguards. The Nam Theun 2 project is likely to be the last large funding package in the Mekong to be conditioned on the environmental and social standards of multilateral agencies (Merme *et al.*, 2014). Despite claims by the World Bank and others that the Nam Theun 2 dam would set an international benchmark and raise the standards on dam building in Laos (Jayasankar and Porter, 2011), the norms, standards and legal frameworks introduced as part of the project were not replicated or followed in subsequent projects, precisely because states and developers believed them to be too onerous or too costly to implement. As a senior representative from the Lao government noted:

> Before, the requirements to mitigate the environmental and social impacts [of dams] were less stringent. The concession agreement … was … altogether around 200 pages. For the concession agreement of the Nam Theun 2, we call it the second generation concession agreement because the contents had been expanded … to meet the requirements of international funders … After the success of Nam Theun 2 we moved to the third generation … The concession of Nam Theun 2 is too complicated and it is very difficult, even for well-trained lawyers, to understand. We have to make it easy … for all project participants contracted – the power takers, the lenders. In the third generation we tried to simplify.
>
> (Interview 46, 2012)

In the face of these changes, international financial institutions and aid agencies have sought to maintain their relevance and influence by supporting targeted programmes in governance and law reform. In 2012, for example, the International Finance Corporation (IFC) of the World Bank Group launched an Advisory Services programme on Environmental and Social Standards in the hydropower sector in the Lao PDR to 'foster a sustainable and commercially viable hydropower, while ensuring that the environment is protected and that local people have access to the water resources they depend on' (IFC, 2015). As part of the programme, the IFC is supporting the Lao government to revise its Water Resources Law (first adopted in 1996), and resuscitate the National Sustainable Hydropower Policy, which has been lying idle since it was introduced by the World Bank as part of a package for the Nam Theun 2 project financing.

The IFC programme seeks to engage government and private sector stakeholders, but also to support their interactions independent of its involvement; the multilateral agency in this context is more facilitator or coach than initiator of change. The IFC has been working, for instance, to build the capacity of Lao government officials to negotiate better concession agreements with independent power producers and to apply and monitor environmental and social safeguards. Furthermore, the IFC is engaging private companies investing in hydropower in Laos to 'improve their risk management by building their capacity to adopt environmental and social standards for their investments, for example, by putting in place environmental and social management systems' (Interview 51, 2012).

In these ways, the former financiers of hydropower development (and would-be overseers of states' transition from command to market economies) have lately assumed the role of advisers or consultants. In the new multi-source funding environment of the Mekong River Basin, in which growth rates have been climbing while those of other regions stall (cf. IMF, 2015), multilateral institutions and donor governments are, it seems, less inclined to prescribe conditions and more inclined to offer counsel to national governments relatively flush with options. Expert knowledge remains influential, but it has taken new forms.

This signifies a new direction in donor and expert inputs in the Mekong being taken in the shadow of China's rise. These are now oriented towards acknowledging and burnishing the strength of central governments, and counselling them towards public–private hybridization, rather than surmounting, undermining or surpassing state power in the name of market competition. The contrast with the tenor of debt-holder-driven austerity and political crisis in Europe could not be more stark; in the Mekong, the scene seems relatively bright and peaceable by comparison. Nonetheless, this latest making of the Mekong River Basin – combining obdurate politics with flexible economics – might be considered a distinct variant of liberal or neoliberal authoritarianism (cf. Wilkinson, 2013; Somek, 2015, discussing liberal or neoliberal authoritarianism in Europe).

Conclusion

Law and development in the Mekong River Basin has a long non-linear history. Law's articulation with power relations and authority in Mekong societies, as we have shown in this chapter, is embedded in layered histories that include colonialism, socialist and post-socialist state relations. Export and transplant of legal reforms based on Western legal models to enable navigation, trade and commerce flourished in French colonial Indochina, albeit with resistance and attenuation. The Communist Party leadership continued the practice of legal borrowing, albeit switching to Soviet models to promote socialist economic development; these too met with countervailing influences and obstructions. It was not until the 1990s, however, that systematic and organized efforts to reform the legal systems of Mekong countries began in earnest, influenced by neoliberal models of development emphasizing free markets, private sector investment and regional economic integration. It is in this context that donors and experts created a complex legal architecture to promote cross-border 'sustainable' hydropower development, with a view to developing competitive energy markets at a regional level.

Donors and experts have played a crucial role in transmitting and retooling legal norms and institutions throughout this development. In part by this means they have contributed to making hydropower into a booming industry, in which the private sector plays an increasingly significant role, backed by government policies. As this chapter has shown, however, donor and expert interventions do not exert a one-directional controlling force. Law reforms have had mixed and unpredictable results. The response to law has been differentiated, converging in some areas, while developing idiosyncratically in others. This has resulted in a mix of both neoliberal and illiberal forms of governance that is, in many respects, quite specific to the Mekong River Basin.

We have seen neoliberal-inspired law reforms result in the creation or entrenchment of monopolistic state utility agencies, as opposed to market competition. Likewise, laws signed and ratified by governments aimed at ensuring poverty alleviation and fair compensation for affected communities have not been implemented, or have been only partially implemented, or have not produced anticipated outcomes when implemented.

These stories of law and development in the Mekong River Basin invite circumspection towards assumptions about the unproblematic transferability of law and the promise of just and equitable development through law (whether that law be rights-guaranteeing or otherwise framed). More fundamentally, it points to the importance of understanding constituents' experiences of legal development and how rule of law initiatives interact with local, regional and global political economies in specific settings – in hydropower, for instance. Certainly, there is a need for greater recognition of the extent, the context and the impact of the plurality of laws.

What constitutes 'good governance' and 'best practice' and what is the 'optimal' mode of law reform for equitable and fair development in the Mekong River Basin will continue to be subjects of much debate. In this context, recourse is often had to models or structural templates for guidance. In the next chapter, we turn to models of 'good governance' adopted in the region – in the organization and work of the MRC, in particular. There, we will show that the 'rule of law' with which the experts discussed in this chapter have long been concerned is not a condition separate from the institutional arrangements designed to promote such rule. Rather, law blurs into governance and structure into practice; institutions designed to ensure good governance become objects of governance attention in their own right. This is certainly the case with regard to the MRC, as Chapter 4 will reveal.

4 Governing a river basin

The work of the Mekong River Commission

In Chapter 2, we presented governance as a practice that entails the blurring of law into societal and political modes of rule-setting and enactment. We also suggested that, applied in the field of resource exploitation and sharing, governance is interpreted in at least three main ways. Managerially, it is seen as encapsulating attempts to navigate and influence a decentred array of institutions and processes surrounding resource-use decisions. Normatively, it is sometimes conceived as a set of arrangements and practices that go beyond the state and invite stakeholder involvement in affairs of public interest. Critically, it is questioned and historicized through the notion of governmentality, as Foucauldian 'conduct of conduct' (Foucault, 1982, pp. 220–221). All three interpretations are evident in the programmatic applications and discursive deployments of governance to the Mekong development landscape. For the most part, however, river basin governance is interpreted and supported through development programmes with reference to a science-based decision support framework of planning and action, and the establishment of mostly procedural rules to guide the conduct of significant players.

Extending our socio-legal purview of the river basin through a governance analysis, we see that the work of law within the wider framework of governance in the Mekong reflects different ways in which the main governance institutions are understood by various state, civil society and international players. In particular, the expectation of some institutions may be for 'harder' regulatory intervention, in the sense of being definitively rule-setting and rule-enforcing, or alternatively for regulating more 'softly' through knowledge-based advice, exhortation and consensus-building.

In this chapter, we examine the role of the basin's key transboundary governance institution: the Mekong River Commission (MRC). Established in 1995, but drawing on a framework for cooperation established under the Mekong Committee as long ago as 1957, the MRC has an ambiguous – hence contestable, negotiable and changeable – governance role (Hirsch 2012et al., 2006). The status of the commission, the agreement on which it is predicated, and the programmes and practices that come under its auspices constitute the subject matter of this chapter. Much has already been written about the MRC as a river basin institution operating on a global plane (Gao, 2014; Myint, ; Cooper, 2012; Hirsch et al., 2006; Razzaque, 2013). The longstanding framework of cooperation surrounding the MRC has given the Mekong a somewhat iconic status internationally, and it has become a point of reference – if not necessarily a model – for other international transboundary initiatives in river basin governance.

Our governance analysis here explores the MRC with reference to the politics of its authoritative knowledge-production, rule-making and policy implementation. We eschew linear assumptions, sometimes advanced in scholarship concerning the MRC, regarding an inevitable or necessarily desirable progression from politics to law, soft law to hard law, anarchic to rule-based regimes, uninformed to informed decisions and so on (see, for example, Dore and Lazarus, 2009). Instead, we present the MRC as an organization whose technical and science-based programmes are part and parcel of a framework of rules and power that governs development outcomes in particular patterned and yet highly contingent ways.

We start with a critical look at ways in which discursive, analytical, material and symbolic practices represent, and programmes largely funded through development assistance support, governance as the work of producing, evaluating and applying science-based rules and practices. The river basin agency as a governance institution receives particular attention, along with the influential policy rubric of an integrated water resources management (IWRM) approach. We then examine the MRC's governance milieu as a series of both tensions and mutual supports between the expert knowledge contained within its technical programmes, soft regulation as represented in its procedures, and the politics of decision-making within and between the MRC's constituent bodies: the secretariat, joint committee and council. We note the tension inherent in the location of the state-centric legal architecture of the MRC amid a normative milieu that is inclusive of a wide range of non-state actors. A case study of the first mainstream dam to have commenced construction on the Lower Mekong River mainstream illustrates the politics of knowledge, embedded power relations and procedures in shaping outcomes. We conclude with a suggestion that soft law is the dominant governance modality of the Mekong River Basin, and that we can see this clearly in the politics of knowledge that shape the MRC.

Beyond law? The politics of the technical in transboundary river basin governance

Transboundary rivers and their basins are increasingly conceived of as entities that are, or need to be, objects of governance (cf. Myint, 2012). This leaves the mode rather than the concept of governance a focus of analysis and debate. Lebel *et al.* (2007), for example, see water governance in the Mekong as something to be democratized, in the sense of broadening public input into decisions, if it is to work effectively and fairly. Scholarly as well as policy-related discussion of governance is often ambiguous with regard to its descriptive (analytical) or prescriptive (normative) intent. Initiatives to bring 'good governance' to river basins often seek to develop a rules-based regime that is informed by science to promote more equitable and sustainable use and management of rivers.

In the following discussion, we explore the institutional deployment of science and associated rule-making and standard-setting in transboundary river basins in the form of river basin organizations (RBOs). While there is a widespread under-standing of the politicized field within which transboundary river basin governance operates, the institutional manifestation of governance in RBOs tends to render that politicized field *technical* through the deployment of science and operational rule-making. That is, RBO governance tends to convert first-order questions (such as whom and what should be valued, how collective life should be lived, how resources and authority should be distributed and so on) into second- or third-order questions (concerning how to act on specific, often narrowly conceived problems in certain specialized, highly routinized modes). We suggest that instead of understanding governance rendered technical as a depoliticizing move, we should rather recognize the politics of the technical. By this we mean to advocate the recovery of first-order considerations and to consider how these have been and are being addressed within technical work.

While governance goes beyond law as commonly understood, legal norms continue to play a critical structuring role in contemporary governance practice. Transboundary river basin governance, in particular, is subject to international law specific to freshwater resources, as well as a range of other legal regimes of general application (listed in an appendix to this book) – those concerning dispute resolution, for instance (de Chazournes, 2013). In practice, however, formal judicial application of law in this area remains quite weak, particularly in the Mekong region. In part this is because the principal international law instrument pertaining to rivers that cross international boundaries – the Convention on the Law of the Non-Navigational Uses of International Watercourses – had not been signed until 2014 by any Asian countries. This is despite the fact that the 1995 Mekong Agreement was drawn up at the same time as the convention was being finalized pre-1997 and that certain norms – for example, reasonable and equitable utilization – are common to both. To the extent that law has previously been invoked concerning the governance of the Mekong, reliance has mostly been placed on legal provisions that are not specific to water. An example is the case (*Niwat v Electricity Generating Authority of Thailand*) of the successful lodging by northeastern Thai communities of a lawsuit in the Supreme Administrative Court against the developers of Xayaburi dam, on the basis that the power purchase agreement signed had not followed Thai law on proper consultation with affected people (Deetes, 2014; see the discussion of this lawsuit in Chapter 7).

In May 2014, Vietnam became the 35th party to the 1997 Convention. Vietnam's accession was significant at a number of levels. Globally, the 35th ratification, accept-ance, approval or accession brought the hitherto latent convention legally into force as of August 2014. Within Vietnam, the accession demonstrated the successful inter-vention of environmental scientists and others concerned to have their government assert and protect the rights of a mainly downstream country. Vietnam's insecurity as a downstream riparian state is exacerbated by the fact that its two major rivers (the Red and the Mekong) have their headwaters in Vietnam's main geopolitical rival, China. Regionally within the Mekong system, Vietnam's accession to a global convention not yet signed by any other riparian Mekong states cast the country as the most multilateralist MRC member with regard to river governance. This resonated widely in a region where non-interference in the sovereign affairs of neighbouring countries is often asserted (usually under the rubric of the so-called 'ASEAN way'). Yet, despite the recent coming into legal force of this relatively 'hard' international instrument, its 1997 promulgation and progressive adoption worldwide in the decades following show little prospect of altering the 'soft' norms that shape governance of the use and management of the Mekong River. It is also likely that the convention will remain a relatively 'soft' hard law instrument, since obligations thereunder are mostly expressed in the forms of procedures and principles rather than prescriptive rules.

The relatively minimal translation of agreed treaty norms into binding legal obligations – a translation of 'soft' norms into 'hard' law, one might say – raises questions concerning the effect of law on governance practice and vice versa. It calls into question, for example, an assumption prevailing quite widely in the Mekong River Basin: that the 'firming up' or 'upgrading' of legal obligations in the region would likely remedy perceived shortcomings in development decision-making and ensure greater accountability surrounding the distribution of riverine resources. Perhaps, on the contrary, the formal status of law is in some respects quite inconsequential with regard to its influence – or lack of influence – upon decision-makers charged with making controversial policy choices in the development sphere. Rather, governance continues to operate in the form of institutional arrangements, the inner workings of RBOs, and the politics of their influence on decision-making. It is to these organizations that we now turn.

River basin organizations and integrated water resources management as governance modalities

While there are various programmatic ways in which governance initiatives operate in relation to rivers, the predominant mode of governance support provided by donors and national governments worldwide has been to establish and build RBOs. Even where they are responsible for river basins within the territory of a single country, these agencies contain an inherently multijurisdictional logic. That logic is based on the river basin's geographical interconnectedness, defined by a unifying natural or biophysical significance – in this case, as an area of land from which water drains to the sea through a common outlet – and hence one within which hydrological interconnectedness and the ecology of land–water relations are assumed to be bound into a common whole (see, for example, Hooper, 2005). The intersection of the natural river basin boundaries by administrative and political borders is what generates the governance imperative, and hence the institutional imperative. The effort to transcend such borders, or render them porous, thereby frames the principle of holistic basin governance. While many domestic RBOs have been established, international transboundary RBOs, and sometimes the sub-basin agencies that are created within them, attract particular attention given the challenges of governing resources and their development across sovereign borders.

Support for international river basin agencies is mainly predicated on the bringing of science and rules to an otherwise un(der)informed and un(der)governed resource-sharing regime. The Asian Development Bank has been one of the main supporters of river basin agencies at various scales. It has established a Network of Asian River Basin Organizations (NARBO), which in the decade following its inception in 2003 saw 79 member organizations join up. Thirty of these organizations are RBOs specific to individual river basins. These include the MRC itself as well as two sub-basin RBOs in Laos (Nam Ngum and Nam Theun-Nam Kading) and one in Vietnam (the combined Cuu Long and Dong Nai RBO) that lie wholly or partly within the Mekong River Basin (Dong Nai is outside), plus another five in Vietnam, one in Thailand and one in Myanmar that are in the Mekong region but lie outside the Mekong River Basin (NARBO, 2013a).

According to its charter, the stated objective of NARBO is 'to strengthen the capacity and effectiveness of RBOs in promoting IWRM and improving water governance' (NARBO, 2013b). IWRM is central to the universalistic and technical framing of river basin governance by intergovernmental institutions, governments, scholars and experts of various kinds. It emerged out of the so-called Dublin Principles at the International Conference on Water and the Environment in 1992. The Global Water Partnership, which emerged out of the Dublin conference and was set up in 1996 with World Bank, UN and Swedish government support, has adopted IWRM as its central governance tenet. The Global Water Partnership defines IWRM in the following way:

> IWRM is a process which promotes the co-ordinated development and management of water, land and related resources, in order to maximize the resultant economic and social welfare in an equitable manner without compromising the sustainability of vital ecosystems.
>
> (Global Water Partnership, 2000)

As the stated objective of NARBO referenced above indicates, IWRM is a core principle of the RBO mission. In this sense, RBOs are the institutional shell within which IWRM gets enacted. The underlying logic of IWRM is that the management of water and other resources within river basins has tended to be fragmented. The fragmentation occurs geographically across jurisdictional boundaries; sectorally across bureaucratic silos; in a scalar sense between centralized decision-makers and water users on the ground; between state, private sector and civil society; and between multiple infrastructure investments considered on a project-by-project basis. In principle, integration addresses a basic governance need arising from this fragmentation. In practice, however, IWRM as a river basin governance principle has come under critique for being ineffective and inoperable (Biswas, 2004) – a 'nirvana' concept (Molle, 2008). It has also been criticized for its overly narrow hydrological orientation and centralist governance bias in dealing with complex river ecosystems (Hirsch, 2012). The following account of the MRC's operationalization of IWRM illustrates these principles and the bases for these critiques.

IWRM calls for both stakeholder involvement in decision-making and the backing up of key decisions by the application of scientific knowledge and methods specific to individual river basins. There is a range of formulations of the relationship between knowledge and stakeholder involvement in IWRM, often emphasizing participation. Molle discusses the 'instrumental rationality' assumptions inherent in the IWRM stakeholder participation model:

> participation features as a key component of 'IWRM in practice' but the approach nevertheless draws more on a concept of instrumental rationality informed by good will and good data (hence, the pivotal roles of the state in empowering people and of experts in providing information) than on the politics of resource management.
>
> (Molle, 2008, p. 133)

One of the more elaborate formulations of the science–society relationship in IWRM is contained in Gooch and Stalnacke's (2010) notion of the science–policy–stakeholder interface (SPSI), which seeks to establish a governance framework that provides the basis for inclusive practice of IWRM as an integrated approach to river basin management. The framework explicitly links natural and social science approaches, and it emphasizes 'the ways in which scientific knowledge must be combined in water management with political and colloquial knowledge' (Gooch and Stalnacke, 2010, p. 150). Gooch and Stalnacke apply the concept of SPSI to a study of four river basins in Europe and Asia, including the Sesan sub-basin within the Mekong River Basin.

In each of these formulations, IWRM is an example of the tendency Tania Li identifies as the impulse to 'render technical' that which is inherently political in the field of development (Li, 2007, ch. 4). In other words, development practitioners tend to seek solutions to problems with little reference to the different sets of interests that may frame the problem, and hence to the possibility or likelihood that what is a solution for some is an exacerbation of the problem for others. The terms of technical solutions lie in 'compromise', 'win–win solutions', 'benefit sharing' and achievement of triple bottom line solutions that simultaneously serve the three pillars of sustainable development declared in the UN's Johannesburg declaration (United Nations, 2002): social equity, economic efficiency and environmental sustainability. But, as Molle suggests:

> the three goals of IWRM are frequently, if not always, antagonistic (hence the conflicts) … [and] trade-offs are necessary and hard to achieve in such situations. This means that all parties must relinquish something and that the outcome of painful political processes in which the different parties rarely wield equal amounts of power will generally fall short of the 'optimal'.
>
> (Molle, 2008, p. 133)

River basin governance in its political setting

While river basin governance through RBOs working according to an IWRM approach presents the management of river basin development as a technical rather than a political challenge, decisions on development of river basins are in fact shaped largely by politics. Recent literature on international river politics has taken a number of directions. Some have characterized global riparian politics using the notion of hydrohegemony, whereby power asymmetries between neighbouring states in a transboundary river system, via their respective abilities to control water, determine who gets what in situations that usually fall short of outright conflict (Zeitoun and Warner, 2006). Sneddon and Fox (2006) prefer the term 'critical hydropolitics'. Critical hydropolitics seeks to understand the politics around the framing of water issues at different scales, and hence how understandings of basin ecology, within particular types of scale-specific institutional arrangements, reflect particular interests. Meanwhile, others have explored the politics of international river governance beyond state-to-state relations, framing governance as 'polycentric' and hence existing simultaneously in multiple institutional forms, and at different locales and levels (Myint, 2012). The politics of knowledge has become a field of analysis and contestation in its own right in the Mekong water governance context (Daniel *et al.*, 2013; Käkönen and Hirsch, 2009). Our governance analysis takes the political nature of scientific and technical framing – that is, the significance of scientific and technical ways of knowing, both as expressions of power and other relations and as means of intervening in or reconfiguring those relations – as a point of departure for a socio-legal analysis of river basin governance.

The inherently political nature of water governance within an international river basin will be shaped by the countries in which an RBO is situated, and the historical and contemporary relationships between them, as much as it is by the platform-specific politics of that particular RBO. This applies clearly to the Mekong, where geopolitical relations between and within countries condition the interplay between explicitly regulatory knowledge (including law) and other forms of technical knowledge (Hirsch, 2015). Geopolitical relations need not, however, be determinative of development outcomes advanced through the work of RBOs. In a transboundary river basin such as the Mekong, the politics of development and resource-sharing are multiscalar, ranging from the regional and global level geopolitics between the countries through which the river flows to the micro-politics of contestation around particular projects and their impacts. Nevertheless, we believe that there is much to be said for probing these dynamics in a relatively flat configuration – as we do in this book – tracking a particular knowledge form through a variety of scale changes (here, tracking regulatory forms, broadly understood), rather than insisting on the determinative force or priority of either the 'macro' or the 'micro' (Latour, 1990; Bourdieu, 1988).

One can, moreover, partially delink these political dynamics from the route and resources of the Mekong River. The political nature of water governance associated with its framing around IWRM and RBOs intersects with geopolitical relations that go well beyond water and the fact of a shared river. Some examples of the non-riparian geopolitics by which water governance is shaped, and which in turn are, in part, shaped by interactions around water governance, include bilateral relations between the MRC's constituent states, and between those states and non-MRC members that are nevertheless riparian countries within the basin. For example, Thailand and Vietnam's competitive relationship with respect to their political and economic influence in the territories that are now Laos and Cambodia has long historical roots. For much of the early period of the Mekong Committee and its reformulation as the MRC (discussed in Chapters 1 and 3), the two countries found themselves on different sides of the Cold War divide (Osborne, 2004).

Similarly, from the 1960s through to the 1990s, Cold War geopolitics shaped regional relations. Until the early 1990s, it was unthinkable that Vietnam, Laos, Cambodia and Myanmar would join ASEAN within the decade. As a result, state-to-state trust-building has been an important part of the regional rapprochement that has shaped development in the Mekong region since the 1990s. A concern with economic development has been central to this rapprochement, whether it be articulated by particular country agendas (such as Thailand's 'battlefield to marketplace' policy of the late 1980s) or through international institutional projects (such as the Asian Development Bank's post-1992 articulation of the GMS programme as an opportunity to 'reap the peace dividend', discussed in Chapter 3) (Hirsch, 2001).

Laos and Thailand have a very particular bilateral relationship shaped by, on the one hand, their cultural and linguistic similarities and, on the other hand, the divergent political and economic paths that they have taken during the colonial and postcolonial periods. The economic relationship between the two countries has been marked in no small part by hydropower development even before the end of the Cold War period. As early as 1971, the Nam Ngum dam in Laos – the only international project built under the auspices of the Mekong Committee – provided for Laos's largest source of foreign exchange other than overseas aid. Even during the Cold War, division between the two countries following the taking of power in 1975 by the Lao People's Revolutionary Party (LPRP) did not prevent the continued sale of electricity from Laos to Thailand. Likewise, during the brief but vicious shooting war of 1988, over three border villages between Uttaradit province in Thailand and Xayaburi province in Laos, Laos continued to sell electrical power to Thailand. Laos and Thailand are also distinguished by being perhaps the most repressive and the most open spaces, respectively, for civil society organizations in the Mekong region. This point of distinction has shaped the decisions by Thai and other investors to seek hydropower development opportunities in the national space of least resistance (see Chapters 3 and 7).

Laos has had a close but unequal political relationship with Vietnam since 1975, based on the links between the LPRP and the Communist Party of Vietnam. For at least three decades, key planks of Laos's national policy followed those of Vietnam, including economic reform. It was long assumed that with respect to decisions affecting the Mekong River, Laos would need to defer to the interests of Vietnam. However, the increasingly close relationship between the Lao and Chinese leadership and the rapid enmeshing of Chinese investment in the Lao economy, including for hydroelectric dams, has shifted the regional geopolitical balance in ways that impinge directly on riparian relations, as we will see in the discussion of the Xayaburi dam that follows.

Likewise, Vietnam has a close political relationship with the Cambodian government of Prime Minister Hun Sen and his Cambodia People's Party, whose origins are in the Vietnamese-installed regime following Vietnam's ousting of the murderous Khmer Rouge. Vietnam's development of hydropower in the upper Sesan, and the relative silence of the Cambodian regime in response to the destruction of lives and livelihoods of mainly ethnic minorities living along the Lower Sesan within its own territory, are in part explained by this political relationship.

China's economic ascendancy with respect to its regional neighbours in the Mekong region, all of which are members of ASEAN, has had a significant bearing on the governance landscape in which decisions about development of the river are made. There are at least three dimensions to this influence. First is the development of hydropower within its own territory by China, based on its non-membership of the MRC and its strength and confidence not only as an upstream country, but also as a regional superpower. Second is China's geo-economic prowess beyond its own borders, nowhere more evident than in hydropower investment in Laos and Cambodia. Third is the role of China in shifting regional political dynamics, so that smaller countries have the ability to offset the concerns of their larger regional neighbours over fears of losing influence to those smaller countries' gain – Laos's confidence to take decisions against the will of its long-time ally Vietnam is a case in point (Hirsch, 2011).

Just as bilateral and trilateral political relationships condition riparian governance, so the subnational politics of civil society, state and business interests within each country also influence, and are influenced by, the politics of technical knowledge with which this chapter is primarily concerned. This is especially stark in the field of hydropower. In Chapter 5, we examine environmental assessment as contested terrain navigated in the register of measurement. In Chapter 6, we explore the influence of transparency norms often associated with regional and multinational corporate interests. In Chapter 7, we turn to the differentiated spaces that NGOs and other civil society players carve out in the governance landscape of the Mekong. For now, we turn our attention once again to an RBO that represents the principal institution associated with governance of the river basin: the MRC.

Governing a basin? The Mekong River Commission

Like other RBOs, the MRC sees its mission largely in terms of an interplay between policy, procedure and science. On its website, the MRC describes its work in the following terms:

> Serving its member states with technical know-how and basin-wide perspectives, the MRC plays a key role in regional decision-making and the execution of policies in a way that promotes sustainable development and poverty alleviation … Since its establishment in 1995 by the Mekong Agreement, the MRC has adopted a number of rules and procedures, such as the Procedures for Water Quality, to provide a systematic and uniform process for the implementation of this accord. It also acts as a regional knowledge hub on several key issues such as fisheries, navigation, flood and drought management, environment monitoring and hydropower development.

The question whether institutions such as the MRC should regulate by more directive or exhortatory means, as opposed to being a 'regional knowledge hub' in the context of decision support and decision-making over mainstream dams, is essentially a political one. This question goes to the heart of the governance dilemmas faced by an international RBO that is charged with managing a resource system shared between sovereign nation-states and utilized by multiple stakeholders within those states. Should the MRC govern by the establishment and monitoring of hard rules, or should the organization seek to advise and consult by a combination of science-based decision support and stakeholder involvement in a 'softer' manner? To what extent can or should rule-making by institutions such as the MRC seek to circumscribe negotiated outcomes and deliberative decision-making by focusing debate around time- and procedure-bound processes? As will already be apparent, this chapter places the MRC's tackling of these questions within a broader institutional, geopolitical, epistemological and historical context than is commonly illuminated in accounts of this organization. The chapter also shows how different actors understand the MRC differently, and hence shows how and why various actors may look, as a matter of reflex, to harder or softer applications of law in governing a river basin.

As discussed in Chapters 1 and 3, the MRC was established in 1995 on the signing of a binding international treaty, the Agreement on the Cooperation for the Sustainable Development of the Mekong River Basin. Only the four Lower Mekong countries of Cambodia, Laos, Thailand and Vietnam are party to the agreement. China and Myanmar therefore remain outside the MRC framework, but they have observer status at meetings.

The agreement was originally to be signed three years earlier, but Thailand's disagreement with Vietnam's wish to maintain veto rights led to a redrafting. As Chapter 3 explained, the commonly understood point of disagreement at the time was whether or not the new agreement would be based on the non-binding 1975 Mekong Committee Joint Declaration of Principles, whereby one country could object to the actions of another if it were deemed to be taking an unfair share of water or related benefits from the Mekong River at the expense of other countries (Makim, 1997; Matthews and Geheb, 2015). The latter was largely based on the international standards set by the 1966 Helsinki Rules on the Uses of the Waters of International Rivers (Interview 25, 2011). Materially, Vietnam was concerned to avoid loss of water to the Mekong Delta, particularly in the dry season, while Thailand was keen to develop its water resources in the dry northeast. Already Thailand had invested in infrastructure components of the Khong-Chi-Mun diversion scheme, but full implementation of this scheme would require taking water off the Mekong and using it to irrigate dry season rice fields, potentially at the expense of downstream farmers in Cambodia and Vietnam.

The primary upshot of the renegotiated agreement of 1995 was Article 5, which greatly limited the circumstances under which one country could object to developments in another country. Under that provision, tributary projects require simple notification and mainstream projects require consultation, but not agreement, among fellow riparian member states (requirements elaborated in the

Procedures for Notification, Prior Consultation and Agreement, discussed below). In other words, the regulatory force of the agreement seemed to be blunt from the outset; it was consciously framed as a binding agreement imposing 'soft' obligations of cooperation and guiding principles. At the same time, however, Article 6 was developed to guarantee dry season flows that would not fall below the average month's minimum. This latter article left open measurement and definitional questions for elaboration under Article 26, which provides for the establishment of Rules for Water Utilization and Inter-Basin Diversions pursuant to Articles 5 and 6 (regarding, for example, what constitutes the dry season). The establishment of such rules would later result in protracted discussions and negotiations over details that would delay effective development of the MRC's key soft regulatory instruments in the form of procedures. Nonetheless, it allowed the agreement to be signed. At the time of its negotiation, neither of the two smaller countries, Laos and Cambodia, had effective bargaining or legal capacity (Interview 25, 2011), which means that the agreement carries a dominant legacy of negotiation between the two larger member countries of the four-nation agreement: Thailand and Vietnam.

As already mentioned, the Mekong Agreement, signed on 5 April 1995, was drawn up in parallel with the drafting of the Convention on the Law of the Non-Navigable Uses of International Watercourses, but the four member countries saw the draft convention as too restrictive in its focus on the river rather than encompassing the resources of the basin as a whole (Interview 25, 2011). On the other hand, those drawing up the agreement took interest in a number of other basin arrangements around the world (Miller and Hirsch, 2002). In the words of the lawyer who assisted with the drafting of the agreement:

> We used the Mississippi, the Columbia, the Colorado [US–Mexico, US–Canada]. Murray Darling [Australia] is not an international river basin, but they acknowledged it's inter-provincial, inter-state. We looked at the Nile, the Rhine and the Danube. I picked out as many as I could.

However:

> [T]hey saw their basin as quite [physically] unique ... From the standpoint of the organisations, we again looked at the Murray Darling river organisation, and many different commissions. They went all the way from investigation to authorities, like the TVA, which can demand things. Again they didn't think that there was a very good model.
>
> (Interview 25, 2011)

Thus, at the time of its drafting, there was a consensus that, while the experience of other rivers and organizations around the world was relevant, the Mekong Agreement should largely be *sui generis*, fitting the specific hydrological and political contexts of the basin and the relationships between the riparian states that share it (Interview 25, 2011). On the other hand, the framing of river basin governance under global principles such as IWRM since the early 2000s sets a universalistic yardstick against which the MRC operates. Events such as the May 2012 Mekong2Rio conference on transboundary river management (hosted by the MRC to bring together international RBOs ahead of the 2012 Rio+20 UN Conference on Sustainable Development) continue to bring out this creative tension between generic and specific governance arrangements (MRC, 2012b). As one of the longer-running frameworks for transboundary river basin governance, the Mekong experience maintains a high profile and serves as a frequent point of reference for those working in or on other basins.

The agreement provided the basis for establishment of the MRC, and it sets out the basic governance framework of the MRC. Unlike the earlier Mekong Committee, the MRC was not established under UN auspices. Rather, it is 'owned' by the governments of the four riparian signatories to the agreement: Cambodia, Laos, Thailand and Vietnam. The MRC is governed by a council, composed of the minister of the relevant government ministry in each of the four riparian member countries that hosts the National Mekong Committee for that country. In each of Thailand, Laos and Vietnam, the Ministry of Natural Resources and Environment takes on this role, while in Cambodia it is the Ministry of Water Resources and Meteorology. None of these is particularly powerful relative to the key energy, finance and planning ministries in each country that have a role in the planning and approval of key projects on the Mekong and its tributaries. The council meets once per year. The joint committee meets twice per year and comprises either the permanent secretary of the relevant ministry or the chair of the National Mekong Committee from each country. The day-to-day work of the MRC is done by the Mekong Secretariat, which is based in Vientiane but also maintains subsidiary programme offices in Phnom Penh. The chief executive officer (CEO) is appointed by the joint committee. The work of the secretariat is largely funded by agencies administering official development assistance from Europe, Australasia, North America and Japan, commonly referred to as 'donors' and designated within the MRC governance arrangements as the 'Donor Consultative Group'. Each member country has a National Mekong Committee that serves as the liaison between the MRC and the relevant governmental agencies of the member nations.

The role of the MRC as a regulatory agency has been much discussed within the framework of 'governance'. It is often assumed in this context that legal or legalistic rules, procedures and protocols, and/or the influence of international law more generally (for instance, the Convention on the Law of the Non-Navigational Uses of International Watercourses), have the potential to introduce a more rational, fair and effective regime of decision-making and stakeholder inclusiveness. Dore and Lazarus (2009) encapsulate this expectation succinctly at the start of a plea to demarginalize the MRC in order to 'assist societies in making better choices about how to share and manage water for the production of food and energy'. They suggest a 'new water governance paradigm' to deal with disputes 'resulting from interventions to natural flow regimes' (without any explicit mention of dams). They go on to propose that:

New regional water governance is vital because these issues have territorial, ecological and political dimensions that need to be managed via regional protocols, rules or benefit-sharing processes.

(Dore and Lazarus, 2009, p. 357)

Each MRC programme is associated with a particular technical area of river basin management and development, and each is supported by one or more of the international donors to the MRC.

The work of the MRC secretariat within most of these programmes is based around two main types of activities. The intersection between these largely defines the governance approach to influencing, and 'lightly' regulating, decision-making around key projects that have potential transboundary impacts within the Mekong River Basin. The first set of activities involves the production of data, science and knowledge, as well as their management and dissemination. The second set of activities is based around planning and rule-making, largely under the rubric of 'procedures', but also in some key overarching programmatic areas.

The institutional and physical setting for these activities matters to their dynamic intersection. Historically, the MRC secretariat maintained some continuity with the earlier Mekong Committee Secretariat, which was set up with an explicit mandate to promote energy production. For its first few years in the 1990s, the MRC secretariat continued to be housed at the same building in Bangkok as the Mekong Committee secretariat, located in the compound of the Thai government's Office of Energy Promotion. Many of the staff continued from the Interim Mekong Committee secretariat to the MRC secretariat in divisional structures that were quite similar to the earlier ones. In the words of a longstanding NGO activist sceptical about the 'new' organization:

> Because its predecessor, the Mekong Committee, was set up to promote hydropower, the sentiment among NGOs that monitor that institution was that it was going to keep doing the same thing. With the transformation from the Mekong Committee to the Mekong River Commission (MRC), the expectation among civil society/NGOs was not high – that it was old wine in a new bottle.

(Interview 23, 2011)

However, the earlier emphasis on planning a cascade of hydropower dams on the mainstream gave way, first to a series of sectoral programmes, and then to a cross-cutting programme structure that would provide the scientific basis for understanding and managing the basin holistically and hence give a more balanced basis for planning hydropower and other resource uses. The MRC considered that a fuller knowledge of the impacts and hence the trade-offs involved in hydropower development would arise from a better understanding of the basin as an integrated resource system.

The decision-support framework of the MRC, based on the principle of technical studies informing strategic and operational decisions, is backed up by a significant investment in the science of the basin through various programme initiatives. These initiatives employ science as a generator of normative standards, rather than as an inflexible judgment tool, in support of better decision-making over hydropower and other forms of development. Much of the donor support for the MRC over the years has gone into programmes, studies, modelling exercises and assessments on which more rational and sustainable decisions are to be predicated. However, these programmes have not stipulated how their outputs are to be used and have laid down few clear pathways for the science to be utilized in project-specific assessments or governmental decisions based on such assessments (see Chapter 5). Nevertheless, enactment of the MRC's procedures has subsequently drawn on knowledge generated in such programmes, for instance in the technical review of projects put through the prior consultation process. This is detailed further below in the case of Xayaburi and Don Sahong dams.

From the standpoint of the riparian governments, the relatively 'neutral' knowledge-making role assumed by the MRC – framed in terms of science and technical assessments – is more palatable than an overtly interventionist stance on projects, the development of which riparian states continue to treat as a matter of sovereign prerogative. Because the work of the MRC secretariat depends on donor funding, the priorities of donors that direct their support to particular programmes has remained significant. By funding the production of scientific knowledge, donors are able to resile from impinging directly on sovereign sensibilities of riparian governments. Scandinavian donors, Denmark in particular, have ensured that the largest portion of the MRC's funding for the years after 1995 went to two programmes in particular: environment and fisheries. These are both programmes seeking to build a body of knowledge of the Mekong as a complex socio-ecological system. The fisheries programme especially has played a spectacular role in raising the level of understanding of the diversity and productivity of Mekong fisheries, increasing the estimates of fish catches based on official figures by an order of magnitude in the decade following the MRC's establishment. Accepted figures on combined capture and culture fisheries from the Mekong increased from about 300,000 tons per year in the early 1990s to up to three million tons per year in the mid-2000s (Hirsch, 2006a). Furthermore, studies of fish migration, spawning and feeding revealed the vulnerability of this fishery to the impoundments required for hydropower development (ICEM, 2010).

The Environment Programme encompasses three elements: the continuance of basic science (for example, through water quality monitoring); the performance of outreach activities (for example, through the production of educational materials including a sophisticated river awareness kit); and the development of assessment principles, through its draft Technical Guidance on Transboundary Environmental Impact Assessment, discussed further in Chapter 5. Other programmes – notably the Sustainable Hydropower Programme – are more directly oriented around the issue of dam construction and are geared at setting (non-enforceable) principles and standards against which individual hydropower projects can be evaluated by public sector decision-makers and by the public at large. While the Fisheries Programme drew for its early assessments on surveys of local ecological knowledge of fishers, there is little engagement between the science produced through MRC programmes and the local knowledge held and shared at a community level throughout the basin. A gap thus exists between the content and application of ecological knowledge produced through the MRC and that inherent in civil society programmes such as Tai Baan, the latter drawing on the experiential knowledge of farmers and fishers to present alternative assessments of projects on the Mekong and its tributaries (Käkönen and Hirsch, 2009; Daniel *et al.*, 2013).

More cross-cutting programmes have been established within and beyond the framework of the MRC with a view to setting basin-wide strategy on hydropower development. An early initiative for an integrative hydrological assessment of the basin's potential and limitations in hydropower and irrigation development was the Water Utilization Programme (WUP), which ran from 2000 to 2008 with support from the Global Environment Facility administered through the World Bank. This programme sought to develop both a real-time hydrological monitoring system and a longer-term strategic assessment of the water available for energy and agricultural development. The real-time system has been set up for flood and drought forecasting, and it is also used for navigation management on sections of the river where freight transport is constrained by water levels during the dry season. The longer-term system has in part been realized in pursuit of the objective of Article 26 of the MRC Agreement: to fine-tune key hydrological parameters and limits that set the basis for key procedures based on definitions of wet and dry seasons, acceptable minimum flows, and so on.

The WUP was elaborated into a more encompassing programme supported by the World Bank under the rubric of the Mekong Water Resources Assistance Strategy (MWRAS). The MWRAS did not come under the auspices of the MRC, but nevertheless employed data generated by the WUP and was understood as a means for the World Bank to become more relevant – even central – to a renewed investment in water resources (mainly hydropower and irrigation) in the Lower Mekong Basin. The MWRAS was predicated on a concern by hydrologists and engineers, long keen to exploit the potential of the river (as discussed in Chapter 3), to show that there was sufficient water for a greater degree of development than was occurring or being envisaged during the early 2000s.

A long-term programme central to the work of the MRC, and one that demonstrates ways in which technical framing may serve particular development agendas, is the Basin Development Plan (BDP). Historically, the BDP can trace its origins to the Indicative Basin Plan of 1970 produced by the secretariat under the Mekong Committee, and the Revised Indicative Basin Plan of 1987 developed under the auspices of the Interim Mekong Committee (Mekong Secretariat, 1970, 1988). These plans were essentially shopping lists of mainstream and tributary dams that represented the culmination of hydro-engineering-driven feasibility studies conducted under the earlier Mekong cooperation framework, as discussed in Chapter 3. The BDP is charged with putting forward a more integrative approach than the earlier Indicative Basin Plans. It expresses a goal of understanding the options, limits and trade-offs associated with water use by the suite of projects proposed under scenario-based basin-wide planning exercises.

One way in which the BDP has moved to a more integrative approach than earlier basin planning initiatives is by employing the rubric of IWRM discussed above; the most recent BDP strategy document is entitled *IWRM-based Basin Development Strategy for the Lower Mekong Basin* (MRC, 2011b). This document invents and uses the concept of a 'development opportunity space' (given the acronym DOS in the strategy) to identify the amount of water that can be impounded and/or diverted without unacceptable impacts on downstream countries. It makes the case for significant future use of the Mekong's tributary and mainstream water for energy and agricultural development, including dams on the mainstream. This way of framing the space to be occupied between existing and planned levels of development, on the one hand, and the levels that could be achieved without imposing unacceptable or unsustainable pressures on water resources and ecosystems, on the other hand, suggests a consensus around a technically calculated room to develop. By framing the DOS at a basin-wide level, the BDP pre-empts or counteracts the critique of specific projects based on their more localized or tributary-specific impacts. It also takes an overtly hydrological approach to identifying 'available' space, neglecting more complex ecosystemic implications of taking up the space with hydropower and irrigation projects that are the main subjects of BDP-driven planning.

In contrast, in 2009, the MRC commissioned a strategic environmental assessment (SEA) to examine the implications of the 11 mainstream dams on the Lower Mekong for fisheries and wider environmental and social impacts, including sediment trapping and its implications for the Mekong Delta. Unlike the modelling behind the BDP, which remained largely in the field of hydrology, the SEA took a more inclusive and multisectoral approach to informing strategic decisions over Lower Mekong mainstream dams. It did so by looking at fisheries, sediment, human livelihoods and a range of ecological, social and cultural values that are not incorporated within the narrower hydrological models.

However, there was a key difference in the status of the two studies in governance terms. The BDP is a core programme of the MRC, which means that its findings carry weight in terms of government buy-in by the riparian members – that is, the governance body's state members (the joint committee and the council representatives of each member state) sign off on key strategic documents, such as the Basin Development Strategy that is a key policy guidance framework that emerged from the BDP. In contrast, the SEA was carried out by consultants, and although these consultants worked for and in close liaison with the MRC Secretariat, the status of the SEA findings remains outside the scope of policy buy-in by the riparian member states. This is the case even though the SEA is based on robust scientific enquiry that in many ways is more comprehensive than that underpinning the BDP.

The modelling of impacts of development strategies and decisions has proliferated further, as whole-of-basin concerns have in part displaced the project-by-project assessments of an earlier era (except in the design and evaluation of financial structures, and associated consultations, as discussed in Chapter 6). A study of the net economic value of Mekong mainstream dams that used an ecological economics approach had far more stark and bleak findings than did the studies of the SEA or BDP (Costanza *et al.*, 2011). That study generated an alternative valuation of the fisheries and other environmental assets of the Mekong affected by hydropower development. On the basis of this, the net present value of the Mekong mainstream dams would be negative US$274 billion, rather than the positive US$33 billion suggested by the BDP. However, as this study was conducted by Portland State University with USAID funding and was even further removed from the MRC governance arrangements, it received relatively little attention. It is clear, therefore, that the status of knowledge in informing key decisions on mainstream development depends to a significant degree on by whom and in what institutional and political settings that knowledge is produced (Costanza *et al.*, 2011). At one level, the profusion of competing studies seems to muddy the governance waters as decision-makers face incompatible advice emanating from the same organization: the MRC in this instance, the source of both the BDP and the SEA. From another perspective, Suhardiman *et al.* (2015) take a more positive view of the multiple assessments as providing political space for deliberation in an otherwise constrained policy environment.

The MRC has also hosted an initiative entitled the Mekong IWRM project that ran from 2010 to 2015. The M-IWMRP, as the MRC refers to it, attributed its provenance to the WUP described above. A key feature of the M-IWRMP was to put 'IWRM principles into practice' (MRC, 2015b), and this it does through support for and implementation of the MRC's procedures and guidelines.

Beyond its knowledge programmes, but supported by them, the MRC has also developed a series of procedures, which are its main regulatory instruments. They differ from the knowledge programmes described above in that they are invoked – in a manner that might be regarded as statutory – to guide and structure deliberation and consultation on specific projects, rather than serving as background decision-support aids. Even so, they are relatively soft instruments in terms of their textual imprecision or open-endedness. As such, they are open to interpretation and experimentation to a degree that gives their invocation a particularly interesting political valence. That is, the procedures and their implementation are negotiable and open to considerable discretion in their deployment (Whitehead, 2011; Suhardiman *et al.*, 2015).

The key procedural provisions through which the MRC governs decision-making on mainstream dam developments with potential transboundary impacts are the Procedures for Notification, Prior Consultation and Agreement among MRC member states, which have become widely known through their acronym PNPCA. The MRC requires riparian member states to follow the PNPCA for any hydropower project or other piece of infrastructure that requires the use of water on tributaries or the mainstream within the Lower Mekong River Basin. For tributary projects, notification is required, which simply means that the riparian member state in which the project is located needs to give relevant information to the MRC for access by other member states, whether or not the project is deemed to have transboundary impacts. For mainstream projects, on the other hand, the PNPCA requires prior consultation between the member states over a stipulated period of six months following submission of the project to the MRC by its host member state. Technically, the PNPCA applies between MRC member states and is not a broader procedure for engaging with affected communities or civil society.

It accordingly preserves the fairly closed model of interstate governance embodied in the Mekong Agreement itself. Nonetheless, the PNPCA has still been the focus of lobbying and criticism by environmental groups from within and beyond the Mekong River Basin and, more discretely, by MRC donors. Moreover, the PNPCA created unanticipated openings for input and contestation, as discussed below.

The MRC plays a role in facilitating this prior consultation in three main ways. First, the data and knowledge held at the MRC allows the secretariat to provide an expert assessment of the project as submitted and to raise issues that relate to potential transboundary impacts. Second, the MRC facilitates consultation within the Mekong River Basin, potentially but not necessarily including consultation with affected stakeholders at various levels. Third, the MRC brings together the results of these assessments, responses to them, and the consultation outcomes, delivering the resulting assemblage to the joint committee to encourage it to reach consensus on the way forward for the project. The following section describes the actual experience with the first two prior consultations that have been carried out for mainstream dams, both proposed by Lao PDR.

Before we explore the MRC's governance practice via a case study of these two Mekong mainstream dams, it is important to consider once more the different interpretations of the MRC's governance role by different stakeholders within and beyond the region. Just as in this book we explore the different expectations and understandings of law at a societal level within the Mekong region, so it is important to take account of the multiple interpretations of the MRC's governance role. This multiplicity is created in part because of the historical and continuing ambiguity surrounding the role of the MRC: as a knowledge hub, a stakeholder platform (with different claims on the breadth or narrowness of stakeholder composition), a rule maker, or a decision-making, policymaking or regulatory enforcement agency (Hirsch *et al.*, 2006).

From the point of view of its riparian member states, the MRC is not a regulatory body in terms that they understand. In other words, the MRC is not perceived as a body that can independently set rules that override or explicitly constrain sovereign decision-making at a national level. In some instances, this view has been made starkly explicit, as in remarks by the Lao vice-minister for energy and mines in an article for Thailand's *The Nation* newspaper on 24 October 2014:

> It seems that what the activists really want is for the MRC to prevent Laos from building dams on the Mekong. Unfortunately this is not something the MRC can do. The Procedures for Notification Prior Consultation and Agreement set forth in the 1995 Agreement are not a mechanism for approving or rejecting any particular project. The MRC is not a building permits office.

In other interpretations, the regulatory capacity of the MRC is more implicit. NGOs have voiced mixed views and understandings of the MRC over time. During the early 1990s, some Thai and international NGOs had a conflictual relationship with the transitional MRC secretariat. Through public statements by its then executive agent, the MRC secretariat appeared to be keen to bring back onto the political drawing board the suite of large dams on the Mekong mainstream that were detailed in the 1987 Revised Indicative Plan, a prospect that riled environmental NGOs (Lohmann, 1990). Throughout the early years of the MRC, NGOs campaigning against dams continued to see the MRC as a promoter of hydropower, locked into old agendas. For example, the first issue of *Watershed*, published by the regional NGO TERRA two months after the signing of the Mekong Agreement, was headed 'New Era, Same Old Plans' (TERRA, 1995). The fact that the first CEO of the MRC was an engineer with publicly stated views that dismissed stakeholder participation fuelled this antagonism. Indeed, the relationship between civil society and the MRC has in part been shaped by the style of the incumbent CEO throughout the MRC's history (Hirsch, 2008).

A different sort of expectation of the MRC among NGOs became apparent in the early 2000s, as the transboundary impacts of Vietnam's Yali Falls dam led to civil society calls for MRC intervention. The expectation voiced in such contexts was that the MRC should act as a supra-national authority and that the agreement should hold upstream countries responsible for downstream impacts. This was met by statements from the CEO and others that the MRC was only as strong as the will of its member governments, that it was not a regulatory agency and that the MRC's role was to facilitate compromise and consensus. The realization of this by NGOs that had earlier tried to call the MRC to account has been accompanied by cynicism on their part regarding the way ownership of the MRC by riparian governments is used by 'experts' to deflect expectations of action by the MRC secretariat:

> [The] MRC doesn't arbitrate anything. It took some 10 years to see this but maybe this is something funny about the MRC. Experts are now very savvy about who owns the MRC and not. It's the governments. I think the leap … to the MRC Council is maybe one reason why there's this division now about what the MRC can and can't do, or in whose interests were they tagged to.
>
> (Interview 20, 2011)

Among these 'experts', donors and scientists have tended to focus on the knowledge-building aspect of the MRC, working within a model of decision support. There has been relatively little direct connection, however, between the MRC's programmatic support for improving scientific knowledge, on the one hand, and expectations that decisions on individual projects would be conditioned by a requirement that such knowledge be applied to the environmental assessments produced for each project, on the other hand (see Chapter 5). It has only been with the review of projects through the PNPCA that the MRC has brought the science that it has produced to bear in the public arena of decision-making on these dams.

A dam too far? Governing projects on the mainstream

On 28 January 2015, the joint committee of the MRC met to consider the results of a six-month prior consultation process on Don Sahong dam, which is the second of the cascade of dams being proposed on the Lower Mekong mainstream (see

Chapters 1 and 5). The prior consultation had been triggered by the PNPCA. Initially, Laos had notified Don Sahong as a non-mainstream project, on the basis that Hou Sahong channel on which the dam is to be built is one of several main braids of the river as it passes over the fault line at the Lao–Cambodian border and hence does not block the entire Mekong mainstream. However, at the request of other members of the Joint Committee in June 2014, Laos reluctantly agreed to submit the project for prior consultation as a mainstream dam. The consultation started officially, but quietly, in July 2014, but the process only got going publicly in October, or halfway through the stipulated six-month consultation period. In January 2015, three of the four riparian members recommended extension of the Don Sahong prior consultation, while Laos as the proposing country announced that it considered the process complete and that it intended to commence construction of Don Sahong dam in the early 2015 dry season. In the absence of agreement, the matter was referred upward to ministerial (or political) levels via the Mekong Council.

The Don Sahong case followed a protracted inaugural prior consultation process that occurred from late 2010 to mid-2011 and beyond in the case of Xayaburi dam. Geheb *et al.* (2015) have documented the Xayaburi case in considerable detail, and the dam has been the subject of several other studies (Whitehead, 2011; Middleton, 2014; Matthews, 2012; Stone, 2011; Matthews and Geheb, 2015). Grumbine and Xu (2011) suggest that the dam represents something of a crossroads for the Mekong and its governance, and that the path forward may lead either to decisions with significant deleterious impacts for the river and communities that depend on it, or to a model process of integrative planning. The subtext of Grumbine and Xu's argument is that the scientific basis for cautionary decisions on Mekong mainstream dams is well established, but the capacity for international cooperation to promote better decisions is the key point of uncertainty or the weak link in the chain.

The NGO International Rivers has provided a critical account of the governance process surrounding Xayaburi, accusing Laos of violating the Mekong Agreement (Herbertson, 2013). International Rivers has similarly been critical of the threat to the food security of the Mekong River Basin posed by Laos's plans to go ahead with Xayaburi dam (Herbertson, 2012). Focusing more specifically on procedural aspects of governance, Jian Ke and Qi Gao (2013) employ the Xayaburi case to highlight the challenges of establishing transboundary environmental assessment procedures in the Mekong (challenges that we explore in Chapter 5). Other analyses include the Xayaburi case as an example of newly politicized governance issues associated with the increased blurring of public and private interests over recent decades (Matthews, 2012; Merme *et al.*, 2014) and what Middleton (2014) refers to as the 'politics of uncertainty' associated with the transposition of diverse normative positions onto a highly unequal terrain of power relations.

Xayaburi has thus become something of an iconic case in literature that links the scientific with the procedural dimensions of Mekong River governance. For many, it crystallizes the politics of the technical as one set of studies generates another, with the project's construction continuing all the while in parallel irrespective of the substantive knowledge produced and the procedures followed. The following account will dwell specifically on the governance role of the MRC through its key PNPCA instrument, focusing on the politics of technical knowledge demonstrated by the Xayaburi prior consultation process and its aftermath. This includes discussion of the precedent set for the next major test of the MRC's governance through the PNPCA in the Don Sahong case.

The timing of the Xayaburi prior consultation was controversial from the start. The Lao National Mekong Committee submitted the project to the MRC on 20 September 2010. In October 2010, the final report of SEA commissioned by the MRC was published (ICEM, 2010). The SEA recommended that Mekong governments adopt a ten-year moratorium on mainstream dams. Its recommendations were already known publicly at the time of the Xayaburi prior consultation submission, but had not yet been officially presented. In the words of an international NGO critic in reference to the lack of deliberative consideration of the SEA on its completion immediately prior to the prior consultation process on Xayaburi:

> There were never any consultations around the final SEA. They jumped right into the prior consultation process for the Xayaburi without looking back at the cumulative impacts and understanding what happened. Even though the countries had been engaged in [the SEA] process for months beforehand and had gone through many consultations. They just kind of pretended the SEA never existed and moved forward. And I think that is a place the MRC could have taken more action.
>
> (Interview 20, 2011)

The test case of the PNPCA thus occurred in the context of a highly charged debate concerning the MRC's position on mainstream dams, exacerbated by the fact that the draft Basin Development Strategy, also produced by the MRC, had in July 2010 identified 'development opportunity space' for at least six of the 11 mainstream dams being proposed, including Xayaburi.

The MRC facilitated two parallel processes for the prior consultation, based on science and stakeholder input respectively. There was limited interplay between these two processes, although the public consultations drew on fisheries and other environmental knowledge that had been produced by MRC programmes over the years. The first was a technical review of the documentation supplied, including the design documents and the environmental impact assessment (EIA) report for Xayaburi. This entailed the setting up of a task group internal to the secretariat to consider the documentation in terms of dam design and operations, hydrodynamic modelling, fisheries, sediment transport, river morphology and nutrient balance, water quality and aquatic ecosystems, dam safety, navigation and the social implications of the dam. External experts were also hired to give special consideration to the two most controversial aspects of Xayaburi: fisheries and sediment passage. The findings of both these groups were presented in a Prior Consultation Project Review Report, which was based in part on the Basin Development Strategy, SEA and the MRC's Preliminary Design Guidance for Proposed Mainstream Dams in the Lower Mekong River Basin (the latter explained below). The report found a number of areas of uncertainty and inadequate data or analysis. It was presented to the joint committee ahead of the final prior consultation meeting in April 2011.

The second process, facilitated by the MRC as part of the prior consultation, was a series of consultation workshops held in Cambodia (two meetings), Thailand (three meetings) and Vietnam (two meetings) to elicit community, civil society and local government views on the dam. No such workshop was held in Laos itself, as the Lao National Mekong Committee took the view that the surveys that had been carried out for the EIA were sufficient for community-level input within the country. In this sense, the environmental impact processes described in Chapter 5 had a direct, albeit counterintuitive, bearing on public engagement surrounding the Xayaburi project, effectively obstructing the Lao public's independent access to information about the project.

A range of both complementary and competing knowledge products, associated in different degrees and in different ways with the MRC, thus informed the consultation process. The technical review process assessed the documentation submitted by the developer for the prior consultation process against the MRC secretariat's Preliminary Design Guidance for Proposed Mainstream Dams. The latter − a 'soft' set of compliance standards, which had been rolled out by the MRC in August 2009 − thus provided an initial benchmark against which the project could be evaluated. The guidelines refer to a number of Articles of the Mekong Agreement. Specifically, they refer to Article 2 (requiring the MRC to evaluate design, operation, impact and mitigation measures for mainstream dams), Article 3 (dealing with environmental protection), Article 5 (concerned with the PNPCA), Article 7 (regarding prevention and cessation of harmful effects) and Article 9 (on freedom of navigation). Despite the MRC basing the technical review on the guidelines, and the guidelines in turn being based on obligations under the Mekong Agreement, the MRC secretariat was not thereby put in a position to make a judgement call on the dam. Rather, the MRC secretariat provided one of a number of inputs into the joint committee's ultimate deliberations on the project.

The prior consultation under the PNPCA also followed particular rules inherent in the Procedures. According to Geheb *et al.* (2015, p. 112), the prior consultation gives the MRC its '[only] regulatory authority in the affairs of the Mekong'. This is an odd statement, in two respects. On the one hand, the prior consultation gives the MRC no power to decide on the dam itself, nor does it allow any one or a combination of riparian countries ultimately to prevent a project from going ahead, or to be compensated for any resulting loss. As such, the prior consultation is clearly not regulatory in a hard, enforceable sense of authorizing a veto, requiring a specific decision, or providing for the imposition of sanctions. On the other hand, if the softer notion of regulation as an exhortatory, goal-setting or framing use of knowledge is taken on board (that which might be identified with analyses of governmentality, introduced early in this chapter and in Chapter 2), then the PNPCA is in fact among a suite of measures by which the MRC seeks to govern: through its knowledge programmes, guidelines and standards. This said, the Xayaburi PNPCA did put to the test the application of procedures through which a member state was required to go under jointly agreed rules, prior to implementing a hydropower project.

The politics around the Xayaburi dam in general have been fractious, and the PNPCA process itself was highly politicized and contentious. There was much criticism of Laos during and following the Xayaburi PNPCA, based on the poor quality of the project documentation. The preparation of this substandard documentation was due in large part to denial by the project developers and the Lao government of any concerns over transboundary impacts – a denial reflected in the EIA's restriction of impact studies to ten kilometres downstream of the dam site. Another focus of criticism was the failure of Laos to release the full EIA report during the public consultations over the six-month period, until shortly before the wrap-up joint committee meeting in April 2011. Also controversial was Laos's continuation of site preparation work in parallel with the prior consultation, which can clearly be regarded as pre-emptive and thus against the whole intent and spirit of the process (Herbertson, 2012, 2013). The procedures require each riparian member state to provide reply forms at the completion of the six-month period, and both Vietnam and Cambodia made it clear in their replies that they regarded the PNPCA as incomplete. Thailand equivocated, but Laos made it clear that it regarded the process as over. The failure to reach consensus over the process, not to mention the dam itself, resulted in the matter being referred upward to the Mekong Council, as noted above.

Conflict over the timing of the PNPCA's commencement, and over the question of whether it was actually complete when Laos claimed it was, transformed basin-wide distributional questions into questions of procedural compliance – activating a politics of the technical along the lines earlier described. Nonetheless, the navigation of the PNPCA procedure by states and the MRC maintained a continuing normative separation between technical and political dimensions of governance, a separation endorsed by many observers. The following interview with an international NGO staff member illustrates the prevalence of this separation in the thinking of many stakeholders:

> [The] PNPCA is a technical process but really it's a political decision. Maybe I would add a nuance to that. Science has had a bearing to an extent on the shape of the political debate, which is important to note, but ultimately, there's a huge body of evidence now that … like, why this project? Why build a project on the mainstream of the river? I'm not advocating tributary projects, but there are less damaging projects at a minimum that could be built. But we also know that there are other alternative solutions to meet Thailand's energy needs inside Thailand. So why is the Xayaburi still going ahead? These are the questions that come back to the political level. It's not based on a rational technocratic process of decision-making.
>
> (Interview 20, 2011)

Furthermore, the apparent rigidity of these power geometries and related distinctions between fields and forms of knowledge seem to have remained even as hydropower financing has moved from the public to the private sector domain (the Xayaburi and Don Sahong projects both being financed largely with private sector financing and without the direct involvement of multilateral organizations). In the words of an NGO observer:

> The law [Mekong Agreement] can be influenced by the weak. Laos can say, 'no I don't care, I want to go ahead on this project'. Before I thought only the powerful countries could do that [referring to when Thailand tried to bend the Mekong Agreement so it could go ahead with the Khong-Chi-Mun water diversion project]. But today I think that actually it's about business, economics.
>
> (Interview 23, 2011)

To the extent that they are regarded as connected, the link between the technical and the political is understood and represented by some in a linear fashion, whereby one gives way inevitably to the other. The CEO of the MRC states the link in the following terms:

> Well … sometimes it's a little bit arbitrary when you draw a line between technical [and political]. I mean, just look at the concept of the Delta Study that they want to do in Vietnam and the Council wants to study the previous scenario work we've done under the BDP and the cases … a decision being made on Xayaburi. We do a lot of technical work, but ultimately what the politicians who are doing the political dealings want to know is answers to certain questions, such as, what's the ultimate impact? Is it going to threaten us in 30 years? Well, there are technical answers to that, although you will have to frame it as, 'if you consider that and that, then it is a threat'. So those technical answers will, should be a basis for political decisions and I think the technical debate does contribute quite a lot. But there's still a bit of dealing in terms of saying okay, well if you go ahead with this dam then that affects other decisions in the future.
>
> (Interview 48, 2011)

The implication here is that the MRC's technical work remains non-political unless and until it is in the hands of politicians. There is little acknowledgement of the politics inherent even within and between core exercises funded by and through the MRC, such as the BDP and the SEA. This is notwithstanding the fact that these drew on the same pool of data (that is, outputs of the MRC's technical programme) to come up with significantly different conclusions, recommendations and strategies for hydropower development on the Lower Mekong mainstream.

The question of the political nature and role of the PNPCA has been addressed in an interesting study conducted during the 2010–2011 process. This study concluded that the experience of the Xayaburi prior consultation was both more and less than it was set out to be under the MRC procedures (Whitehead, 2011). This is because, on the one hand, the rules set up for the prior consultation resulted in a messy outcome and a petering out of the process without resolution. No agreement was reached on the project, or even on the ultimate status or adequacy of the consultation process in decision-making about the dam. On the other hand, the process itself provided openings for a more inclusive and robust political deliberation on the dam than would have occurred without the processes facilitated by the PNPCA. It also facilitated the formation of alliances, such as that noted by Vietnamese NGOs as having formed between civil society, the scientific community and government in the case of the downstream country (Interview 35, 2011).

Ultimately, and as extensively documented in Matthews and Geheb (2015) and elsewhere, the construction of the Xayaburi dam has gone ahead. The PNPCA and its aftermath resulted in a commitment by the Lao government to address significant concerns by requiring investment in a redesign of the fish passage facilities, the addition of an extra navigation lock, and provision for sediment flushing through the radial gates of the dam. Initially, the cost of these improvements was stated at US$100 million (International Rivers, 2013b). In January 2015, it was reported that the improvements – the redesign for which still had not been made public by the time the construction of the dam was reported as 40 per cent complete in early 2015 – would actually increase the overall 115 billion baht (approximately US$3.5 billion) cost of the project by another ten billion baht, or more than US$300 million (Polkuamdee, 2015). It was not clear whether the developer or the Lao government has been made contractually responsible for this extra investment, but the same report indicated that the developer would seek compensation from the Lao government for these costs (Polkuamdee, 2015).

For many, the prior consultation process for the Don Sahong dam took a similar trajectory to that of Xayaburi. Initially, as indicated above, the Lao Ministry of Energy and Mines was reluctant to submit the project for prior consultation, but considered that acceding to requests to do so might be less complicated than seeking arbitration on whether Don Sahong was in fact a mainstream dam. In an interview given a year prior to the initial attempt by Laos to submit the project through a simple notification, a senior government official stated:

> I ask the developer to go to the law firm in Singapore … and the two lawyers in the same firm have different ideas, different interpretation, and I said then 'stop'. I [would] probably spend more time deciding what should it be before I submit it, I say I go ahead with prior consultation six months and that's it but because if they start to decide … it will probably take a few years and spend a lot of money. We decided to do it that way.
>
> (Interview 52, 2012)

The change of mind was temporary, however, and the prior consultation commenced on 28 July 2014. The Thai, Cambodian and Vietnamese reply forms submitted six months later all requested extension of the prior consultation, while Laos once again stated that the process was complete. Once again, the matter was raised to the Mekong Council level, but with an indeterminate timeframe for decision-making. Once again, site preparation work was being conducted all the way through the prior consultation process. Once again, the donors to the

MRC issued Development Partner statements of concern, highlighting that the organization and its procedures had been supported with donor funds in order to promote a more cautious path of hydropower development on the Mekong mainstream and urging Mekong governments to follow such a path.

Conclusion

In this chapter, we have examined the main governance framework for the Mekong River Basin. This centres on an RBO – the MRC – the programmes of which are largely geared to the production of technical knowledge and procedures. These outputs are, in principle, geared towards better decision-making around development projects that make use of the basin's resources. We have shown that by framing its authority around technical scientific knowledge-making and knowledge-gathering, and orienting the production of such knowledge, as well as related information-sharing and procedural compliance, around second-order considerations of mitigation, design adaptation and so on, the MRC sidelines or forecloses first-order deliberation – such as consideration of whether or not to build a dam. The experience of Xayaburi dam in Laos, and the repetition of that experience (of Laos engaging in deliberation on design and mitigation while contractual and physical preparations for dam construction proceed apace) in the case of Don Sahong, indicate that this governance modality fosters unilateralism in addressing first-order considerations, despite the cooperative framework through which it is carried out.

The case of the prior consultation over mainstream dams exemplifies the limits to governance that has been rendered technical through a rules-based regime informed by science and thereby made agnostic to the political context in which it is embedded. Yet it also demonstrates how a procedural, technical arena provides some political openings for more inclusive governance. In its strictest interpretation, the letter of the PNPCA only requires interstate consultation. In practice, however, and subject to the discretion of the leadership of the MRC – and particularly the secretariat that manages the process – it has progressively evolved into a practice of consultation involving the public, civil society, local government and other stakeholders, even though such broader consultation is not mandated by the PNPCA. Despite their deficiencies (explored in Chapters 5 and 6), public consultations nevertheless represent an example of flexible governance, potentially open to strategic intervention by a range of actors. This move towards greater flexibility and accessibility has occurred in part because of prevailing public expectations concerning the need for greater openness and participation – in other words, this partially reflects the impact of transparency norms, discussed in Chapter 6 and 7.

It is clear, then, that soft law is indeed the governance 'modality' currently prevailing in the Mekong River Basin, at least in institutional settings of and surrounding the MRC. This has come to be, however, for more complex reasons than those that might be suggested by the intuitive assumption that this is a stage on an upward-inclined journey from no law to softer law to harder law. The experience of the PNPCA shows that soft law may also be a wedge in a door that sometimes opens further because of social pressure and prevailing normative expectations (cf. Sneddon and Fox, 2007). At the same time, the PNPCA draws on and gains credence from the hard law of individual riparian countries, most notably in the field of environmental and social impact assessment. It is to this field that we now turn.

5 Assessing a river basin

The politics of the technical

Prior chapters have examined how understandings of the Mekong River Basin have been crafted over time, and how the challenge of trying to govern the basin, largely by technical means, has been handled by the Mekong River Commission (MRC) since 1995. In Chapter 4, governance of the river by the MRC under the 1995 Agreement was subjected to critical analysis. This chapter elaborates on these governance issues by examining how Mekong River Basin development activities and conditions have been evaluated and measured through environmental assessment (EA) processes. It explores some examples of these EA processes and comments on the politics of measurement and assessment in this field. As the Mekong region becomes 'the scene for one of the most intensive hydropower developments globally' (Keskinen *et al.*, 2012, p. 320), the legally required use of EA processes gains in importance as part of development decision-making. EA as experienced in the Mekong has come to reflect a complex and at times confusing network of international, regional and national legal norms, practices and institutions. In the Mekong River Basin, EA processes, now more than ever, have become part of the interconnected stories of law. That is, they are embedded in practices of legal claim and counterclaim and involve a litany of instruments and institutions, proponents and opponents, allegiances and disaffections. Collectively, these enliven a sense of law – including law regarding EA – as unwieldy and yet still filled with potential, as we noted in Chapter 2.

The chapter begins by explaining briefly the current EA processes used in the Mekong at the national and regional level. The forms of EA include environmental impact assessment (EIA), social impact assessment (SIA), cumulative impact assessment (CIA), strategic environmental assessment (SEA) and, in some Mekong jurisdictions, health impact assessment (HIA). Also discussed is the increasingly significant area of transboundary environmental impact assessment (TbEIA). The chapter then moves to an investigation of the technical politics of EA, building upon our discussion in Chapter 4.

Light is shed on the practice of technical politics in the EA context by examining three very different hydropower development projects: A Luoi dam in Vietnam and Xayaburi and Don Sahong dams in Laos. The chapter grapples with the divergent interests with which the relevant decision-making processes must deal and how these are expressed in a register of measurement and evaluation through EA. We ask *who* participates in EA processes and who does not, or does not adequately, *what* is assessed and what is not in the course of EA, and the determinants of the *who* and the *what*, once again raising the concept of a river as a 'contestable resource', as canvassed in Chapter 1.

Environmental assessment legal requirements at the national level

To grasp what is intended to be assessed by EA in the Mekong River Basin, one needs to have careful regard to EA legislation enacted by each of the Lower Mekong countries and the EIA guidelines that each has developed. For this reason, Table 5.1 provides a snapshot of the main elements of the relevant laws, draft laws and guidelines. The table only records whether or not the particular element is mentioned, not whether the provision is 'adequate' or 'properly implemented' by reference to international standards and comparators; original instruments and documents should be referred to in order to gain a full picture. Briefly, the distinct EA processes referred to in the legislation and guidelines included in the appendix can be described as follows.

Transboundary environmental impacts are defined by the MRC's *Draft Technical Guidance* document for conducting TbEIA (MRC, 2012a) as:

> significant environmental impacts/changes originating within the territory of one Member Country which potentially affect other Member Countries. The environmental impacts/changes include effects on water quality, quantity, flow regimes, river morphology, biodiversity or further consequent impacts to people's livelihoods depending on the Mekong River and its tributaries.

The EIA process that addresses transboundary impacts is defined in the *Technical Guidance* as 'a process for assessing the potential transboundary environmental impacts/changes of the proposed project/activity'. Expert interpretations of 'impact', 'potential' and 'significant' are thus critical in any determination of what a TbEIA is to assess (MRC, 2012a).

SIA, as a more specialized aspect of EIA, involves assessment of the social and often the economic aspects of any particular development on communities potentially affected thereby. SIA is generally focused on 'analysing, monitoring and managing the social consequences of development' (Vanclay, 2003, p. 1) or, in the words of Cambodia's draft EIA Law of 2015, it is:

> the process of study and prediction of positive and negative impacts on the socio-economic aspects of society from projects and different development activities together with determination of appropriate measures to protect, mitigate, and compensate project-impacted-persons for the impacts on their lives, livelihoods, welfare, and health.
>
> (Cambodia, 2015, Annex 1)

As in TbEIA, practices of anticipating, investigating, detecting and documenting 'impact' serve important gatekeeping functions in any SIA process. There will almost invariably be socio-economic factors that can never register as such – for example, the long-term psychological or spiritual effects (and associated physical manifestations) of the displacement of peoples from their traditional lands. In the SIA context, however (unlike in TbEIA), the calibration of positive impacts is expressly invited in this analysis. This suggests that SIA may be less a practice of risk aversion or defending socio-economic rights than one of cost–benefit analysis or pragmatic weighing-and-balancing.

CIA refers to assessment of 'the impact on the environment which results from the incremental impact of the action when added to other past, present, and reasonably foreseeable future actions' (Lao PDR, 2012). It may also imply that 'all EIAs must analyze and evaluate the cumulative impact caused by existing and future projects in the surroundings of the Project, which may trigger significant environmental or social impacts' (Cambodia, 2014). In other words, CIA suggests a requirement 'to evaluate the cumulative impacts of multiple activities' (Dore *et al.*, 2010, p. 33) or to engage in 'assessment of the multiplier or cumulative impacts of more than one project or action across a geographic area or in the same area over time' (Carew-Reid, 2014, p. 4).

SEA is generally understood as a practice of long-term strategic planning, as opposed to an assessment focusing on a particular project or plan, or even a cumulative series of projects. In contrast to project-based and site-based EIA, SEA is by definition both strategic and conceptual, 'entailing larger scales and longer time horizons. It is more qualitative, more uncertain, vaguer and more distant from the public than EIA' (Lawrence, 2013, p. 25). In general terms, SEA entails assessment of 'the likely negative effects of policies, plans and programmes on the environment' (Montini, 2013, p. 250). It also assesses the impacts of alternative directions for the development through the construction of models and scenarios, thus supposedly broadening the scope of the assessment. What an SEA assesses thus differs quite markedly from what is assessed by an EIA, SIA, CIA or even TbEIA.

A more detailed elaboration of what may be assessed under the rubric of SEA is provided by Cambodia's draft EIA law, which states:

> SEA refers to a range of analytical and participatory approaches that aim to integrate environmental considerations into policies, plans and programmes and evaluate the inter-linkages with economic and social considerations. The principle is to integrate environment, alongside economic and social concerns, into a holistic sustainability assessment.
>
> (Cambodia, 2015, Annex 1)

It is not clear from the Cambodian draft what 'holistic sustainability assessment' is intended to mean, but according to Montini the term refers to the *merging* of SEA and EIA processes (Montini, 2013, p. 253; see further below).

The Vietnamese legislation defines SEA as 'analysis and forecast of existing or potential impacts on the environment, which have been described in the development strategy, planning and proposal, in order to provide measures to control and reduce adverse impacts on the environment, and to serve as a ground for and to be incorporated in such development strategy, planning and proposal with the objective

Table 5.1 Main elements of environmental assessment laws and guidelines in the Lower Mekong

Jurisdiction	Latest enactment or draft law or guideline	SIA	SEA	TbEIA	CIA	Public participation	Access to information
Cambodia	Environmental Protection and Natural Resource Management Law 1996 Sub-decree on EIA Process 1999 Draft EIA Law 2015	Yes	Yes	Yes	Yes	Yes	Yes
Laos	Law on Environmental Protection 2013 Decree of EIA 2010 EIA Guidelines 2012 Ministerial instructions 2013	Yes	No	Yes	Yes	Yes	Yes
Thailand	Enhancement and Conservation of National Environmental Quality Act 1992 EIA in Thailand 2012 (new EIA law contemplated)	Yes	No	No	No	Yes	Yes
Vietnam	Law on Environment Protection 2014 EIA/SEA Circular 2008 Decree on environmental protection, planning, strategic environmental assessment, environmental impact assessment and environmental protection plans, 2015	No	Yes	No	No	Yes	Yes

of ensuring the sustainable development. (Law on Environment Protection 2014, Article 3 para 22). Ke and Gao record that Vietnam has applied SEA a number of times in the context of power development planning (2013, pp. 989–990; see also Carew-Reid and Roop, 2013, p. 5).

The possible governance implications of national legal guarantees of 'public participation' and 'access to information' in the Mekong River Basin, highlighted in Table 5.1, is a theme that we take up in Chapter 6.

On examination of the legislation and guidelines on EA of the four Lower Mekong countries, we see that none of them directly refers to any specific obligations under the Mekong Agreement, although all of them refer to hydropower as a specific category for assessment, with electricity-generating capacity being the main determinant of the level of assessment required. Given the development pressures and continuing debate on hydropower projects and their effects, it is not surprising that national legislation and policymaking surrounding EA in the Mekong River Basin continue to be regularly reviewed and remain in a state of flux.

Taking account of this flux, we set forth below a brief overview of the content of EA laws in each of the four riparian states upon which we focus in this book. Cambodia, Laos and Vietnam have redrafted their legislation and policies over the past few years, and continue to amend them, while changes to the Thai provisions continue to be contemplated.

Cambodia

Cambodia's Environmental Protection and Natural Resource Management Law, drafted in 1996, included as one of its stated objectives 'to assess the environmental impact of all proposed projects prior to the issuance of a decision by the Royal Government'. A Sub-decree on Environmental Impact Assessment Process passed in 1999 makes clear that an Initial EIA (IEIA) or EIA is required for every private and public project or activity (including hydropower development with a capacity of one megawatt or greater), with detailed requirements contingent on the type and size of the project. The stated goal of this legislation is to 'encourage public participation in the implementation of EIA process and take … account of their conceptual input and suggestion for re-consideration prior to the implementation of any project'. A general guideline on IEIA and full EIA was published in 2009 and confirmed the previous enactments, but made exceptions for 'urgently needed' projects that are decided directly by the Cambodian government.

Despite these requirements, it has been reported that 'only about 5 per cent of the roughly 2,000 major development projects, such as dams, roads and bridges, approved by the government between 2004 and 2011 carried out environmental impact assessments' (AECEN, 2012). Ostensibly to redress this, as noted above, Cambodia, has now drafted a specific law on EIA that contains detailed provisions for EIA, SIA, TbEIA and SEA (Cambodia, 2015).

In an interesting instantiation of public–private hybridization upon which we have remarked elsewhere in this book, Cambodia's new EIA Law is being drafted in collaboration with an NGO law group (Vishnu). The latter group has stated that 'the key priority for Vishnu and the Ministry [of the Environment] is that the Law be developed in the most transparent and participatory manner possible' and that it 'will serve as an important regional model both in terms of process and substance' (Vishnu Law Group, 2015).

While it does not directly mention the Mekong Agreement, the Cambodian draft EIA legislation includes the term 'Transboundary Agreements', defined as

'Agreements made amongst riparian states about how shared (water) resources will be utilized by the parties involved, and the processes that will be followed to sustain these understandings' (Cambodia, 2015, Annex 1). Article 49 states: 'A Project that has potentially significant trans-boundary environmental impacts is required to conduct a trans-boundary Environmental Impact Assessment (TbEIA).'

With regard to assessment reports previously issued under Cambodian law, our interview material indicated that Cambodian NGOs have been wary of their quality, in part because of the outsourcing of compliance of EA requirements to paid consultants incentivized to facilitate the clearance of proposed projects. For example, with regard to the Sub-decree on Environmental Impact Assessment, one of our informants remarked:

> Businesses are required to conduct this. Unfortunately, they hire consultants to do these assessments. There is communication between consultants, they copy each other's information, or even use NGO sources. So you never expect good assessment reports because the company owner pays the consultant to do the report and they check the report. The projects are always feasible, always feasible.
>
> (Interview 6, 2011)

Laos

The Lao government issued a Regulation on Environment Assessment in 2000 under the Law on Environmental Protection 1999 and a Decree on EIA in 2010. The decree was replaced in 2013 by a Ministerial Instruction on the Process of Environmental and Social Impact Assessment of the Investment Projects and Activities and a Ministerial Instruction on the Process of Initial Environmental Examination of the Investment Projects and Activities.

Laos also introduced EIA Guidelines in 2012 for a trial period of two years, during which time all stakeholders were called on to submit comments for improvement. Given the experience of the Xayaburi hydropower development and the significance of an MRC-commissioned SEA in shaping the debate (although not in informing the decision) surrounding that project (see Chapter 4), it is curious that these guidelines do not contain any specific provisions on SEA. Nevertheless, TbEIA is required for investment projects beyond national borders (Articles 4(5) and 6(2)).

Thailand

Thailand has used the EA process as a screening device since the early 1980s, with legal provisions introduced in its Enhancement and Conservation of National Environmental Quality Act 1992. Thailand issued a comprehensive set of EIA guidelines in 2012 (Thai Office of Natural Resources and Environmental Policy and Planning, 2012), which specifically include provisions on social and health impact assessment; these do not as yet have the force of formal law.

Vietnam

The Vietnamese Law on Environmental Protection of 1993, revised in 2014, includes detailed provisions on EIA, especially concerning the range of projects that require an EIA, as well as how the process should be carried out. It also includes SEA requirements (see Articles 15–23). In 2008, the Vietnamese government issued a Circular Guiding Strategic Environmental Assessment, Environmental Impact Assessment and Environmental Protection Commitment, and in 2011 it issued a further Decree Providing Strategic Environmental Assessment, Environmental Impact Assessment and Environmental Protection Commitment. In early 2015, yet another decree was issued on environmental protection, planning, strategic environmental assessment, environmental impact assessment and environmental protection plans.

Environmental assessment as a practice of technical politics

Consistent with its prominence internationally, EA has become an increasingly important aspect of the practice of governance in the Mekong region, informing and tempering development activity. The requirement to conduct one or more forms of EA in connection with scrutinizing, evaluating and authorizing development has gained acceptance in the Mekong River Basin, largely under the influence of international standards and external agents such as donors, investors and NGOs (see Chapter 3).

While these assessment processes are generally characterized as technical, the techniques in question are as much political as they are practical or scientific. That is to say, they concern first-order issues (what is, and what is not, arguable or knowable, and what ought to be done) as much as they concern second-order questions (those commonly understood as matters of 'implementation', 'adequacy', 'usage' or 'compliance'). The initial choice of who drafts the EA's terms of reference, for instance, determines the scope of what is assessed and, by implication, what is not assessed. Contestable norms surround the identification of what is measured – that is, what counts and can be made countable – and how. Further choices must then be made as to who conducts the assessment, who should be consulted, and at what stage, and who is allowed or invited to participate, and in what capacities (for example, as local governments, local communities, NGOs, individuals and so on). The way in which the resulting EA report or statement is published or otherwise distributed, and whether or how it is used to promote the development activity, are also relevant political considerations.

The first-order emphasis upon 'voice' in the EA context – and associated assumptions about rational planning – are among the EA features that some have taken issue with (Fischer, 2003). Implicit or explicit awareness of such political dimensions of EA among design and construction professionals has been confirmed by scholars of engineering:

> Engineers are not neutral bystanders in these political, economic and environmental struggles but rather – such as with damming projects along the

Mekong – active players. Whether they recognise it or not, the decisions they make with respect to projects they work on means they are political players and complicit in outcomes that lead to the curtailment or infringement of affected people's rights.

(Barrington *et al.*, 2012, p. 34)

A significant feature of the politics of EA in the Mekong River Basin is its mediation by highly standardized contractual structures and a fairly limited line-up of repeat-player consultants. The tasks of scoping, data collection, arrangements for public participation, technical analysis and the preparation of the EA report or statement remain largely dominated by international consultancy firms. Such firms are necessarily in the employ of governmental and private sector development interests. However, subsequent distribution, usage and interpretation of, and negotiation surrounding an EA report rarely involve these consultants directly. Consultants '[a]lmost never see the results of their work, which discourages their sense of responsibility and encourages a formulaic approach' (Fisher, 2013, p. 8).

The practice of outsourcing EA to consultants, often responsive to a project-specific, financier-approved brief, creates a sense of EA as both ubiquitous and rootless. According to Fisher, confusion surrounds by whom, exactly, EA requirements are imposed or whose objectives they are designed to serve:

While few project promoters go so far as to completely deny the necessity of environmental analysis, virtually all complain about the increasingly formalistic and bureaucratic procedures, and agree with the impression that the primary objective seems to be to satisfy the environmental departments of the financing banks.

(Fisher, 2013, p. 8)

Due in part to the nature of EA processes themselves – in requiring a wide variety of choices to be made at both macro- and micro-levels and sequestering these, one from the other, in a predetermined procedural sequence – EA almost invariably elicits a divergence of views and priorities. This divergence may, in some instances, be made explicit. However, unlike at the national level in many other countries, where EA processes can be legally challenged for failing to meet legislative requirements, there is no regional tribunal or body that is empowered to resolve any particular contest between project proponents and opponents in transboundary contexts. As established in Chapter 4, the MRC is clearly not such a 'policing' organization, (and see Hirsch *et al.*, 2006). In the theatrical analogy, the MRC may at times be a producer of an EA drama, but it is not a director.

Partly responding to the lack of control over how environmental impact assessments are conducted and how the results are used, a range of alternative tools for measuring and assessing environmental impacts of hydropower dams has been developed by the hydropower industry, international development organizations and civil society groups. Many of these assessment tools have involved a multi-stakeholder approach to get buy-in from multiple sectors, such as the

World Bank-initiated World Commission on Dams, the Hydropower Assessment Sustainability Assessment Protocol and the Rapid Sustainability Assessment Tool. Civil society has also responded to the limitations and inadequacies of conventional EA processes through grassroots research initiatives that seek to mobilize local knowledge to produce 'people's EIAs' (Manorom, 2007; Tai Baan Research Team, 2004).

Who decides?

The tension between the various interests at stake in EA raises questions as to who makes the decisions involved in EA processes, on what grounds and for whose benefit. The various forms of EA practised in the Mekong, as varied, multilayered and imperfect as they are (and perhaps in part because of their imperfections), furnish a significant case for characterizing the region as a socio-legal arena, with the attendant 'drama' we mentioned in Chapter 1.

In these assessment processes, despite the formal requirements for consultation found in national legislation, and persistent efforts by communities to interject, the cultural, social and economic interests of local communities affected by these activities are often given less emphasis than other interests. In most Mekong River Basin EA processes, financial compensation and/or land replacement decisions are dictated by national governmental authorities. On the other hand, government promises – for example, regarding the provision of alternative settlement land – are, on occasion, also not met, thus disappointing both communities and private companies that have placed reliance on such undertakings.

As noted above, each of the Mekong jurisdictions has enacted legal provisions and has generated policies on EA pursuant to which decision-making authority will be allocated. In addition, the MRC has issued its own draft guidelines. These provide some answers to the 'who decides' question, at least in the first instance. Nonetheless, a factor conditioning this distribution of decision-making power – and the consequent realization of national law and policy objectives, including those of EA – is the relative status of the government departments charged with making decisions with regard to an EA and acting upon its findings. This is a more nuanced aspect of *who* decides. In the Lower Mekong states, the environment ministry that carries out reviews of the EA processes is generally of a low political status within hierarchies of government (Wayakone and Makoto, 2012, p. 1663). Often, it is a higher status economic department that will have the final say over whether a development project will go ahead.

What is assessed?

Irrespective of who is understood to be driving the EA process in a particular instance or how that process is framed (that is, as EIA, SIA, CIA, SEA or TbEIA), its orientation and content will always be directed towards assessing 'impacts'. Precisely what range of actions (or inactions) will be reviewable for impact, and how those actual or potential impacts will be measured and documented remain, nonetheless, unresolved as general region-wide concerns. Given the emphasis that has now come to be placed on EA processes as part of the practice of planning and managing major development projects in the Mekong River Basin, the provision concerning EA under the Mekong Agreement is, perhaps surprisingly, brief. Under 'Functions of the Joint Committee', Article 24D states: 'To conduct appropriate studies and assessments for the protection of the environment and maintenance of the ecological balance of the Mekong River Basin.' Among the specified functions and duties assigned by the council and joint committee, the agreement requires the MRC secretariat to '[f]ormulate the annual work program, and prepare all other plans, project and program documents, studies and assessments as may be required' (Article 30C). These are then taken into account to a greater or lesser degree in subsequent decisions and policy formulations.

The MRC's policy development in the past few years with regard to EA has focused primarily on TbEIA and more recently on SEA. With regard to TbEIA, the MRC made an initial attempt in 2005 to spell out an appropriate framework, with a revised version in 2010, and then, as noted above, produced a *Draft Technical Guidance* on transboundary EIA in 2012. That last document observes that:

> it is rather difficult to define common quantitative standards that could deter-mine significant environmental impacts for a specific type of proposed project/ activity. National EIA legislation[s] of the Member Countries var[ies] with regards to EIA requirements according to list, size, magnitude, nature and loca-tion of proposed projects/activities. The criteria for conducting a national EIA therefore differ among the Member Countries.
>
> (MRC, 2012a, p. 5)

Beyond the broad umbrella terms of 'impacts', 'the environment' and 'ecological balance', precisely what is to be assessed within the scope of EA thus tends to be deferred to later in the decision-making process. It is a matter to be worked out in the context of 'project and program documents' with reference to 'proposed projects/activities'; it is not a matter for a priori or project-independent debate. As a consequence, attention commonly comes to rest – in the context of EA – more on project delivery than on the evaluation of alternatives (including the 'no-go' alternative). The 'technical' issues around which EA commonly revolves include the evaluation and mitigation of the impact of dam building on deforestation; the destruction of agricultural plots, cash crops and fruit trees; the building of access roads to the development site; water quality and hydrology; arrangements for moving communities out of the path of the dam; and the dam's accoutrements, such as the barrage, powerhouse and transmission lines.

Insofar as the breadth or depth of a particular version of EA is contested, relatively infrequent recourse is had to formal legal avenues of complaint in the Mekong River Basin. In other regions, where an EIA process is judged to be inadequate for technical reasons, or for not considering and analysing all statutorily required fac-tors, the resulting environmental impact statement can be declared invalid through an administrative appeal or by a court or tribunal. In those circumstances, the assessment process may then need to be recommenced or its scope expanded, and the project may be delayed or even halted. However, in the Mekong, such legal challenges to EA processes are rare (as noted with regard to the *Niwat* case below and in Chapter 7) because of factors such as lack of information and inadequate procedures, or no procedures at all, for complaint through institutional or judicial processes. Civil society tends not to look automatically to the law to air or address EA-related concerns, largely because of distrust of its institutions. For example, one NGO representative stated:

> I think the people, they understand about the law, but they disbelieve about the mechanism because the … judicial system is corrupt and most of the judges are appointed by the party … The system is corrupt from the grassroots to the top.
>
> (Interview 6, 2011)

Several interviews confirmed that NGOs and communities try to address their problems in other ways, including working both formally and informally with government ministries and local authorities (Interview 6, 2011; Interview 8, 2011; see also Chapter 7).

The question of what is assessed in an EA thus devolves, frequently, into a question of who may participate in the framing, conduct and contestation of EA processes.

Who participates?

A key predicate and overarching rationale for EA requirements is the apparent virtue of 'participation' in politico-legal processes of decision. Priority is given, in this context, to participation by collective agents standing outside the government – civil society actors – with a view to the latter exerting some check upon the power of the former. As we discuss in Chapter 6, the political ideas underpinning such a commitment to participation (and its close correlate, transparency) may not be uniformly shared across the Mekong River Basin. Nonetheless, the question of who participates in EA remains a recurrent concern, thanks in part to its regular airing by NGOs at a range of scales. This may, sometimes, be in lieu of a more provocative (and thus politically risky) airing of distributional or abuse-of-power-related concerns.

Participation by civil society in an EA is first conditional upon knowing that such a process is, or is about to be, underway. From the point of view of the government department or the developer involved in a hydropower project, however, EA is often seen as a narrow means to a predetermined end – namely, that the hydropower project be approved. Accordingly, EIAs are sometimes not published, since publication may not be required by applicable national law but depends on the will of the government or the developer, and it is not necessarily in their interests to do so. As a result, members of civil society may not even know if an EIA has been conducted. As one of our informants stated: 'When they build a dam or conduct an EIA, they don't inform communities' (Cambodia NGO, Interview 8, 2011).

Without intervention from NGOs, then, the public at large often does not become aware of – let alone get invited to – meetings designed to ensure participation in EA. Who is actually allowed to be involved is determined to some extent by who is arranging the consultative meetings. As one Lao official observed, hydropower development is 'more about engineers, not about people as much' (Interview 49, 2012). A consultant in Vietnam, talking about a meeting with the Power Trade Coordinating Committee, put it more bluntly:

> The whole room was filled with black suits. Even the women were wearing black suits. And there's nobody from any other sector there. These guys are in a platoon against the real world, power people, all in black suits, are on a mission to save their countries and are 'cocooned' in the power sector.
>
> (Interview 32, 2011)

Thus even where commitment to public participation is manifest, the standardization of participatory rituals under the rubric of EA has the potential to hollow out the political significance of participation for those who do gain access. On being asked about Cambodian regulatory requirements demanding community consultation, one Cambodian informant observed:

> Yes, always this is the problem. There is a legal requirement for that, but they don't do it, or they don't do it properly. They just tick the boxes.
>
> (Interview 6, 2011)

Other interviews in Cambodia (for example, Interview 8, 2011) likewise revealed that legally mandated participation requirements are not well followed, whether in their letter or in their spirit. As the commune chief of Sre Kor in Stung Treng province of Cambodia – affected by the Lower Sesan 2 dam – stated: 'we have been living like a frog in the well – not knowing anything' (quoted in Di Certo and Titthara, 2012).

Moreover, where broad participation in a process of EA consultation is in evidence in the Mekong, this does not necessarily ensure that people's concerns will be taken seriously. As a senior Lao official stated, 'we take legitimate concerns, not rubbish one[s]'. Civil society input is regarded as 'good if they give legitimate comments', but the government will ignore too many negative comments against hydropower development. The official argued that 'hundreds of hydropower projects are developed around the world, and asked why Laos should be unfairly singled out'. Complaints that Xayaburi dam 'will stop fish, kill 60 million people' were said to be unhelpful, since at other dams 'people still enjoy fish' and some even supply all the fish to Vientiane (such as ten tonnes a day from Nam Ngum). This official blamed overfishing and overpopulation, not hydropower, for declining fish stocks. He also stated that 'our leaders do not know why they [civil society] exaggerate' (Interview 52, 2012).

Accordingly, no matter how pervasive the discourse of participation, the professional culture of hydropower protagonists – and the governance styles that they help to foster – tend to collide, recurrently, with demands for greater civil society participation. With regard to Xayaburi, for example, our interviewees indicated that public consultations were not what they had expected:

> We thought that the community consultative meeting was about communities joining the meeting to get information and to voice their concerns to the government. But no communities joined that consultative meeting, only the government. Commune councils were there. I checked with them and they didn't have any information about Xayaburi. Some of them thought Xayaburi dam was in Cambodia, not in Laos! It seemed they didn't know anything about it. That was strange for us. In terms of NGOs, they only invited NGOs working in that area. We wrote a letter to ask CNMC [Cambodian National Mekong Committee] if we could join the meeting and they gave us permission to participate. They allowed six NGOs who are part of RCC [Rivers Coalition in Cambodia] to join the meeting. Most of the participants, especially local government authorities, didn't know much about Xayaburi. So we shared information about Xayaburi at the meeting. We shared documents we had brought with us with all the relevant stakeholders and we joined in the discussion.
>
> (Interview 8, 2011)

Striking an optimistic note, Ke and Gao maintain that public involvement in the EIA process 'is taking a leading role in promoting legal reforms towards deliberative democracy' in the Mekong River Basin. They argue that the growth of environmental NGOs is the most vibrant aspect of civil society development in the region, and that public participation in the EIA process is 'a breakthrough point for a wider or expanded application of this mechanism at different levels of decision-making in the future' (Ke and Gao, 2013, p. 995). Our study of EA participation in the Mekong does not, however, support the view that wholesale breakthroughs along these lines are imminent. Instead, our sense of the openings and possibilities that EA participation may occasion is more interstitial.

By now, it may be apparent that the responses to the questions that we have posed regarding EA in the Mekong River Basin – who decides, what is assessed, who participates – depend to a large extent on the expert practices of measurement or evaluation that are central to EA. The particular mandate, training and expertise of the party or parties charged with carrying out EA in any one instance will inform the types and scope of the impacts that they anticipate and identify. This, in turn, will determine the range of people identified as stakeholders who may, on that basis, be entitled to participate in the EA process. Beyond these general tendencies and beyond the framing by legislators and policymakers of 'best practice' in the EA field, however, it remains important to consider the range of ways in which EA has been conducted in particular instances of Mekong River Basin hydropower development. The following section will do so in relation to three of the dam suites introduced in Chapter 1: A Luoi, Xayaburi and Don Sahong.

Analysis of environmental assessment: A Luoi, Xayaburi and Don Sahong

This section demonstrates that the EA processes utilized for the A Luoi dam, the Xayaburi dam and the Don Sahong dam differ considerably. Examination of the technical politics at play in each instance reveals somewhat different features of the socio-legal landscape of EA, as this has taken shape – and continues to take shape – in the Mekong River Basin.

A Luoi

Vietnam's law and policy on environmental issues – a snapshot of which was provided above – has changed considerably since the introduction of the Doi Moi or 'renovation' policy from the mid-1980s. With respect to public participation, Dao comments that:

> once the country opened its economy and began pursuing neoliberal develop-
> ment policies, pressure began to mount from international and environmen-
> tal groups promoting better accountability, transparency and participation in
> the development process. The government in general has begun to pay more
> attention to these issues. As a result, there has been a clear shift over time in
> the policy and planning process for the dam-associated displacement of local
> populations.
>
> (Dao, 2010, p. 327)

A particular attenuation of these 'neoliberal development policies' is in evidence in the EA related to the A Luoi dam.

The A Luoi dam is situated in Thua Thien Hue province some 70 kilometres west of Hue. The dam, one of three in the A Luoi district, is the largest in the province. Its construction is unlike the majority of storage and run-of-river dams in the Mekong River Basin, owing to its function of diverting water across a major escarpment, which involved the excavation of a 14-kilometre-long, five-metre-wide headrace tunnel to link a large reservoir to the power station several hundred metres below. More than 30 kilometres of roads were constructed and more than 20 kilometres of transmission lines were put into place (International Waterpower and Dam Construction, 2009).

The dam is built on the upstream of the A Sap River, which is the headwater reach of the river that becomes known as the Sekong as its waters cross into Laos and then Cambodia. As noted in Chapter 1, the dam represents an inter-basin diversion, diverting water to the Bo River, thereby increasing capacity of the Huong Dien hydro dam on the Co Bi River, with most of the water that would otherwise have flowed into the upper Sekong in Laos being lost to that system. As recognized in the environmental impact statement, the dam site is only a short distance from the Vietnam–Lao border, and as the dam was recognized as potentially causing some transboundary impact with respect to Laos, province-to-province governmental consultations were held across the border, with a Lao delegation going to Hue province (Interview 41, 2011).

However, despite the dam's proximity to Laos, the perception of the A Luoi District People's Committee was that '[u]ntil now, Laos has no concerns ... A Luoi district is seen to have a good relationship with Kaleum district and Salavan province in Laos' (Interview 41, 2011). Thus, in contrast to the centralized national decision-making that appears to be typical of most other hydropower developments, provincial governments in this case appear to have had some influence with respect at least to the assessment of potential transboundary impacts – although not on whether the development itself should go ahead. In the case of A Luoi, no formal TbEIA was carried out. In addition, the development was not notified by Vietnam to the MRC pursuant to Article 5A of the Mekong River Agreement, which requires that '[o]n tributaries of the Mekong River, including Tonle Sap, intra-basin uses and inter-basin diversions shall be subject to notification to the Joint Committee'.

The area affected by the A Luoi dam is around 4,200 hectares, involving six communes consisting mainly of two ethnic minorities, the Pa Ko and the Ta Oi. Pham Huu Ty et al. (2013) record that the dam and reservoir displaced 218 households in total (more than 870 villagers). Many of the households moved to a newly built resettlement village some 15 kilometres from the dam site. The storage of the water in a reservoir has resulted in the flooding of agricultural areas, while the building of the roads has affected access to forests and coffee crops. Community incomes have reduced considerably, and the participation of local people in decision-making has been limited (derived from Hoang, 2012 and Pham et al., 2013).

Disputes concerning resettlement have centred on the level of compensation for affected communities. The A Luoi District People's Committee Chair indicated that some 190 billion VND (US$9 million) was allocated for compensation to the communities. Some of the residents whose land was not flooded initially refused to move from the old area to the new area because of the living conditions and the poor quality of the soil in the new area. Although the local communities petitioned both the company and the government about their situation, little action was taken to address the complaints (drawn from Hoang, 2012 and informal interviews with a small number of residents in 2011).

To place the A Luoi development into a broader national Vietnamese perspective, up to 2007:

> The construction of hydropower dams has displaced 44,557 households or about 200,000 people and expropriated 133,930 hectares of land. Although hydropower dams have the potential to bring benefits in terms of power supply, flood control, and irrigation, their construction can harm ecosystems and uproot local communities.
>
> (Pham et al., 2013, p. 679, internal references omitted, citing Ministry of Industry and Trade, 2007)

A UNDP study that analysed the environmental and social costs and risks of hydropower found that, regarding costs, 'citizens are generally worse off after resettlement' (UNDP, 2013, p. v). Nevertheless, according to Vietnam's national master plan for power development, 'Vietnam continues to give priority to hydro-power station development, bringing its capacity from 14,000MW as in the present to 21,300MW by 2020', and it 'continues to develop small and medium-scale hydro-power stations' (Ministry of Industry and Trade, 2014). Recent reports indicate that Vietnam's hydropower plants now number 284, with 204 hydropower projects under construction. However, 12 planned projects have recently been scrapped because of perceived 'low economic benefits and high social and environmental risk' (Ngan Anh, 2014).

In order to address concerns of displaced communities, the prime minister of Vietnam issued a detailed decision, setting out general principles for compensation, support and resettlement (Decision No. 34/2010/QD-TTg: see Vietnam, 2010). In spite of these stated principles, however, Pham *et al.* argue, in relation to the prime minister's decision, that:

> The compulsory acquisition process implies that all activities of compensa-
> tion and resettlement are designed and carried out by the responsible agencies
> of local government and then presented to affected people with details on
> losses, compensation values, and resettlement plan. There is no policy to put
> the people to be displaced in the centre of decision making process.
>
> (Pham *et al.*, 2013, p. 681)

During our field visit to the A Luoi dam, we were unable to gain a clear picture of whether affected communities broadly supported or opposed the A Luoi project, or were satisfied with their resettlement and compensation or with the EA process. Our visit to the resettlement village was arranged by Vietnamese officials and we had to be accompanied by provincial authorities. They identified the villagers with whom we spoke, and we were forbidden to take photographs of the area. The incident illustrates that the space for civil society actors to examine and assess hydropower projects in Vietnam is rather constrained, even (or perhaps especially) for foreign academic researchers. However, other research (Pham *et al.*, 2013) has established that there is a good deal of dissatisfaction surrounding the A Luoi dam.

We were, nevertheless, able to gain some insight into the EA and related consult-ation processes that were carried out for the A Luoi dam. The development project consultants retained for the A Luoi project talked with the People's Committee and villagers about resettlement locations. It was explained to those who initially did not wish to move that their houses would be flooded, and that the 'social and eco-nomic development of the province necessitated' the project (People's Committee, Interview 41, 2011). The People's Committee (effectively the local government) and the A Luoi Central Joint Stock Company (the developer) both claimed that the 'consensus' of the people was received. 'There was no great difficulty in resettlement because it was relatively easy to get the consensus of the people' (Interview 42, 2011).

As the Hue Foreign Affairs Department put it, the people 'feel sad about leaving home, but happy about their new wealth' (Interview 39, 2011). One resident of the resettlement village we visited agreed with this description, saying that he was 'excited' to be living in the new village. In summary, he stated that there was better land, many jobs, more money, better houses and furniture, electricity and piped water, schools and a medical centre. He said that he was 'not very sad to leave the old village – we wanted to leave to develop the economy' and that a lot of information was provided beforehand and the company helped him to move. The resident's spouse was more ambivalent: 'I can't say I'm happy; not right to say I'm sad; it is too far from uncle and relatives in the old place' (Interview 43, 2011).

The A Luoi People's Committee claimed that many documents were issued to the people, and there were government lawyers to 'guide people to understand' the land law and the compensation law: 'every month or quarterly the lawyers would go to the communities to explain the law – for example, on resettlement' (People's Committee, Interview 41, 2011).

On the other hand, provincial policies for resettlement and compensation have been described as 'inconsistent, overlapping and leading to prolonged complaints by the local community (Hoang, 2012). As noted by Tran Van Ha in his case study of involuntary resettlement for the Son La hydropower project, participation is a relatively new concept in Vietnam. Even so, Tran contends that, in the context of hydropower resettlement over the past few decades, Vietnam has seen 'significant improvements in the supporting institutional framework under which participation is carried out' (Tran, 2011, p. 40).

According to our own interviews, as well as research by others, the residents displaced by A Luoi had little or no direct participation in any level of the decision-making or project evaluation process, including the EA. The environmental impact statement produced for the A Luoi development did not include any specific SIA study, while the scientific studies undertaken in that context appeared to downplay the impact of the building of the dam and the power plant.

The EA process nonetheless seems to have been crucial to the consolidation of the A Luoi dam into an object of 'consensus'. Regardless of whether EA participation can or should be judged 'adequate' against external or internal measures, the EA process seemed to have been critical in this instance in creating prospects for political and legal resolution around the project. Movement through the sequenced steps of an EA process seemed to encourage, at once, a sense of progress and a sense of inevitability surrounding the A Luoi dam. Measurement, in an EA mode, thus served as an end in itself. The mere fact of an EA being conducted may have been more or less enough to set aside questions otherwise attending, such as inter-basin diversion of water.

Any such sense of resolution has, in contrast, been far harder to generate in relation to the Xayaburi and Don Sahong projects. This is undoubtedly a consequence of their scale and clear transboundary dimensions, but it may also be (in part) a result of the broader range of EA methodologies in contention surrounding these projects.

Xayaburi and Don Sahong hydropower projects

In Chapter 4, discussion about the governance framework for the Mekong River Basin included an analysis of governance arrangements around the Xayaburi and

Don Sahong hydropower projects. We argued there that 'soft law' is the prevailing governance modality in this setting. That discussion is continued here, but we now focus particularly on the contrasting EA aspects of the two projects.

As stated in the 2010 SEA commissioned by the MRC:

> The 2003 PNPCA [Procedures for Notification, Prior Consultation and Agreement] protocol and its 2005 procedural guidelines require Member Countries to notify the MRC in the event they wish to engage in any major infrastructure developments (such as hydropower schemes) on the mainstream Mekong or tributaries, particularly if those developments may have significant trans-boundary impacts on people or the environment downstream.
>
> (ICEM, 2010, p. 41)

As noted in Chapter 4, the Xayaburi proposal was the first hydropower development to trigger the prior consultation process under the PNPCA, as it was the first mainstream dam on the Lower Mekong. The Don Sahong dam project in southern Laos is planned to be the second mainstream dam after the Xayaburi project. The PNPCA process for Don Sahong was initiated in 2014 (MRC, 2015a).

While the Lao EIA provisions, briefly described above, mandate 'the participation of local administration, mass organisations, and population likely to be affected by the respective development project or activity', participation in relation to the Xayaburi EIA was limited – nowhere more so than in Laos. The environmental impact statement that resulted from the Xayaburi EIA was prepared outside Laos by Bangkok-based firm TEAM Consulting Engineering and Management Co. (2010). The conclusions and recommendations of that statement summarize its four main environmental aspects: physical resources, ecological resources, human use values and quality-of-life values. The definition and breakdown of environmental impacts relate to each of the 'project features' – namely, the barrage, powerhouse, transmission line and access road during both the construction and the operation periods. The flavour of the EIA as a project justification document (as we discussed earlier in this chapter) is apparent in its first few lines: 'The Mekong River possesses a huge hydropower potential, both in its mainstream and its tributaries. A large part of this potential is in the Lao territory due to the country's favourable climatic and geographic conditions' (TEAM Consulting Engineering and Management, 2010, p. 1.1). The Xayaburi EIA did not include reference to any attempt at involving communities with regard to impacts, even though the EIA report reiterated the statutory participation requirements outlined above (TEAM Consulting Engineering and Management, 2010).

While space for participation by civil society in countries such as Laos is limited, international NGOs have nevertheless been able to influence assessment and planning in relation to various hydro development projects, particularly when those projects have involved international public financing. A regulation on NGOs was enacted in 2009, allowing for domestic NGOs to register with the government. Nevertheless, as Wayakone and Makoto (2012, p. 1661) record, international NGOs have been able to influence assessment and planning in relation to various hydro-power development projects in Laos.

International Rivers, for example, criticized flaws and gaps in the Xayaburi EIA, including the lack of key technical information and analysis on fisheries and aquatic resources, water hydrology, sediment transport and earthquake risks; and that the report only included a downstream area of ten kilometres, with transboundary, basin-wide and cumulative impacts not addressed (Trandem, 2011). The fact that the Xayaburi EIA did not assess downstream impacts beyond ten kilometres – thus not taking into account effects on Cambodia and Vietnam territory – was seen as a major gap. One NGO interviewee with whom we spoke noted that the organi-zation in question 'tried to make a statement … saying the dam should not be approved without having evidence that it will not have impacts on downstream countries' (Interview 8, 2011).

Failure to assess downstream impacts has its antecedents. For example, the EIA for Yali Falls did not consider downstream impacts in Cambodia (Mean PowerPoint presentation in Interview 10, 2011). With respect to the A Luoi dam, the attention paid in the EIA to downstream impacts was minimal. According to the scientific evidence canvassed in that EIA, 'impacts of the decrease in the river downstream in Laos are trivial' (Central Region Hydropower Stock Company, 2006).

The level of public consultation surrounding the Xayaburi EIA was similarly the subject of criticism. The Xayaburi EIA's consultation process was restricted to 252 households located in the project site – less than 8 per cent of the people who would be directly impacted by the project (Trandem, 2011, p. 2; see also BankTrack, 2015b). Despite its shortcomings, this EIA consultation was cited by the Lao gov-ernment as a justification for not conducting any public consultations in Laos as part of the PNPCA process, as discussed in Chapter 4.

A novel development in the Xayaburi case – and an expression of the dis-satisfaction surrounding the related EA process – was the filing of a petition by the Network of Thai People in Eight Mekong Provinces in the Thai Supreme Administrative Court (*Niwat v Electricity Generating Authority of Thailand*). The peti-tion asked the Electricity Generating Authority of Thailand 'to respect community rights and comply with the 2007 Constitution by arranging transparent public hearings, as well as health and environment impact assessments before signing power purchase agreements'. The case appears to be the first time that a civil soci-ety organization in the region has directly challenged in court any of the processes by which a transboundary hydropower development has been decided. Regardless of the outcome, opening the court doors in this case may arguably lead, in the longer term, to further actions to access the decision-making processes – not only in Thailand, but also in the other Lower Mekong countries (see further Chapter 7).

Increased insistence on a transboundary EA approach among policymakers and donors in the Mekong River Basin, as recorded by the MRC's 2012 draft Technical Guidance on Transboundary Impact Assessment, has called into question the legiti-macy of unilateral decisions, such as those made in relation to Xayaburi and Don Sahong. The imperative of developers and governments taking into account the broader concerns of social and economic impacts on local communities on the one hand, and downstream and transboundary communities on the other hand, was also promoted by the 2010 SEA, as discussed in Chapter 4. The SEA (as would be expected) urged the taking of a broader view than EIA would typically entail, leading to its finding that the dozen or so potential mainstream projects 'could have profound and wide-ranging socio-economic and environmental impacts in all four riparian countries' of the Lower Mekong River (ICEM, 2010, p. 9). As one independent consultant based in the region observed:

> [I]n this region where you have very little integrated planning, SEAs are often forced back into trying to fill gaps and voids that should have been conducted in a proper integrated planning process.
>
> (Interview 32, 2011)

Certainly, it can be said that the SEA carried out in the shadow of the Xayaburi project illuminated the highly complicated relationship between realizing national economic aspirations for development and trying to achieve ecological and social sustainability on a regional basis. It revealed, too, unresolved tension within the politics of the technical in this context, as different actors have championed differing EA methodologies and outputs. Measurement has proven to be an intensely negotiated practice: a matter of representation and re-representation at different scales, and translation between them, from the broad timescale and spatial reach of an SEA to the truncated space and time of a project-specific EIA and back again.

It is also clear from the Xayaburi EA experience and the subsequent SEA process, as compared to that surrounding A Luoi, that the imperative of conducting or commissioning some mode of public participation is now broadly affirmed in the Mekong region, at least with respect to large-scale projects with acknowledged cross-border dimensions:

> [O]ne reason [for this] may be the inherent value of consultations and transparency, in terms of providing information and to planners. Another reason may be that if stakeholders are not provided with an official channel to influence decision-making – and therefore the likelihood and severity of the impact – social unrest may ensue.
>
> (ICEM, 2013, p. 43)

Another way of interpreting this endorsement of 'participation' is that development interests, including governments, feel obliged to be more transparent and consultative in order to placate disaffected communities and groups, in an attempt to remove bothersome barriers, or to demonstrate 'good governance' practice vis-à-vis international financiers. A further possibility is that calls for 'participation' – and actions taken in this context – serve as proxies for other types of grievance or intervention. Consider the interviewee we quoted above, reporting on 'community consultative meetings' held in Cambodia in 2011 in relation to the Xayaburi project. Despite finding the arrangements 'strange' and concerning, our interviewee did not resile from political action:

Most of the participants, especially local government authorities, didn't know much about Xayaburi. So we shared information about Xayaburi at the meeting. We shared documents we had brought with us with all the relevant stakeholders and we joined in the discussion.

(Interview 8, 2011)

Faced with a foreclosed process, our interviewee seized an opportunity to try to prize that process apart, recognizing that to join in the discussion is also (potentially) to change it in some way.

Further to this theme, there was a discernible shift in tone between the Xayaburi and Don Sahong EIAs. While in the Xayaburi environmental impact statement of 2010 there was no mention of the 1995 Mekong Agreement, and very limited mention of any notions of sustainability, the Don Sahong EIA of 2013 referred to the Mekong Agreement having 'guided the development of the Project and the associated environmental and social documentation'. 'This process is ongoing', the Don Sahong EIA stated. It noted, too, that the general principles for planning hydropower development include that the development must be 'equitable and sustainable'; that it 'should be seen in [the] context of the regional energy sector, in particular realistic future energy demands'; and that '[f]isheries and navigation are integral elements of hydropower dams, [so that] it is necessary to find optimal solution[s] to conjunctive hydropower generation, navigation lock operation and fish migration' (National Consulting Company, 2014, p. ix). It remains to be seen whether this change in the rhetoric of EA over recent years will be a harbinger of any significant change in deliberative, planning and decision-making processes surrounding hydropower development in the Mekong River Basin, or whether the renewed salience of the Mekong Agreement might open up new occasions and institutional settings for political action.

Conclusions

What then do the processes of EA, as practised at the regional and national levels, say about the socio-legal conditions of the Mekong River Basin? Can we accept, for example, that the river and its waters have, in some senses, become primarily economic commodities 'neglecting its other uses and benefits for livelihoods such as capture fisheries' (Lazarus *et al.*, 2011, p. 248)? Practices around the initiation and conduct of EA in the Mekong River Basin do seem to be geared more to promoting the interests of developers than to other considerations. To what extent might the governance role of the MRC, discussed in Chapter 4, exert some countervailing force in this regard?

Despite the Mekong Agreement's prescription for member states to 'cooperate in all fields of sustainable development, utilization, management and conservation of the water and related resources of the Mekong River Basin', and the MRC's mandate to facilitate this, there appears to be little that the MRC can do to *determine* the actual practice of EA concerning hydropower development. At a regional level, nevertheless, analysts have called for existing EA procedures to be revised, and have looked to the MRC to spearhead this process, calling it 'the central institutional actor responsible for transboundary environmental governance in the region' (Sneddon and Fox, 2006, p. 188) or 'a key institute leading such transformations' (Keskinen *et al.*, 2012, p. 323). Others have thrown some doubt on the capacity of the MRC to take on such an explicitly 'regulatory' role (Hirsch *et al.*, 2006, p. iii; Interview 32, 2011). Recent policy discussions in the Mekong indicate that these issues – including the MRC's ongoing role in EA processes – are a live aspect of debate: witness the detailed report produced by the MRC for a 2014 conference promoting the 'benefits of transboundary cooperation': *Cooperation for Water, Energy, and Food Security in Transboundary Basins under Changing Climate* (Bach *et al.*, 2014).

Divergent expectations of the MRC, EA processes and the Mekong Agreement notwithstanding, actors concerned with hydropower development in the Mekong River Basin do continue to look for legal 'fixes' and principled templates to address the dilemmas they confront. One international NGO active in the field, EarthRights International, has observed, for example, with regard to obligations under the Mekong Agreement to cooperate and promote sustainable development, utilization, conservation and management: 'the Agreement's legal procedures promoting cooperative decision-making under the auspices of the Mekong River Commission (MRC) are insufficient to carry out the spirit of the Agreement' (King, 2011).

Mekong River Basin governments, too, have been voicing dissatisfaction with the current 'framework' and calling for changes, some of which could potentially be addressed by EA-related law reform. At the 19th MRC Council meeting, for example, Cambodia called for 'increasing attention with practical action from the riparian countries in a more effective framework' (Sin, 2013). In addition, the joint development partners expressed concern 'about the social impacts and environmental risks associated with the construction of the Xayaburi hydropower dam … also given that alterations to the initial design have not yet been formally communicated through the auspices of the MRC' (MRC Council, 2013; see also Rieu-Clarke, 2014, p. 3). From Vietnam, a thinly veiled critique was levelled at the Xayaburi process at the 19th Council meeting:

> [E]*ach* riparian country should show their responsibility by assuring that any future development and management of water resources proposed in the basin should be considered with due care and full precaution based on best scientific understanding of the potential impacts.
>
> (Nguyen, 2013, emphasis added)

Legal scholars have reform ideas at the ready that could be responsive to at least some of these concerns, including a number that relate to EA. In his work on holistic impact assessment, for example, Montini has suggested the merger of the SEA and EIA processes into a single process, in order to achieve a broad paradigm shift in decision-making away from a neoclassical economic model, as reflected in the 'green economy' approach (championed by the Rio+20 outcome document, *The Future We Want*' (United Nations, 2012)), towards one that is based on the principle of sustainability. Montini defined that principle as 'the duty to protect and restore the integrity of the Earth's ecological systems' (Montini, 2013, p. 244). Given the shortcomings of EIA and SEA, Montini argued, merging them into a single 'holistic impact assessment' process in order to promote a 'true vertical integration framework inspired by the same kind of holistic approach' might result in a more effective tiering of SEA and EIA, and 'reduce the lack of coordination presently existing between the two procedures' (Montini, 2013, p. 255). Analysts such as Rieu-Clarke (2014, pp. 23, 33) have likewise called for greater consistency among legal requirements for EA and more general harmonization between the laws of different national jurisdictions.

There may be merit in exploring these and other ideas circulating for legal and institutional reform. Our study of EA processes in the Mekong River Basin suggests, however, that the rearrangement of principles and the implementation of law reform programmes can only take people so far through the unresolved controversies surrounding hydropower development. To navigate this socio-legal terrain, one needs to acquire fluency, too, in the politics of the technical: to grasp what may be at stake in banal and routine practices of measurement, modelling and participation-management. It is through such practices that affected humans and non-human life, conditions and phenomena come to count; it is by these means that 'impact' comes to register upon those wielding governance power, whether 'hard' or 'soft' in form.

To render a hydropower development project assessable through various modes of EA practice is thus a powerful form of knowledge-making. Although many identify EA practice with heightened transparency and accessibility, its effects might be better grasped in terms of layering. In the Mekong River Basin, for instance, EA offers the prospect of arranging a project according to several different evaluative schemes: from the policy- and future-oriented arrangement of an SEA to the relatively tight, project-oriented configuration of an EIA. Laid one upon the other, these different renderings of a development project may reveal mismatches and shortfalls that can afford a basis for political action: consider, for instance, the incommensurability that arose between the 2010 SEA (founding a recommendation that all mainstream dam development be suspended) and the 2010 Xayaburi EIA (clearing the mainstream dam development to proceed). They can also, however, have a dulling, insulating, cushioning effect in combination, turning 'impacts' into quantifiable, compensable lumps (ready to be smoothed away) and making the underlying project seem, in some respects, harder to reach, question or assess – as appears to have been the case in relation to the A Luoi dam. In the chapter that follows, we reflect further on some of the counterintuitive effects that 'participation' and 'transparency' can have, turning from assessment and measurement to the practice of making things publicly visible.

6 Disclosing a river basin
Transparency and its discontents

Preceding chapters have depicted actors at work in and around the Mekong River Basin in a range of postures: planning, governing, managing, evaluating. This chapter will focus on a common dimension of those operations, namely the practice of making the Mekong River Basin – and relations among its people – visible. More precisely, the chapter will highlight the proliferation of an expectation and rhetoric of transparency in the Mekong River Basin and explore some of its rival understandings and implications.

A focus on transparency in the Mekong River Basin is timely and important for two main reasons. First, the invocation of transparency correlates strongly with expectations of, or demands for, financial information circulating in global capital markets to be valid, uncorrupted and reliable (Salvioni, 2002). Accordingly, the prominence of norms concerning transparency has increased in the Mekong River Basin alongside growth in private sector financing of hydropower development (Middleton et al., 2009). A focus on transparency affords us a way of engaging with some of the ways that these shifting financial dynamics are reshaping practices and experiences of governance.

Second, close analysis of governance techniques dedicated to the production and maintenance of transparency helps to explain frustration or fatalism apparent among many in the Mekong River Basin in relation to hydropower development. Commitments to transparency seem to have climbed higher and higher on the reported policy agendas of governments and of intergovernmental and non-governmental institutions active in the Mekong River Basin (see, for example, Menon and Melendez, 2011). Nonetheless, the very factors contributing to this – a prevailing focus on electricity market liberalization and regional energy trade as critical drivers of development (described in Chapter 3) and the proliferation of civil society actors (described in Chapter 7) – have supported a turn towards private sector funding, planning, expert analysis and management in Mekong River Basin hydropower since the 1990s. This 'turn' has made relevant decision-makers harder to 'see', interact with or pursue grievances against, even compared to public sectors that have been (outside Thailand) relatively impervious to community input. This chapter details this shifting economy of perception – that is, the way capacities for insight of various kinds are unevenly distributed in the Mekong River Basin with regard to hydropower development.

As will become clear from the interview material presented below, 'stakeholders' in Mekong development-related conflicts have cultivated (to varying degrees) fluency in transparency language and associated modes of action, as well as in a related vocabulary of public participation and information sharing. The framing of claims, conflicts and engagement in terms of transparency in the Mekong River Basin has helped to foster particular understandings of commonality, and to encourage movement away from others. That is, transparency's proliferation has engendered particular ways of understanding people's shared space, politico-legal agency and collective capacity at the expense of others. This is the case even as demands for transparency remain markers of (often unresolved) disaffection and disagreement.

Among the senses of commonality that transparency policies have helped to sustain in the Mekong River Basin is a commonality-for-development. As discussed below, this is sometimes expressed with reference to a market economy goal (in relation to which transparency is often a euphemism for decentralized planning purged of corruption) and sometimes in terms of progress towards enhanced democracy (in relation to which transparency is again often used euphemistically). In both formulations, recourse to transparency in the Mekong River Basin has tended to elicit *ex ante* commitments to particular distributions of resources and agency at conflict sites. This chapter will elucidate those commitments, and the role of transparency policies in supporting commonality-for-development, even while recognizing that this is not all that transparency has meant or could yet mean in the Mekong River Basin.

The chapter begins by sketching some potential intellectual and political underpinnings for the positive valence that the term 'transparency' is commonly given, without venturing any comprehensive 'world history' for the concept. It continues with an account of the rise of transparency as a 'principle' shaping conduct and expectations in the Mekong River Basin, oriented initially around its somewhat belated emergence in the paper trail of the Mekong River Commission (MRC). This will open out into a sketch of two rival understandings of the term that have gained particular currency among those engaged, in various capacities, with Mekong River Basin hydropower development: transparency as development-oriented 'public participation' and what we term here 'remedial transparency'. Attention will then turn to effects that transparency rhetoric and ideas appear to have had in decision-making, analysis and community engagement surrounding the Nam Theun 2 hydropower project in Laos, the features of which have been outlined in Chapter 1.

Throughout the account that follows, the place and impact of 'law' – including international law – will often be difficult to pin down. As foreshadowed in Chapter 2, this book works with a pluralized understanding of law, yoking together 'hard' and 'soft' elements. At times, the story that this chapter tells will be a story of 'hard law' – international environmental treaties, for instance – exerting 'soft' influence, that is, shaping conduct beyond the reach of their formal effectiveness or enforceability. Elsewhere, this will be a story of the 'hardening' of 'soft' norms, such as the translation of certain expectations concerning 'good governance' into contractual preconditions for multilateral financing (under relevant financing agreements) or for government approval (via the relevant concession agreement). At other points, this chapter will track dynamics of policymaking, argument and analysis, and examine their effects, more or less without regard to the formal status of law as such.

The claim embedded in this account is that, in the context of Mekong River Basin development, distinctions between law and policy, or between 'hard' and 'soft' law, are now frequently folded and fused under the rubric of elastic umbrella-concepts circulating on a global scale, such as the 'rule of law', 'governance', 'market practice' or indeed 'transparency', as Chapter 4 has already shown. Differentiation – between law and policy or law and politics, national and international laws, formally enforceable and 'merely' exhortatory forms of law – remains important, but practices of differentiation along these lines typically occur en route to some other objective, rather than being the subject of prior consensus or ultimate resolution (Fischer-Lescano and Teubner, 2004). Analyses that proceed on the basis that soft-to-hard or hard-to-soft transitions in law will have certain predictable effects are, accordingly, unreliable – perhaps increasingly so.

Also advanced in this chapter is an argument that international norms and institutions have particular salience amid the understandings that orbit and inform Mekong hydropower development. Accordingly, deficiencies and difficulties in the governance of development in the Mekong River Basin are as attributable to the international realm as they may be to national spheres, notwithstanding the preference of many for treating international law and policy as exogenous to, and necessarily curative of, such deficiencies (Thomas, 2011, p. 1023; Rittich, 2006, p. 210).

Possible underpinnings for commitments to transparency in the Mekong

Transparency – broadly, the condition of making decisions, laws or policies plain and comprehensible to those impacted by them (that is, their terms and the process of their making) and exposing them, in turn, to some prospect of question – is widely identified with better quality decision-making. That connection is assumed as a matter of reflex in some circles, although it has also been the focus of specific argument and empirical analysis (Creighton, 2005; Fenster, 2006; Islam, 2003; Stiglitz, 2002).

Instrumental rationales for transparency are not, however, the only reason that it may be embraced; transparency may be valued as an end in itself, often as a permutation of the right to freedom of expression, or a necessary feature of the infrastructure of rights protection. For instance, discussing ramifications of the obligation to respect and ensure the right to freedom of expression under Article 19 of the International Covenant on Civil and Political Rights (to which Cambodia, Laos, Thailand and Vietnam are parties), the UN Human Rights Committee observed that 'the principles of transparency and accountability [are] … essential for the promotion and protection of human rights' (Human Rights Committee, 2011). The link that the committee draws between transparency and accountability – that is, to the prospect of making individuals, governments and institutions answerable for any shortcomings in their performance of duties, discharge of responsibilities or regard for rights – is one frequently made, although not necessarily realized. This association is one to which we will return in discussing 'remedial transparency' below.

However it is expressed, the sense that transparency is a 'good thing' often seems indebted to political commitments beyond those enshrined in human rights law explicitly, which debts may or may not be made explicit. In the context of hydropower development in the Mekong River Basin, these normative commitments seem more a matter of presumed consensus or dissensus than explicit argument. Amitai Etzioni (2010) has observed that '[t]ransparency is viewed as a self-evident good in Western society'; the reference to 'Western society' hints at one or more countering views, but Etzioni does not name or engage with those views. Similar assumptions will be discernible in some of the quotations presented later in this chapter. This section does not attempt to unravel this 'self-eviden[ce]' diagnostically. Rather, its purpose is to convey a sense of the very different bundles of political investment or inheritance that actors may carry into their enactments of 'transparency' amid Mekong River Basin hydropower development.

Transparency for state security, public peace and social well-being

In pan-Atlantic (Anglo-American and European) policymaking and pedagogy, in which some actors engaged in Mekong hydropower development will have been directly or indirectly immersed, transparency is identified especially with the thought of Jean-Jacques Rousseau, Jeremy Bentham and John Dewey, among others. Put another way, framings of transparency are often traced to the political traditions of republicanism, utilitarianism and/or pragmatism, and the instincts that each implants. Assumptions informed by these traditions seem especially prevalent among those experts described in Chapter 3 becoming increasingly active and authoritative in Mekong River Basin development since the end of the Second Indochina War (the Vietnam War) in 1975.

In Rousseau's eighteenth-century writing, transparency was associated with a human experience of 'really know[ing] with whom one is dealing', an experience reminiscent of a pre-Enlightenment state in which 'men found their security in how easily they saw through each other' (Rousseau, 1997, p. 8). Creating political conditions that were favourable to some regeneration of this experience – in however compromised a fashion – was vital to Rousseau's republican vision (Starobinski, 1988; Marks, 2001).

In a very different political setting, a century later in Britain, Jeremy Bentham lauded transparency – or, as he put it, 'publicity' – for its public utility. Transparency was useful, he argued, for its propensity to 'constrain [political leaders] to perform their duty', to 'secure the confidence of the people' in their so doing, and to enable the public 'to form an enlightened opinion' and to share that with their leaders (Bentham, 1843, pp. 310–311).

In the context of early twentieth-century American pragmatism, John Dewey made an argument somewhat comparable to Bentham's, although aimed at 'evolv[ing] and perfect[ing]' social policy tools in operation, through 'observing, reporting'. 'Whatever obstructs and restricts publicity', Dewey wrote, 'limits and distorts public opinion and checks and distorts thinking on social affairs' (Dewey, 1927, p. 167). A 'good state', in Dewey's account, 'relieves individuals from the waste of negative struggle'. It 'confers upon [individuals] positive assurance and reinforcement', characteristically in a problem-solving mode (Dewey, 1927, pp. 71–72). Transparency or publicity was, in Dewey's account, a means to that end.

Unsurprisingly, references to Rousseau's, Bentham's, Dewey's or related writings do not appear in the paper trail generated in connection with Mekong hydropower development, nor are such references dotted throughout our interview records. Nonetheless, postures taken in relation to transparency in this context – by donors and NGO staff especially – may be informed, to some degree, by republican, utilitarian or pragmatic ideals or impulses. Derivations of these distinct traditions seem to animate at least some of the tensions and divergences discernible when people talk and think about transparency in the Mekong River Basin.

Those of republican inclination, for instance, may approach public transparency primarily in terms of the realization of 'dominion' or full citizenship and the 'non-domination' of individuals (Pettit, 1997; White, 2011). Those of a more utilitarian bent might regard a commitment to transparency as contingent upon that principle's delivery – through institutional and doctrinal expression – of 'security against misrule' (Kelly, 2001). Pragmatists may regard transparency foremost as a mechanism to foster experimentation, peer comparison and learning, to facilitate problem-solving in social life (Dewey, 1927; Simon, 2004–2005, pp. 192–198). People may disagree about what transparency could or should mean in Mekong River Basin hydropower development in part by reference to these contending lines of thought.

Transparency in Marxist thought

Given the political history and two current governments of the four riparian states on which this book focuses, it is important also to consider whether Marxist–Leninist thought yields a distinctive sense of transparency with which some engaged in Mekong River Basin hydropower development may be working, or in which some in this field will have been schooled.

Insofar as the ways of thinking about transparency, and the rationales for promoting it, briefly outlined above, concern lines of sight and communication between, on the one hand, those responsible for governing and, on the other hand, the governed, there is no counterpart to this in Marxist thought. As Marx famously wrote in 'Critique of the Gotha program', the 'government machine' of the state was, for him, but a vehicle of 'political transition' (Marx, 1875). After the revolutionary transformation of society, Marx envisioned, '[t]he entire nation will be governors and there will be no governed ones' (Marx, 1926, p. 545).

The Marxist critique of ideology, the 'common sense of capitalism' and 'false consciousness' have, nonetheless, sometimes been taken to evoke an 'alternative consciousness' whereby 'scales fall from people's eyes or as they wake up, as if from a dream, and, all at once, see the light, glance directly through the transparency of things immediately to their essential truth, their concealed structural processes' (Hall, 1986, p. 33). Revelatory expectation along these lines has been a feature of some Marxist political thought within the region. Craig Reynolds and Hong Lysa have written of the long shadow cast upon Thai progressive thought, for example, by Jit Poumisak's mid-twentieth-century scholarship, evidencing concern with unmasking the 'real face' of institutions of dominance (Reynolds and Lysa, 1983).

Some appeals to transparency in the Mekong River Basin do, accordingly, convey a sense that market-oriented development programmes entail some 'illusion, a trick, a sleight-of-hand' (Hall, 1986, p. 37). This sense – and an associated preoccupation with disabusing people of these illusions – seems especially to animate demands for remedial transparency, discussed below. Many characterize difficulties surrounding Mekong hydropower development as matters of exposing falsehood and eliminating myth, especially those identified with the interests of global capital (see, for example, Chachavalpongpun, 2011).

Louis Althusser, however, led the way away from such readings of Marxist thought; he was scathing of the 'naïve[ty]' of Engels's return to a 'transparency' that presupposed that 'intersecting forces' of 'individual wills' make history (Althusser, 1977, p. 124). The problem to which Marx gave the name 'ideology' was not, Althusser showed, one of correctible error, want of information, or the distortion or concealment of underlying truths; it was not, accordingly, a problem amenable to redress through transparency. Rather, what proves most potent, in Marx's writing, is precisely 'what we are always seeing, what we encounter daily' (Hall, 1986, p. 38). As Althusser, Hall and others have shown, the opposition of falsity to truth, or opacity to transparency, is not one on which Marxist thought depends, nor one that it aspires either to puncture or to sustain. There is, accordingly, no Marxist equivalent to transparency along the lines outlined.

If there is any arguable point of resonance between Marxism and demands for transparency prevailing in the Mekong River Basin, it is in the expectation of public participation. This could perhaps be taken to overlap with Marxism's emphasis on collective political agency or 'species being' (Marx, 1843, pp. 33–34). As Marx explained, however, calls for 'the so-called sovereignty of the people' under the rubric of a '*democratic republic*' by no means correspond with the '*revolutionary dictatorship of the proletariat*' of Marxist imaginings, democratic government having been characterized by Marx as 'a police-guarded military despotism, embellished with parliamentary forms' (Marx, 1875, p. 538).

Transparency in Buddhist teachings and the impact of 'sacred nationalist' thought

It is similarly difficult to identify any plausible point of comparison between the foregoing traditions and ideas of transparency advanced in Buddhist texts. The Buddhist notion of *śūnyatā* has sometimes (atypically) been translated into English as transparency, although a far more common translation is 'emptiness' (Corless, 1989, p. 27). However, 'transparency' in the Buddhist context implies a relation to the self – or to the illusion of self – rather than a relation of governed-to-governing or vice versa. Notions of transparency that have come to be expressed in law, policy and development practice in the Mekong River Basin do not, accordingly, appear indebted to Buddhist teachings.

Beyond the range of Buddhist thought, some scholars of Southeast Asian politics have written of the prevalence and significance of 'sacred nationalism' – in Thailand, for example (Fong, 2009). Such traditions may have particular implications for transparency, Fong suggests, arguing that:

> [Thailand's] nation[al] and historical construction occurs in secretive monarchical institutions where political manoeuvrings are kept from the public, "helped by the fact that royal actions were above public criticism" due to the country's enforcement of *lèse-majesté* laws – laws that prohibit any criticism of the monarch and members of the royal family.
>
> (Fong, 2009, p. 674; quoting McCargo, 2005)

Be that as it may, any suggestion that Southeast Asian nations and cultures have a particular disposition in relation to transparency – that they are, for instance, chronically non-transparent – risks fuelling a 'self-orientalizing discourse' that 'Asian economies are dominated … by exclusive and unassailable ties between small communities of people': discourse often mobilized to shortcut inquiry or reject reform (Morris, 2000, p. 474). If the sorts of political ideals outlined above do not resonate widely or consistently among peoples of the Mekong River Basin, this may have more to do with the fraught and disjunctive history of 'democratization' in the region than any necessary tension with indigenous thought or practice (for contending accounts of Thai democratization, by way of example, see Kasian, 2006; Winichakul, 2008).

Market transparency

One variant of transparency that does appear to have gained significant ground in the Mekong River Basin in the context of hydropower development – arguably more than notions of transparency traceable to the political traditions of Rousseau, Bentham or Dewey as such – is the idea of transparency as a requirement for the efficient operation of markets. Where a paragon of 'market equilibrium' is associated with an ideal of 'complete information', as it has been in neoclassical economic thought and at least some of its derivations, then informational transparency and market liberalization become natural bedfellows (Blyth, 2003; Kopits and Craig, 1998, p. 7). Accordingly, arguments for transparency in market terms often serve as proxy arguments for government deregulation. The apparent opening of regulatory processes to a supervisory public (through disclosure requirements and reporting practices, for instance) may be seen to justify dispensing with other forms of supervision or constraint.

Friedrich Hayek can perhaps be credited most with emphasizing the importance of information dispersal to an economic system founded on market competition (Hayek, 1944, pp. 51–52). In Hayek's account, transparency is a principle of visibility: valued because it improves price discovery, enables market participants to 'watch' and respond to one another competitively, and 'provide[s] a favorable framework for individual decisions' (Hayek, 1944, p. 52; 2011, p. 332). Yet it is, at the same time, a principle of invisibility or blindness (Foucault, 2008, pp. 278–280). The need for unfettered transmission of information among individuals – and for legal and institutional arrangements that encourage that – is understood to arise from the growing 'range of [human] ignorance' (Hayek, 2011, p. 77). 'When we reflect how much knowledge possessed by other people is an essential condition for the successful pursuit of our individual aims,' Hayek wrote, 'the magnitude of our ignorance of the circumstances on which the results of our action depend appears simply staggering' (Hayek, 2011, p. 75). In Hayek's account, transparency is cast as necessary in light of the presumed failure of central planning in a modern economy, given the 'synoptic view' it would demand – at least in the case of 'planning against competition' (Hayek, 1944, pp. 43–44, 51). Market transparency presumes the invisibility of the economic system as a whole and the blindness of the central planner, as much as it anticipates the transformative effect of information freely distributed among individuals.

Worries about the limits and dangers of central planning that may be traceable to the pervasive influence of neoliberal economic thought – Hayek's influence especially – are relatively widespread in the Mekong River Basin, even among those who are by no means proponents of neoliberalism. Those more concerned with democratic decentralization than enhancing competition often share Hayek's lack of confidence in expert oversight and a commitment to transparency as its corrective.

Criticism of the MRC-introduced Decision Support Framework (DSF) may be read at least partially in this light, for instance. In use since 2004, the DSF was 'set up to assist planners to assess the magnitude of changes brought about through natural and man-made interventions in the water resource system, as well as the impacts that these will have on the natural environment and upon people's livelihoods' (MRC, 2014a; Sarkkula *et al.*, 2007). One finds an unlikely echo of Hayek's emphasis on the limits of human knowledge and the dangers of 'synoptic' planning in a thoughtful 2010 joint report of the Australian Mekong Resource Centre and Oxfam Australia, which observed that 'few if any existing impact assessment processes in the Mekong are able to encompass the magnitude of cumulative or combined impacts of development in different parts of the basin and at different scales' (Lee and Scurrah, 2009, p. 32). In light of the 'uncertainties and shortcomings of existing models and assessment tools', the authors argued for the models concerned, 'accompanying planning processes', and the 'methods and assumptions built into the models' to be 'placed in the public domain' (Lee and Scurrah, 2009, p. 34). In this instance and elsewhere, a confluence seems to arise between competitive capitalist or neoliberal rationales for transparency and democratic or human rights-informed arguments for knowledge and decision-making decentralization. Thanks perhaps to this confluence, suspicion of central planning and basin-wide oversight has become embedded in the law and development common sense by which many concerned with hydropower development in the Mekong River Basin are guided.

Law and development common sense

The shaping of law and development common sense in and around the Mekong River Basin fills a larger canvas than this chapter can depict. Indeed, this book as a whole takes only a partial glance at this common sense, focused, as it is, on a selection of hydropower development projects in four riparian nations. The rise of expectations of transparency in the Mekong River Basin from the time of the 1995 Agreement on the Cooperation for the Sustainable Development of the Mekong River Basin (the Mekong Agreement) onwards is part of a larger story of the turn to institutions in economic thought and renewed focus on the 'rule of law' in development policy during that period, a story told in Chapters 2 and 3 and elsewhere (for example, Thomas, 2011; Kennedy, 2006a). It is as a necessary element of promoting the rule of law or 'good governance' that transparency has often been championed in the Mekong River Basin, as is evident in the characterization of the term, as a central feature of 'good governance', by the UN Economic and Social Commission for Asia and the Pacific (UNESCAP, 2009).

Experienced law and development practitioners have, nevertheless, conceded that 'there is a surprising amount of uncertainty about the basic rationale for rule-of-law promotion' (Carothers, 2006, p. 17). This is as true of transparency promotion as it is of any other recurrent feature of rule of law rhetoric and programming. To some, transparency is expected to aid economic growth and the transition to, or entrenchment of, market economies, in ways outlined above. To others, transparency helps to engender democracy. Neither of these causal assumptions has been well substantiated but, as Thomas notes, this 'does not appear to have affected the momentum behind programs' that are designed to give effect to them (Thomas, 2011, p. 1016).

The prevalence of the former view – identifying transparency with market liberalization and economic growth – was made apparent, for example, at a ministerial conference on public–private partnerships (PPPs) in the Asia Pacific, organized by UNESCAP in 2012, at which the executive secretary of UNESCAP, Dr Noeleen Heyzer, spoke with hydropower very much on the agenda. Dr Heyzer emphasized a link between the adherence by states to 'rules and regulations' ensuring transparency and their prospects for economic prosperity, observing that 'mechanisms to ensure transparency and accountability' are necessary features of the successful implementation of PPPs in order to eliminate 'infrastructure bottlenecks' (including '[e]rratic power grids'), close 'development gaps' and build 'more inclusive prosperity' (UNESCAP, 2012). As Dr Heyzer's focus on power grids suggests, foreign currency-generating and foreign investment-attracting sectors (such as the energy sector) are those in which greatest emphasis has been placed on transparency, in contrast to other areas of policymaking and governance where foreign financial stakeholders are less in evidence. Transparency mechanisms typically emphasized in such contexts are concerned primarily with risk reporting and auditing.

At the same time, many promote commitments to transparency in the Mekong River Basin as political way stations en route to democracy or, in Thailand's case, democracy's fuller flourishing (on Thailand's tradition of 'semi-democracy', see

Pathmanand, 2008, p. 139; Samudavanija, 1989). Asked about the meaning of transparency in the literature and practice of the donor institution in which this person worked, one interviewee commenced a response in terms of personal democratic convictions, before shifting into a more typical diplomatic register, concerning the benefits of orchestrated 'dialogue' and 'constructive discussion':

> I'm a bit of a deliberat[ive] democrat so I believe, some would say unrealistic[ally], that if you were a little more open, a little more transparent, there are implications … because your mandate is known, your decisions are known and there is opportunity for some public discussion and/or push back … So we have been trying to modestly contribute to some simple normalization, openness and understanding of mandate and then actually, you know, discussion … opportunities where the development partners or citizens … are having discussions with their own ministers around tables … We're funding quite a bit of that through round tables, dialogues, as in multi-dialogues … designed and led by people who have the competence and credibility to … have a constructive discussion.
>
> (Interview 50, 2012)

To the extent that transparency is framed as a precondition for democracy in the Mekong River Basin, this also serves as a point of resistance or distinction. It is precisely this association of transparency with democracy that prompts some in the Mekong River Basin to chafe at its invocation. Transparency tied explicitly to democratic governance is seen by some as problematic because of the difficulties, costs, delays and 'competitive disadvantage' associated with the ensuing encumbrance – and likely questioning – of decision-making power. As one interviewee from a Thai NGO observed with regard to hydropower development in Cambodia and Laos:

> They keep saying China or Vietnam is a model that can make decisions in one week. The most competitive advantage in their view is if you take one day for an investment decision and Thailand is bad because people demand participation; that is what they say. So when we say, why don't you look at Australia and Europe, [they say] how long do they take to build a dam?
>
> (Interview 23, 2011)

A senior government official in Laos expressed precisely the views anticipated in the foregoing remarks, albeit without attributing these to any influence of democratic ideals or polities:

> The development of hydropower projects in Laos has become more and more complex … We have so many parties involved in the decision-making and that's why lately a large project takes a long, long time to be developed … Before you would have one or two meetings but today they require you to do at least two village meetings, and then we have a district meeting and then we have a provincial meeting and then we have a [Ministry of Natural Resources

and Environment] site visit and technical meeting with them, four or five meetings. For one provincial consultation, it costs us more than some projects, US$20,000, for all participants to stand and sit down for half [a] day. Why so expensive?

(Interview 52, 2012)

Transparency in national constitutional law

One possible expression of this sometimes controversial focus on promoting transparency in law and development work since the 1990s (described in Chapters 2 and 3) has been the writing of direct or indirect transparency commitments into the positive law of Mekong nations – specifically, into constitutional law – in ways potentially applicable to hydropower development.

Under the 1997 Thai Constitution, for instance, provision was made for public involvement in development and environmental and natural resource decision-making. This was maintained in the 2007 Constitution, under which a 'person' has 'the right to receive information, explanation and justification' from the government before any activity is permitted which may affect 'the quality of the environment, health and sanitary conditions, the quality of life or any other material interest concerning him or a local community, and shall have the right to express his opinions on such matters' (section 57). Citizens also enjoy the right to participate in public decision-making affecting their rights and liberties (section 58), to present a petition and be informed of its result (section 59) and to sue the government (section 60). Protest action taken in Thailand concerning hydropower development, partially on the strength of such guarantees, is described in Chapter 7. However, the 2007 Constitution was suspended on 20 May 2014, under martial law, and an Interim Constitution was adopted in July 2014. At the time of writing, the constitutional parameters for information flow between government officials and their constituents in Thailand remained, accordingly, in flux.

In contrast to the 2007 Thai Constitution, Cambodian constitutional law does not guarantee rights of information, access to the reasons of public decision-makers, or public or community participation in decision-making about development, the environment or natural resource exploitation. Rather, the 1993 Cambodian Constitution guarantees citizens generic rights to participate actively in political, economic, social and cultural life (Article 35) and to denounce or file claims in relation to any breach of law by the state (Article 39), without providing for public access to information to facilitate the realization of those rights. In addition, the constitutional protection afforded freedom of 'expression, press, publication and assembly' (Article 41) is conditioned by a requirement that no one exercise this right 'to infringe upon the rights of others, to affect the good traditions of the society, to violate public law and order and national security'.

Positive commitments surrounding transparency in the Lao Constitution are similarly generic, yet somewhat more elaborate. In 2003, the 1991 Lao Constitution was amended to guarantee Lao citizens a right to lodge complaints and petitions and to propose ideas to relevant state organizations in connection with public interest issues or their own rights and interests (Article 41). The same Article provides that complaints, petitions and ideas of citizens shall be examined and resolved by the state in accordance with law. The guarantee of a 'right and freedom of speech, press and assembly' was a feature the 1991 Constitution maintained in 2003, without the right-specific caveats of the Cambodian Constitution (Article 44). Nonetheless, overarching requirements that Lao citizens 'respect the Constitution and the laws … observe labour discipline, [and comply with] the regulations relating to social life and public order' condition its exercise (Article 47).

Vietnam's 1992 Constitution (both originally and as amended in 2001) guaranteed citizens a right of access to information, with its exercise to be prescribed by law (Article 69). This right, including the 'prescribed by law' conditioning, was maintained in the 2013 Constitution (Article 25). Also common to both the 1992 Constitution (Article 53) and the 2013 Constitution (Article 28) is a right to participate in the management of the state and society, and to discuss and propose to state agencies issues concerning certain localities or the whole country. New to the 2013 Constitution, however, is a requirement that the state create conditions for such participation and 'publicly and transparently receive and respond to the opinions and petitions of citizens' (Article 28). At the same time, in a 19 March 2013 speech on constitutional revision, the Vietnamese prime minister, Nguyen Tan Dung, is reported to have called upon the Communist Party, the state and 'every single citizen' to 'fight against unconstructive speeches and actions that sow division and harm solidarity in the Party and society' (Bao Moi, 2013). Accordingly, some 'opinions and petitions' fare better than others under these constitutional provisions, as Chapter 7 discusses.

However ambivalent the commitment of political leaders to transparency, and however conditioned the means of enlivening these constitutional rights (conditions explored in Chapter 7), it has clearly become an expectation in recent decades that constitutional law among riparian nations of the Lower Mekong River Basin will provide some positive assurance of the entitlement of citizens to access information concerning – and/or to petition the state regarding – public decision-making. This assurance, at least on its face, extends to public decision-making about hydropower development.

Becoming transparent: a principle of MRC work

Against the backdrop of these contending loyalties and translations, transparency has come to have particular resonance in the Mekong River Basin. This section examines some of its site-specific operations in this regard, taking the MRC as a potential gauge of broader shifts in Mekong River Basin language and practice.

In and around the MRC, hydropower development in the Mekong River Basin is increasingly chaperoned by explicit appeals to, and affirmations of, transparency. This need not, however, have been the case; indeed, there were early indications to the contrary. For instance, no reference to transparency, no mechanism for disclosure or scrutiny of decision-making, and no provision for public access to information featured in the 1995 Mekong Agreement.

Of course, the entry of four states into that international agreement, the text of which was made publicly available, did entail some commitment to transparency in general terms. The Mekong Agreement demanded accession to a mode of decision-making that would, to the extent of its coordinated nature, potentially be reviewable by and/or involve parties outside each participating state's national government. The agreement generated expectations – among its states parties, but also more broadly – that decision-making within its scope would at least arguably correspond to 'rules and regulations' stipulated therein or contemplated thereby, however permissively expressed. The agreement provided, for instance, for 'notif[ication] and consult[ation]' among member states in emergency situations, as well as 'cooperat[ion]' in a range of other matters, effectively contemplating the sharing by states parties of information with the MRC, other states parties and at least some circle of persons with expertise in these areas (Mekong Agreement, 1995, Articles 1, 2, 4, 6, 10). More generically, as José Alvarez has observed, 'treaties facilitate transparency … [by] provid[ing] access to stabilizing or reassuring information about others' level of compliance, or provid[ing] methods to clarify expectations' (Alvarez, 2005, p. 339).

Nonetheless, it remains noteworthy that neither the language of transparency, nor any explicit procedural framework for governing-to-governed interaction or information-sharing with 'the public' (in any version) appeared in the textual record of the MRC's establishment or the early years of its operation. Tellingly, the Procedures for Data and Information Exchange and Sharing, published by the MRC on 1 November 2001, were concerned exclusively with 'the reciprocal transfer of data and information among the member countries' (MRC, 2001). Initial formulations of what might be called 'proto-transparency' in the Mekong River Basin thus revolved around a closed circle of governance, in which both decision-makers and parties affected by decisions were conceived of as national governments, and information was envisaged moving between them via the channels of the National Mekong Committees and the MRC, with no overflow of, outflow from or inflow from beyond those channels expressly anticipated.

It did not take too long, however, for transparency to emerge as a 'principle', suggestive of a broader and looser array of meanings, to which the MRC, member states of the Mekong Agreement and others concerned with Mekong River Basin development became expected to have regard. In Article 3 of the Procedures for Notification, Prior Consultation and Agreement under the Mekong Agreement, developed by the MRC in 2003, 'transparency' was the fifth of five guiding principles identified (MRC, 2003a). Since then, the invocation of transparency as a generic, guiding 'principle' has become a recurrent feature of MRC procedures documents. It features as such in Article 3 of the Procedures for Water Use Monitoring, published by the MRC in 2003 (alongside five other principles guiding their implementation), Article 3 of the Procedures for Maintenance of Flows on the Mainstream, published by the MRC in 2006 (alongside a commitment to the principle of 'reciprocity') and Article 3 of the Procedures for Water Quality published by the MRC in 2011 (alongside an explicit encouragement of 'public participation') (MRC, 2003b, 2006, 2011a).

Transparency has also become a principle by reference to which MRC performance has come to be evaluated by key sources of funding and support, as well as by critics. Sensitivity has grown, for instance, to the 'black-boxing' effect of scientific and financial modelling in the Mekong River Basin. Scholars and activists alike are attuned to challenges associated with opening the contingent assumptions of such models to public or third-party scrutiny when they serve as a basis for significant decision-making, a concern often in tension with the insistence by investors and developers on protecting confidential business information (Sarkkula *et al.*, 2007; Costanza *et al.*, 2011). The MRC has increasingly been called upon to conduct or oversee modelling in ways responsive to transparency demands. For example, the World Bank, reporting on the outcome of a US$11 million grant from its Global Environmental Facility to the MRC to develop an agreed methodology and set of rules surrounding water utilization among Cambodia, Laos, Thailand and Vietnam, observed that:

> A better communication strategy to the stakeholder[s] regarding the application of the hydrological models and water utilization procedures could have been considered. The DSF is now used as the main tool for planning water resources investments. However, sources of the models need to be disclosed to the general public for independent review and examination, at least through a step-by-step approach to confirm the accuracy and robustness to make the model more credible and trustworthy; otherwise, the model would be considered as a black box that cannot be evaluated objectively. Further, the progress and final outcome of the water utilization procedures would have been disclosed and disseminated not only to the line ministries, but also to the riparian communities and local governments which were directly affected by the procedures.
>
> (World Bank, 2009, p. 18)

Also significant with respect to the proliferation of transparency-talk in the work of the MRC over the past decade are changes in leadership and leadership style that have occurred within the organization. In its early years, MRC work practices evidenced 'the MRC leadership [having] little time or interest in interacting beyond … the donor community and the riparian member governments' (Hirsch, 2008, p. 39). According to one interviewee at the MRC, the organization was initially 'very technical', unwilling to foster or broach explicitly 'political discussions', an approach that changed with the appointment of Joern Kristensen as CEO in 1999:

> Then we had Joern Kristensen coming in, restructuring it a bit, coming in from a diplomatic background, he opened it up a bit more in terms of how we really put forward basin development ideas and start[ed] to deliver more in [terms of] procedures and implementation.
>
> (Interview 48, 2012)

If Kristensen's tenure as MRC CEO, from 1999 to 2004, inaugurated 'a more transparent science-based understanding of the implications of various development options' within the MRC, this was reversed somewhat under the subsequent leadership of Olivier Cogels. The MRC's strategy from 2004 to 2007 has been described as one of 'selective engagement', including 'holding back results of its publicly funded scientific analysis if those results indicated adverse impacts of development' (Hirsch, 2008, p. 41). In 2008, however, Cogels's initial successor, Jeremy Bird, brought to the organization 'a reputation as a much more open and inclusive leader than his predecessor'. This is an institutional disposition that subsequent leaders of the organization appear to have worked to maintain (Hirsch, 2008, p. 43).

However uneven the MRC's historical practice in releasing information publicly and interacting with NGOs, academics and community groups, a broad trajectory of increasing concern with transparency over time does seem to be in evidence in and around the MRC. By the time of our interviews with MRC secretariat and environmental programme staff in 2012, detailed accounts of transparency tripped quite readily off the institutional tongue (Interview 47, 2012).

MRC efforts towards ensuring greater standardization, clarification and opportunities for review in Mekong River Basin development are exemplified, in particular, by the decade of work that the MRC has devoted to trying to facilitate agreement upon a Framework for Transboundary Environmental Impact Assessment (TbEIA Framework) among parties to the Mekong Agreement, discussed in Chapter 5 (see Environmental Law Institute, 2009; Ke and Gao, 2013). Proposals for a TbEIA Framework in this context have recurrently been benchmarked against international treaties to which the riparian states of the Mekong are not currently parties, invoked as indicators of global expectation rather than as sources of legal obligation: the 1991 Convention on Environmental Impact Assessment in a Transboundary Context (Espoo Convention, 1991) and the 1998 Convention on Access to Information, Public Participation in Decision-making and Access to Justice in Environmental Matters (Aarhus Convention, 1998). Also influential as benchmarks for the TbEIA are the World Bank's Operational Policy on Environmental Assessment and the Asian Development Bank's Environmental Assessment Guidelines, the broader import of which was discussed in Chapter 5 (World Bank, 1989; ADB, 2003; for an example of such benchmarking, see Environmental Law Institute, 2009).

The development of operational understandings of transparency in the context of environmental impact assessment (EIA), with reference to international treaties and standards in this area, has lent the term a particular tenor. As Chapter 5 made clear, EIA regulation promotes an *episodic* understanding of the demands of transparency, oriented around particular events or projects conceived as 'triggers' to notice requirements. It encourages a focus on *formal channels* for public notice, participation and consultation, involving *mediation by experts*. It directs effort and attention towards *prediction*, especially the prediction of harms and threats, and the distribution of information following the scope and trajectory of those predictions. It encourages *segmentation* of the population into particular categories and subcategories, or the eliciting of certain *subjectivities* (such as those of 'stakeholders' and 'affected groups'). We will return to these EIA-related aspects of transparency below.

Less tied to EIA practice, however, was the multidimensional account of transparency offered by a senior member of the MRC secretariat:

> [Transparency] means may different things. On an administrative level ... it means that we should operate like a good international organization who should have international tendering, we should fight against corruption, we should be clear about how we will work internally. When it comes to information dissemination and working with both our member countries and others, it's trying to go in there and be clear about what we are talking about, what our decisions are, what we are working with ... and be open about where this position that we take in a particular context comes from – [from] a technical standpoint or from a consensus that we are achieving [among] the countries ...
>
> (Interview 48, 2012)

This account encapsulates a sense of transparency '*of* governance': an imperative to make plain the operations of governance to those entities and people who are to be governed, as a matter of 'principle' and 'good ... organization[al]' conduct. It also expresses an idea of transparency '*for* governance': as a means of maximizing regulatory effectiveness by informing thinking, modulating behaviour, engendering allegiance and fostering legitimacy (Mitchell, 2011, p. 1882; Brunnée and Hey, 2013, p. 25).

In both these senses, it is apparent that transparency has become an important feature of practices of authorization within and surrounding the MRC, whether the decisions in question present as technical, political or regulatory in character. Indicia of transparency have become significant to the MRC establishing and maintaining a sense of its operations being publicly authorized, and to others – both public and private actors – establishing authority vis-à-vis the MRC.

At the same time, even approaches to transparency concerned more with enhancing rather than challenging governmental rule remain controversial among the National Mekong Committees and at the level of the MRC's joint committee and council. A former member of one of the National Mekong Committees explained how varying degrees of disclosure among Mekong nations remain a source of tension:

> We have procedures [regarding development with transboundary impacts] but we would like more detail. However, that is difficult to attain because it depends on the capacity of different countries. The different mechanisms in each country make it difficult. For example, say we want more data or information, but in Cambodia and Laos the monitoring is weak compared to Thailand and Vietnam. That is the reason why sometimes one country gets annoyed saying that: 'we supply a lot already, now we are waiting for others to supply data. We have provided a lot already, but other countries not much. Let them contribute.'
>
> (Interview 33, 2011)

Rivalry and tension evidently persist, with respect to cross-border information-sharing, regardless of the prevalence of transparency principles among participants in the Mekong Agreement. As for transparency vis-à-vis the public in a hydropower development setting, the view of the World Commission on Dams regarding developments in that regard has been less than sanguine. In its influential *Dams and Development* report of November 2000, the commission observed: 'While there has been a growing emphasis on transparency and participation in decision-making involving large dams, especially in the 1990s, actual change in practice remains slow' (World Commission on Dams, 2000a, p. 176).

Rival transparencies: contending accounts from beyond the MRC

As will already have become apparent from the benchmarking of TbEIA proposals to international instruments, the trajectory of MRC policy and practice with regard to transparency outlined above likely reflects the impact of broader trends and normative practices. In one commentator's account, such trends and practices have invested transparency with global ubiquity, notwithstanding the 'chameleon-like' properties of the concept:

> We demand transparency of our partners, our colleagues, the city council, the government, international institutions and even the objects we use … Transparency [has become] a standard of (political, moral and, occasionally, legal) judgement of people's conduct. A narrative of transparency permeates our daily life … Transparency is all around us.
>
> (Bianchi, 2013, pp. 1–2)

Global policymaking and law-making have almost certainly contributed to a sense of transparency being 'all around' in the Mekong River Basin, albeit in variegated ways. It is, however, beyond the scope of this chapter to survey such global initiatives and their differentiated effects. Overviews of law-making and policymaking activity on this theme are presented elsewhere (for example, Bianchi and Peters, 2013). Accounts are also available of specialized understandings of transparency that have developed in particular fields of law and policy: in international economic law (Zoellner, 2005–2006); in law and development (Mock, 1999–2000); in international environmental law (Ebbesson, 2013); and in the related spheres of international relations and political science (Grigorescu, 2007). Scholars have also surveyed the adoption of transparency policies within international organizations and by national governments (Grigorescu, 2007; Donaldson and Kingsbury, 2013; Relly and Sabharwal, 2009). This section of the chapter will not replicate or summarize any of these more comprehensive or in-depth accounts. Rather, the section will focus on two rival yet related accounts of transparency put forward in connection with hydropower development in the Mekong River Basin: transparency as development-oriented public participation and remedial transparency.

Transparency as development-oriented 'public participation'

Perhaps the most prevalent idea of transparency amid law, policy and debate in the Mekong River Basin is one tied to EIA processes and standards. As noted above, EIA regulation envisages transparency in terms of information-flow concerning foreseeable risks or impacts of proposed development activity. Central to this idea of transparency is the expectation of its realization through 'public participation': a standardized process oriented around the specification of 'publics' and the convening of hearings, meetings and other sites of 'consultation' or 'discussion' for the distribution of information and the receipt of 'feedback'. Information-sharing at such sites is typically envisaged as a process of exchange, albeit often a lopsided one (Ebbesson, 1997; Mostert, 2003).

This is the main sense in which transparency has featured in intergovernmental statements of policy concerning water management: for instance, in the Dublin Statement adopted in 1992 (under the auspices of the UN Administrative Committee on Coordination Inter-Secretariat Group for Water Resources) and the Ministerial Declaration on Water Security in the 21st Century, adopted at The Hague in 2000 (under the auspices of the World Water Council) (ACC/ISGWR, 1992; WWC, 2000). Both these statements endorse a 'participatory approach' or 'participatory process' of water management (Mostert, 2003). Likewise, the World Commission on Dams has advocated strongly for transparency in decision-making surrounding dam development, with an emphasis on 'informed participation in decision-making processes' such as through the convening of 'stakeholder forum[s]' (World Commission on Dams, 2000a, p. 217; Dore and Lebel, 2010). The basis for this advocacy is the Commission's finding that 'the most unsatisfactory social outcomes of past dam projects are linked to cases where affected people played no role in the planning process, or even in selecting the place or terms of their resettlement' (World Commission on Dams, 2000a, p. 176).

Perhaps under the influence of this international standard-setting and advocacy, and its national and community-level echoes (discussed in Chapters 5 and 7), or perhaps because of a lack of alternative prospects for engaging with key decision-makers, EIA provision for project-oriented 'public participation' is a prominent way in which transparency has come to be understood and enacted in the Mekong River Basin in connection with hydropower development. The importance of EIA processes in this sense is apparent in the following quote from a representative of a Cambodian NGO:

> Before it wasn't easy to take action on environmental issues, but now they have environmental policies and regulations, for example, sub-decree on EIA … We comment on the EIA process, and we provided comments on the Lower Sesan 2 EIA. Even though NGO comments were not really taken into account by consultants, the company or the government, we nevertheless had a chance to talk. Actually they invited us, even though they didn't give us enough space, they still invited us to give comments. Even though the government doesn't make any information public, in this type of meeting they make themselves

visible, sit in the front and provide some comment. Before, [we] could not say anything directly about the dams – about Sesan or the Yali Falls dam. But now it's a little bit easier.

(Interview 10, 2011)

The significance of national and international EIA policymaking in shaping these expectations of transparency is also evident in the following quote from the representative of a Thai NGO, in which the opening of government to scrutiny is identified, initially at least, with 'mak[ing] use of EIAs' in processes inspired by practices in 'other countries':

> Public participation comes from strong demand and education. For energy planning, an expert behind closed doors has been doing that for a long time. Only recently has the public started to think we should disclose this process, in other countries they are doing it like this. It depends on how the public demands … But of course when the public is strong it tries to make use of EIAs, especially when they provide public participation, the public tries to use that space to make something different from the framework of EIA. Then that is something. If the EIA can't solve that, the public raises questions so then that delegitimizes the EIA.
>
> (Interview 23, 2011)

A sense of transparency that abounds in the Mekong River Basin, with regard to hydropower development, is one in which 'public participation' of a kind that EIA processes contemplate is axiomatic.

Remedial transparency

Lodged within the preceding accounts of transparency, but also a stand-alone idea exerting influence in the Mekong River Basin, is a further notion of transparency that expectations of 'public participation' do not encapsulate. This is as a means to remedy grievances, address injustice or exclusion and, at least potentially, bring wrongdoers to account; hence the recurrent link drawn between transparency and accountability, remarked upon above. Bound up in this sense of transparency is a concern with 'access to justice' (and its obstruction), whether through direct or mediated contact with legal professionals and institutions, or through the promotion of a more generalized awareness of impacts and 'rights'. Explaining the work of an international NGO based in Thailand, one interviewee observed:

> [T]he Mekong mainstream dams and large hydropower dams on the tributaries are potentially serious because so many people rely on the rivers, fisheries for food security and livelihoods. The people who live along the river have very little rights. There is little consultation with people. They don't have a say in decisions that affect their lives. At the local level there's a disconnect with decisions which are made at higher levels … Our campaign is trying to get the issue out to people, helping them understand the different studies have been

done so they understand what kinds of impacts can be expected from this dam. The Mekong River Commission put up a little bit of information but a lot of it is not accessible to the majority of people. So we try to make that information more accessible to local groups and communities.

(Interview 20, 2011)

The promotion of debate or the fostering of public pressure through the local or international media is also an important feature of remedial transparency. One representative of an international NGO based in Thailand remarked, for example, that:

[O]ne of the important things [our organization] can do is to work with international media to create pressure that will make it more likely that governments will respond to.

(Interview 20, 2011)

Among those affected or potentially affected by hydropower development, transparency is a recurrent feature of languages of claim and grievance within the riparian nations of the Mekong, whether explicitly or implicitly. Concern is expressed about concentrations and abuses of power and the difficulty of attributing responsibility for or against these, with transparency conceived as an antidote. Worries about lines of sight and their obstruction are, for instance, implicit in the following remarks from a Cambodian NGO concerned with hydropower:

People … are afraid to use the court, and that's not just concerning environmental cases, in other cases as well … If you want to get a copy of the documents, you have to pay. It's not a free distribution of documents for a public hearing.

(Interview 17, 2011)

What this interviewee did not go on to say, but seems to imply, is that greater visibility around the basis for court decisions, including access to relevant documentation, would assuage peoples' mistrust of the legal system and make them more willing to pursue such remedies as may be available to them under Cambodian law. In this sense, these remarks endorse a remedial sense of transparency. So too did a February 2014 letter, championing the interests of 'affected communities', that the Save the Mekong coalition of NGOs sent to the Cambodian, Lao, Vietnamese, Thai and Chinese heads of government regarding the Lower Sesan 2 dam (discussed in Chapter 7). In that letter, the coalition elected to emphasize 'access to information, meaningful consultation and participation in decision-making' and an expansion of the EIA area, rather than the choice of governments to pursue hydropower development per se, relative to other alternatives (Save the Mekong, 2014a). Likewise, the Lao 'social enterprise' known as CLICK observes on its website: 'We believe in the power of information. The more information that is available and accessible, the more choices for people to development' (CLICK Laos, 2014). However scantly elaborated the connection, in practice, between 'more information' and 'more choices for people' affected by hydropower development, the notion that information can remedy wrongs and augment political capacity is apparently a matter of quite widespread conviction in the Mekong River Basin. This is despite the same organizations' recognition that access to development information frequently has not been empowering; in a June 2014 letter highlighting the threat posed by mainstream dams, the Save the Mekong coalition observed:

> Over the past six years, widespread public opposition to the Mekong main-stream dams has been expressed nationally, regionally and internationally through petitions and letters to regional governments and the MRC, yet con-struction on planned projects has continued unabated.
>
> (Save the Mekong, 2014b)

These divergent versions of transparency remain in dynamic relationship. Sometimes development-oriented transparency seems ascendant. Asked about interactions with NGOs airing concerns about the adverse impact of hydropower development (invoking remedial transparency, in other words), a senior Lao government official suggested that concerns of this kind could readily be dispensed with when 'negative to development':

> [I]t's good if they give legitimate comments, that maybe we overlook it or … it's good in that aspect. But there are a lot of comments and activities that are negative to development of hydropower, I think that is not acceptable and we ignore it.
>
> (Interview 52, 2012)

Elsewhere, however, the opposition between these views of transparency, and contending versions, seems not so easily overcome, as the following remark, from a representative of a Thai-based NGO active on hydropower issues, implies:

> So when you ask, what is people's perception of law … I think [that the World Bank and the Asian Development Bank] see law as a truth to promoting the economy. But civil society expect[s] that the law is something that will help to protect and benefit of the majority of the country, or [mitigate] the social and environmental impacts of development. I think they are opposite attitudes.
>
> (Interview 23, 2011)

This sense of confrontation is apparent too in the same senior Lao government official's account of decision-making around hydropower, in which our interviewee's tone became far less strident or assured:

> I don't know how many laws we have but then we … Even the Ministry of Natural Resources [and Environment] are so afraid of whatever they say now, whatever they agree to or whatever comment they make… What happened? Before we could develop anything, now we cannot do anything, it has to be approved by other member countries … so everybody is so afraid to give any

firm opinions and what does it mean? What can we do? This is becoming very hot issue in Laos I think because … I mean not only the Mekong mainstream now, even the tributaries we need approval from the member country, what can you do then, you sit and do nothing.

(Interview 52, 2012)

Even as notions of transparency embedded in law and policy at multiple scales have helped to undergird a regulatory infrastructure and delimit a field of action oriented around and towards hydropower development in the Mekong River Basin, the operationalization of transparency norms in this context does seem to have generated, or be generating, a sense that those in power in any one jurisdiction can no longer 'develop anything' to the full extent of their financial capacity. Consideration must, rather, be given to the 'agree[ment]' or 'comment' of those across national borders, perhaps even to obtaining the 'approv[al] by other member countries', or so our interviewee above suggests (with frustration).

The contending versions of transparency highlighted here are, of course, not exhaustive of significance given, or appeals made, to transparency in the Mekong River Basin. The notion of a right to information (drawing upon constitutional guarantees outlined above) may, for example, be implicit in some of the human rights advocacy discussed in Chapter 7, but a right-based notion of transparency is not one on which we focus here (see, generally, Klaaren, 2013). The next section of this chapter will elaborate on the distributions and effects of the rival notions of transparency that we have outlined, with particular reference to the Nam Theun 2 dam.

Transparent action

The interplay between these accounts of transparency in the Mekong River Basin – transparency as a principle of MRC work, transparency as development-oriented public participation and remedial transparency – has had distributive and constitutive effects in and around the Nam Theun 2 project in Laos. None of these effects is fully anticipated by the accounts of transparency's value (or irrelevance) outlined above (that is, those associated with the republican, utilitarian, pragmatic, Marxist, neoliberal, 'sacred nationalist' or law and development traditions).

The basic features of the Nam Theun 2 project, including its financing structure, were outlined in Chapter 1. This section will highlight some of the ways in which notions of transparency have helped to direct energy, distribute resources and shape engagement in and around the project, including to foster a commonality-for-development among the people of Laos and those otherwise concerned with the project, and to place that commonality in question.

The 'publics' of Nam Theun 2

In order to deliver on the promises and requirements of 'public consultation' demanded by multilateral lenders to the Nam Theun 2 project, one or more versions of the 'public' had to be mobilized and made visible (Mirumachi and Torriti, 2012; Singh,

2009). A common starting point for this endeavour – enlivened as such by the World Commission on Dams Framework mentioned above, among other instruments – has been the concept of 'stakeholders' (World Commission on Dams, 2000a).

Historically, the stakeholder implied custodianship or trusteeship: specifically, the stakeholder was one with whom money was deposited, for a time, for some gambling or transactional purpose ('stakeholder, *n*.', OED Online, 2014). The stakeholder of the Nam Theun 2 project, however, owes less to eighteenth- or nineteenth-century gambling and financial practices than to a management strategy that emerged, for the most part, in the 1980s, known as a stakeholder approach (Freeman and McVea, 2001). The focus of this approach was on equipping business managers strategically to manage relationships with 'any group or individual who is affected by or can affect the achievement of an organization's objectives' (Freeman, 1984, p. 46). The 'stake' by which a stakeholder may be identified has thus been conceived for the most part in terms of impact or effects: effects *upon* a business enterprise and effects *of* that enterprise. Additional criteria have been proposed for determining 'who and what really counts'. Some have, for instance, differentiated salient effects according to influence, relationship and claim (Mitchell *et al.*, 1997). Others have framed 'stakes' in terms of risk (Clarkson, 1994). Nevertheless, in management literature, the goal of a stakeholder approach in relation to those so identified has been framed in terms of balance: 'stakeholder management is a never-ending task of balancing and integrating multiple relationships and multiple objectives' (Freeman and McVea, 2001). The influence of this managerial approach was apparent in the following remarks from a senior staff person at a Thai commercial bank involved in hydropower development financing:

> [A]dopting integrity principle[s] doesn't mean that you can shy away from other people, so communication at [the] right time with the right content is something that is a part … [of] a framework for reputation management … Talking too much sometimes is not good especially for the reporters these days, sometimes they just keep whatever they would like to hear …
>
> (Interview 29, 2012)

So managed, a 'public' to which the Nam Theun 2 project was to be made transparent was a public configured always around and towards the project, according to predicted effects of the project and anticipated capacity to affect the project. The project was cast as a fulcrum upon which 'balance' among stakeholders was expected to come to rest.

It was on this basis that the Nam Theun 2 project's 'stakeholders' were further categorized: as '[p]eople directly affected by the Project' (Project Affected Peoples, or PAPs); 'GoL [Government of Laos] officials at the district, provincial and national levels'; the 'broader interested regional and national community'; 'NGOs operating in the Lao PDR and particularly those in the Project area'; and '[i]nternational NGOs, international organizations, and the local, regional and international media' (Nam Theun 2 Power Company, 2005, vol. 1, ch. 4, p. 2). According to the project's Social Development Plan, this followed 'extensive scoping of issues and review of findings' (Nam Theun 2 Power Company, 2005, vol. 1, ch. 4, p. 2). This scoping or review did not, however, appear to extend to the merits or demerits of a stake-holder approach, or alternatives to that approach. Rather, from the beginnings of public consultation around the project in the mid-1990s, the exercise was framed as one of identifying and allowing participation by 'stakeholders' (McPhail and Callieri, 1998; Nam Theun 2 Power Company, 2005, vol. 1, ch. 4, pp. 1–6).

In addition to breaking project stakeholders down according to anticipated effects, institutional type and indicia of interest – as the preceding paragraph high-lighted – the Nam Theun 2 Social Development Plan categorized stakeholder groups on a scalar and geographic basis with regard to 'the social and resettlement components of the NT2 Project' (Nam Theun 2 Power Company, 2005, vol. 1, ch. 4, p. 2). Among 'local' stakeholders, the public was organized by household, village and community and by location: specifically, whether they were households or villages 'on the Nakai Plateau'; communities 'living along the Xe Bangfai' River or downstream of the dam; or communities 'with assets or land under the Project (construction) Lands'. The closer the 'public' was to the project construction site, the smaller the scale of the units into which they were divided – namely, household or village. Those located on the Nakai Plateau were also aggregated and disaggre-gated according to 'ethnicity' based on '[d]etailed anthropological studies' (Nam Theun 2 Power Company, 2005, vol. 1, ch. 4, p. 5).

The further the 'public' was positioned (geographically) from project construc-tion, the more its configuration in relation to the project depended on pre-existing, formal political and legal arrangements for representation or delegation, as well as on generic enterprise or institution type. At the 'regional' level, for instance, the cat-egory 'stakeholders' extended to 'community leaders', 'GoL agencies' at the district and provincial levels, and '[b]usinesses and contractors'. At the 'national' level, the 'public' became the 'People of the Lao PDR' and, alongside them, recognized stake-holders included 'GoL Ministries' and the '[n]ational media'. At the 'international' level, the notion of stakeholders seemed to return to its financial roots; stakehold-ers at this scale included 'international power utilities', in particular the Electricity Generating Authority of Thailand, the World Bank, the Nam Theun 2 Project Company, 'investors and Financial Institutions', international NGOs, international media and '[o]ther hydroelectric dam developers' (Nam Theun 2 Power Company, 2005, vol. 1, ch. 4, p. 5). To some extent, this progression towards ever-greater medi-ation seems logical and sensible. At the same time, however, there are aspects of this categorization that seem quite counterintuitive, even questionable.

It is striking, for instance, that among those pre-existing institutions envisaged as 'stakeholders', bodies representative of 'workers, farmers and [the] intelligentsia' do not appear – despite these groupings being characterized as 'key components' of the Lao polity under the Lao Constitution (Article 2). Likewise, it is noteworthy that only some of those 'social organizations' that feature in the Lao Constitution as 'the organs to unite and mobilise all strata of the multi-ethnic people to take part in the tasks of protection and construction of the country' – the Lao Front for National Construction (LNF), the Lao Federation of Trade Unions, the Lao People's Revolutionary Youth Union and the Lao Women's Union (LWU) – appear as vehicles or sites of 'public participation' in connection with the Nam Theun 2 project (Article 7). The LWU and the LNF did register in the project's Social Development Plan: these were two 'mass organisations' with which '[m]eetings, discussions and seminars' were held 'from 1995 onwards' at a local, regional and national level for '[t]raining and capacity building … in consultation techniques and participation as members of PCPD [Public Consultation, Participation and Disclosure] Teams' (Nam Theun 2 Power Company, 2005, vol. 1, ch. 4, p. 7). It is clear that these social organizations identified with Lao communist organizing could only be admitted to the Nam Theun 2 project's 'publics' once pre-qualified for that purpose through 'capacity building'. 'Local GoL Organizations' and 'Provincial Organizations' were similarly the targets of such training in 'consult-ation techniques' (Nam Theun 2 Power Company, 2005, vol. 1, ch. 4, p. 7).

The enactment of transparency – in the mode of 'public consultation, participa-tion and disclosure' – was thus among the 'techniques' that needed to be acquired by those pre-qualified as 'stakeholders' of the Nam Theun 2 project. A 'public' that would suffice for international standards on public participation was something that certain segments, leaders and institutions of the 'People of the Lao PDR' had to learn to become. This becoming public was inserted into a developmental trajec-tory envisaged in parallel to the Nam Theun 2 project's realization: a trajectory of 'capacity building', based on an assumption of 'low' political capacity on the part of Lao government institutions, social organizations and bureaucrats (UN Department of Economic and Social Affairs, 2005).

A prevailing expectation was that the project would serve – with regard to assur-ances of transparency and in other respects – as a 'standard setter' for future polit-ical and economic development in Lao PDR (Laking, 2008, p. 3; see generally Triantafillou and Nielsen, 2001). The 'Decision framework for processing the pro-posed NT2 project', published by the World Bank in 2002, asserted that 'it is critical to the success of the project that sustained progress on reforms be maintained over the long-run' (World Bank, 2002). Enactments of transparency in and around Nam Theun 2 also helped to reinforce a developmental narrative surrounding the World Bank itself, which was, during the period of the project's negotiation, preoccupied with the development of metrics around governance and information flow, as well as with its own institutional reform on such issues (Kaufmann *et al.*, 2003; Bellver and Kaufmann, 2005; Bello and Guttal, 2006). Yet it remained just as conceivable that this useable knowledge assembled for the Nam Theun 2 project's realization – the tailored configurations of the public and techniques identified with their consulta-tion – would be exhausted in its manifestation as such. Arguably, that is what has occurred vis-à-vis the project's mainstream successor, Xayaburi. Around Xayaburi, few remnants of the audit, consultation and reporting architecture erected around Nam Theun 2 are still in evidence (Rieu-Clarke, 2014; Middleton, 2012).

Transparency to underwrite project regulatory and financial infrastructure

Indicia of transparency – enactments of 'public consultation, participation and disclosure', for instance – were taken to ensure for the Nam Theun 2 project 'broad-based' support, both within Lao PDR and internationally (World Bank, 2002). Yet these also served to underwrite the project's bespoke legal regime. Part of the regulatory architecture devised for the project was a multipart supervisory mechanism involving expert panels, revenue management arrangements, audit institutions and advisers, some of whom were introduced in Chapter 1. The ensuing sense of the project being continually monitored – albeit by an institutional 'gaze' generated, for the most part, solely for that purpose – helped to deflect criticism and assuage demands for remedial transparency by other means. The structures of internal and external monitoring put in place in connection with Nam Theun 2 were deemed 'invaluable' by World Bank staff in answering the demands of '[i]nternational stakeholders' for 'open, transparent, independent mechanisms to monitor project performance', even though 'not all parties agree on what this entails' (Porter and Shivakumar, 2011, p. 71).

Another implication of transparency-ensuring mechanisms being tailored specifically for the Nam Theun 2 project may have been to engender or reinforce a sense of development information as proprietary and metered out, ordinarily, by contractual arrangement. Information generated by the project's International Social and Environmental Panel of Experts (the Expert Panel) was made generally available on the project's website (albeit predominantly in English) and was at times critical of the project's management (see, for example, POE, 2007, p. 9). Yet this information was still clearly aligned with the project's investors and responsive to their demands, especially those of the Lao government to which the Expert Panel reports. Our interview with a senior official in the MRC secretariat suggested that this is typical of the way that development information has come to circulate in the Mekong River Basin:

> We have some difficulties in some areas where we would like to share more data, more information, and we're constrained by two particular issues. One issue is that some of the countries don't think that that's information that should be shared too widely, but shared in a smaller government-to-government group. The other one is this issue of the value of the information and although we have a disclosure policy, if a consultancy company comes and asks for particular information we will have to say, well what are you going to use it for? Are you going to use it for commercial purposes? Are you going to sell it? Are you going to do this and this and this and this … If a member government comes and asks for information then it's easier to disclose that information because they are owners of it also and we have a system.
>
> (Interview 48, 2012)

Information surrounding hydropower development in the Mekong River Basin is typically made transparent, it seems, by permission, special agreement or proprietary right, rather than as a matter of public entitlement or legitimate public interest. This understanding may be traceable to the contract-dependent nature of most of the transparency mechanisms arrayed around Nam Theun 2.

Transparency as distancing

Information-sharing and consultation about the Nam Theun 2 project and review of its progress have, nonetheless, brought a wide range of people concerned with the project – especially those identified as 'stakeholders' – into regular contact with one another. Informal conversations conducted during a visit to one of the resettlement villages on the Nakai Plateau left an impression that village inhabitants were regularly inundated with visitors and subject to periodic surveys (site visit to a Resettlement Village, 2012; see, for example, Souksavath and Nakayama, 2013).

At the same time, a number of sources suggest that disaffection and complaints concerning the project among those resettled rarely reach beyond the village headman and that some village headmen actively discourage feedback (Souksavath and Nakayama, 2013, p. 83; Singh, 2009, p. 500). Given the retribution to which those perceived as critics of the Lao government have sometimes been subject, this is understandable (UN Working Group on Enforced or Involuntary Disappearances, 2013). However, risks associated with the people of Lao voicing their views about the Nam Theun 2 project seem to have been given relatively slight attention among those charged with amassing 'broad-based support' for the project, as the following quote from an anonymous source of Guttal and Shoemaker's, responding to an invitation to attend a meeting with World Bank officials, suggests:

> About the [World Bank] have a meeting with NGOs on February 6. I am truly appreciated in your sincere to the socio-environmental impacts because of Nam Theun 2. However, about my conduct, as I am Lao, I am living in Laos. As you know, the political in Laos is really different so far, government very pro in this project. So, if me or someone of Lao speak out in tend to against, we cannot live, especially tell to the foreigner of important person which concerning to the project, even if we are truly identify to the deficiencies, and truly wanted to improve. By the reason I am quiet, and better I should be find another way to help the poor in other parts … I am very sorry, but I can not directly tell to some person such kind of important meeting under dangerous of situation political, is really feel unsafely, if I say anything against, everything finish for me, I can not work.
>
> (Guttal and Shoemaker, 2004)

The proliferation of opportunities for direct or mediated encounter among project stakeholders, within and among nations of the Mekong River Basin, thus seems, somewhat counterintuitively, to have maximized a sense of distance and estrangement among them. One interviewee from a Thai-based NGO remarked that '[l]ack of political space, especially in Laos [is growing]. In Vietnam it's growing slightly but still quite closed. In Cambodia and even in Thailand it's closing' (Interview 20, 2011).

This distancing effect of project-related transparency efforts may be most apparent in the Nam Theun 2 project's visitor centre, where one can view annotated photographic displays and watch videos about the project in air-conditioned comfort (see Chapter 1 and, more generally, Otomo and Eslava, 2010). But for some Lao architectural flourishes, the design and atmosphere of the visitor centre could not feel more remote from the conditions to which most people are accustomed in the surrounding area, even those occupying the relatively salubrious dwellings constructed in the resettlement villages. The outflow of information through such an imposing architectural medium may do more to block engagement with the project than invite it. This, in turn, may help to fuel appeals to remedial transparency, as described earlier.

Transparency as commonality-for-development – prospects for its retooling

Transparency is, as observed earlier, a term commonly carrying a positive inflection. In many accounts, it is associated with the righting of wrongs, the empowerment of the disempowered, the inclusion of the excluded, the correction of asymmetries and the promotion of economic growth. As this chapter has shown, however, transparency elicits a range of expectations and understandings, these often in tension with one another.

In the case of the Nam Theun 2 project, contending versions of transparency seem to have coalesced, for the time being, around an idea of transparency for which the project itself is the primary touchstone. It is in this project's shadow that a sense of Lao 'publicness' has been made newly visible. It is by reference to this project's development that many in Laos, and elsewhere in the Mekong River Basin, have been encouraged to re-envision their pasts and tailor their futures.

So framed, the public made transparent to itself and to others, in connection with the Nam Theun 2 project, is both fragile and isolated: tethered to the physical site of the project, conditional upon project-specific agreements and circumscribed by expert prediction of project impacts. Yet this sense of commonality is, at the same time, uneasy, contested and amenable to strategic intervention. As shown in the somewhat cynical redeployment of 'transparency' to anoint Nam Theun 2's successor – the Xayaburi project – one of the world's most 'transparent' dams (on the basis of its navigability by migrating fish), transparency remains open to creative retooling (Al Jazeera, 2012). Moreover, despite its apparent frustration time and time again, a sense of transparency as potentially remedial persists among the peoples of the Mekong River Basin.

Given transparency's differing permutations and combinations in the Mekong River Basin, movement upwards on one or other transparency index should not be taken as unequivocally positive. Even so, the foregoing account is put forward less with an eye to condemning reliance upon or appeals to transparency, or replacing transparency-related measures with something else, than with an eye to more astute navigation of the work of transparency norms and the commonalities they elicit. The story of Nam Theun 2 is not a story of transparency's failure; it is a story of transparency – or transparencies – most productively at work. For all its developmental routinization, transparency remains a basis for challenging hydropower development, as the following chapter will show.

7 Contesting a river basin
Civil society's legal strategies

While civil society actively challenges hydropower development in the Mekong River Basin, the role of law in these struggles has seldom been examined, or made visible as an instrument or locus of contestation. In part this is because of common assumptions in the region about the limited utility of law as a tool of resistance. First, there is a view (expressed here by a donor representative in Cambodia) that hydropower is 'not governed by law' at all, but determined by the political economy of 'who has the most power and resources', in a region where governments are 'completely committed to physical development' and the 'transformation of the landscape' (Interview 2, 2011).

Second, to the extent that law does have traction, it is believed to instrumentally enable the powerful to get what they want: 'the most powerful will invoke law' where it suits them (Interview 2, 2011). Some in the Mekong River Basin (such as the Thai NGO quoted here) see law as an instrument of corporate interests: 'the aim of law is to promote the economy, not about protection or mitigation' (Interview 23, 2011). According to this viewpoint, law enables rather than constrains development. Legal rights *against* governments or companies are not taken seriously in weakly institutionalized states driven by pragmatic, transactional political cultures, personalities and patronage.

In turn, those adversely affected by hydropower development are thought to understand law as being repressive and constraining rather than enabling and remedial. Thus, some villagers affected by hydropower development in the 3S rivers region of Cambodia's Stung Treng province reported that law is about 'what it is forbidden to do' rather than about rights and protections (Interview 18, 2011). One regional NGO agreed that 'law tells people where their position is, not their rights for protection' (Interview 9, 2011); another Cambodian NGO said that 'law is used to control, and to protect the powerful' (Interview 7, 2011). Among Thai NGOs, too, there is a sense that law is 'not giving rights to people, but taking customary rights from people' (Interview 23, 2011).

Third, disputes are said to be resolved not by formal legal institutions or through Western notions of the rule of law, but by ad hoc negotiation – 'through rice wine' (Interview 2, 2011) – which suits those with power. 'Law in practice' is different from law on the books and comprises 'informal or shadow institutions behind the formal … regime' (Interview 7, 2011).

Fourth, even if law potentially offers redress, communities affected by hydropower development – typically rural, indigenous and poor – are believed to have a limited understanding of law and little capacity to invoke it to vindicate their rights. In the Cambodian village mentioned earlier, a villager stated that 'we do not know what the law is' and hence 'we are blind' (Interview 18, 2011).

This chapter challenges these assumptions about the role of law in the governance of hydropower in the Mekong. It shows that law is often strategically mobilized by actively engaged civil societies as a form of resistance and argument to challenge the developmental agendas of governments, corporations, investors and development banks. The frequent resort to law by civil society suggests that it is seen as a useful mode for challenging dominant assumptions about the political economy of development – namely, that money talks, politics rules and law is irrelevant or marginal. Law has a curious resonance and appeal to civil society, even when many know that often it does not work as they would hope or expect.

In particular, the chapter shows that law is central, not marginal, to decisions about hydropower development. It also shows that while law often facilitates development, its plurality of norms and processes also enables civil society to resist development, or influence its shape, or temper decision-making. Law is not only a tool of domination, but also 'provides space for resistance', 'a terrain of contestation' (Rajagopal, 2005, p. 348), and 'resources for people to argue for their "rights"' (Pirie, 2013, p. 12). Sometimes this even happens through the courts, although more commonly law is utilized elsewhere. Further, awareness of law among affected communities is not as low as is often thought, but rather a critical (and plural) consciousness of law (Trubek, 1984, p. 592) permeates civil society thinking and action on hydropower in the Mekong River Basin.

This chapter accordingly examines how and why law and legal institutions are used by civil society to contest hydropower development. It unsettles the common cause for development around which transparency norms, considered in the previous chapter, encourage many in the Mekong River Basin to gather. It is true that civil society rarely seeks to challenge hydropower in the courts, except in Thailand. The challenge for civil society is how to effectively use law where a key mechanism for its enforcement – the courts – is effectively unavailable. This chapter's account of how different actors take up this challenge may seem to return us to a more 'classical' notion of law than was in the foreground in Chapter 6. Less focus is placed here on the work of 'umbrella concepts' in eliciting subjectivities and shaping aspirations, and more on the explicit mobilization of legal rules. Nonetheless, this chapter's story remains one in which law is socially constructed and contested and operates as both a vehicle and an engine of hope.

As will be demonstrated, law is often drawn upon as a language for articulating claims and grievances in formal and informal interactions with state administrative and political authorities. The normative authority of the law is invoked to strengthen the bargaining position of those affected by projects, and to persuade authorities to abandon or mitigate dam proposals.

Civil society has also used law in attempting – often not very successfully – to engage and persuade private actors. While hydropower companies and investors insist that they follow national laws, more proactive policies of corporate social responsibility (CSR) are not well developed. Civil society has had difficulty even in contacting companies. Civil society has, however, used a non-binding complaints mechanism of the Organisation for Economic Co-operation and Development (OECD) with modest success. Some civil society organizations have also been involved in the development and trial of the hydropower industry's Hydropower Sustainability Assessment Protocol, which aims to measure the environmental and social performance of projects.

A further core strategy of many local NGOs and INGOs, particularly in Cambodia and Thailand, is to provide legal education, training and 'empowerment' to affected communities. Such activities are a response to limited legal knowledge in the region, and aim to enable people to articulate legal claims against states or developers. The dissemination of legal ideas also provides a shared vocabulary for civil society networks to draw upon in organizing regionally to contest hydropower development.

As in every society, law is not the only or even the main frame of resistance. Grassroots social movements, political advocacy, social justice discourses, mass mobilization, direct action and media campaigns are some of the other means of contesting development. Many of these are used in concert with legal strategies or to advance legal arguments. Even so, law is surprisingly pervasive. Some of the most successful mass movements, such as the opposition by the Assembly of the Poor to the Pak Mun dam in Thailand, have been both facilitated by law and creative of it.

In terms of what 'law' is considered in this chapter, in practice civil society's legal engagement in disputes about hydropower development tends to be centred in *national* environmental law, water law and land law, and, increasingly, private law. This is inevitable because national laws primarily govern hydropower decisions and project implementation, whereas regional and international laws are more distant and less institutionalized. As noted in Chapter 2, however, distinctions between national, regional and international law are not as sharp as is sometimes supposed (Rajagopal, 2005, p. 346). Rather, law is often plural, hybrid and 'transnational' (Koh, 1996) and invites a 'politics of resistance that is neither purely local nor global' (Rajagopal, 2005, p. 348). According to an international NGO, international standards are not seen as 'radical' in the region (Interview 22, 2011). As elsewhere (Dupuy and Vierucci, 2008; Boyle and Chinkin, 2007, pp. 41–97), NGOs in the Mekong region have increasingly sought to influence the making, implementation and enforcement of international norms in domestic legal orders.

The discourse of human rights, however, has thus far taken a more marginal place in strategies of legal resistance in the Mekong River Basin. In part this is because human rights are often seen as more 'political' than the more 'technical' or 'scientific' discourses of environmental law that prevail in hydropower development in the region. It is also because human rights can be perceived as challenging state political authority in a region where none of the Mekong countries is genuinely democratic. For example, according to one domestic NGO, Vietnam 'thinks outsiders use human rights against it so is resistant to the idea' (Interview 34, 2011). Even so, there are formal openings for human rights claims; all of the Mekong countries provide constitutional protection for rights, and Thailand has a National Human Rights Commission empowered to consider rights violations.

In terms of the actors examined here, the notion of 'civil society' encompasses a variety of groups pursuing common interests outside the family, state and market (CIVICUS, 2005). This chapter focuses only on domestic, regional and international NGOs concerned about the adverse environmental, social and economic impacts of hydropower development, and 'grassroots' community groups affected by it. Some of the key actors were introduced in Chapter 1.

Law in administrative and political engagement

The most common way in which civil society invokes and engages with law in Mekong hydropower governance is through the state's administrative and political processes. As discussed later, in part this is because, as observed by a Cambodian legal NGO, many people do not trust the courts (Interview 6, 2011) or because the courts are inaccessible or risky to use. In engagements with authorities, law (alongside other discourses) often serves as a language of claims, entitlement and resistance framing a dispute. Because it is external to the subjective demands of affected persons, and validated as authoritative by the constitutional structures of the state itself, law can empower community demands, lend credibility and legitimize claims beyond the self-interest of those invoking it. Law serves as a form of strategic argument in interactions with state officials, regardless of whether it can be enforced through binding court or tribunal decisions.

The profuse civil society engagement with authorities both evidences its familiarity with law and stimulates its acquisition of further legal knowledge. Law may originally issue from the state, but once unleashed the state cannot exclusively control it and social forces can work to reconstitute it. As Merry argues, 'talk about the law by ordinary people contributes to what law is and what it is not ....These understandings, which can be called legal consciousness, include people's expectations of law, their sense of legal entitlement, and their sense of rights' (Merry, 1992, pp. 209–210). Despite the lamented weakness of the rule of law in Mekong countries, these 'interpretive acts by ordinary people', even when 'unauthorized and illegal' (Rajagopal, 2005, p. 353), often contest and refashion statist visions of law.

Diverse modes of legal consciousness are evident in communities affected by hydropower development. Villagers in Cambodia expressed familiarity with natural resources law that had a direct impact on everyday life, such as fisheries, land or forestry laws (Interview 18, 2011). Communities also often described their interests in normative language, even if it is not pegged to knowledge of particular laws. Thus, according to a local NGO in the 3S rivers region of Cambodia, villagers sometimes speak of their 'right' to water (Interview 10, 2011). Sometimes such understandings align with law on the books; sometimes they are based on formally unrecognized traditions or customs; and at other times they are claims beyond any law. It is not unusual for social legal consciousness to involve expanded notions of what the law actually says (Merry, 1992, p. 214).

While affected Cambodian villagers professed little knowledge of state law, their statements in fact evidence a strong awareness of the technical legal concepts that dominate development planning, mitigation and remedies. Villagers stated that the government did not research the impacts of development: 'before development in the area, they should study the impacts and consult first, negotiate, pay compensation, reach agreement' (Interview 18, 2011). This is unsurprising: villagers do not live in an isolated, pre-legal, rural bubble. Most villagers have mobile phones; many have access to diverse media; quite a few have been visited by NGOs and some are members of NGOs; and some have had contact with hydropower companies, environmental impact assessment (EIA) consultants, and multilateral development bank staff.

Local authorities

For many rural and indigenous people living in the Mekong River Basin, customary law (albeit dynamic and often shaped by the state or other external influences, including colonial histories) is typically their foremost frame of legal reference. Customary law, however, normally only applies to settle local disputes between community members and thus is inoperative (or subordinate) in addressing disputes over land or resources with outsiders (UNDP Cambodia, 2007, p. 3). Communities are, accordingly, forced to go outside customary law to contest state-driven hydropower development, typically escalating through the hierarchy of state authority from the village chief through to the commune, district, provincial and national levels (UNDP Cambodia, 2007, p. 4).

The first point of administrative contact with the formal state is local authorities. Local authorities can often address, regulate and take enforcement action concerning complaints or disputes relating to natural resources, such as illegal fisheries or logging, or pollution, as a Cambodian provincial government officer noted (Interview 15, 2011). There are also examples (as in the Voen Sai district of Ratanakiri province in Cambodia) where NGOs have successfully advocated with district and provincial authorities for the return to the community of land illegally encroached upon by private companies (Interview 12, 2011). As one Cambodian NGO observed, NGOs often play a facilitative role in organizing dialogue and meetings between villagers and local authorities (Interview 17, 2011).

While local authorities formally apply state law, they may be sympathetic to the concerns of local people and even base decisions 'on concepts and norms of traditional law, as much as or more than they are on the application of national laws' (UNDP Cambodia, 2007, p. 5). As a senior provincial government official in Cambodia observed, a 'compromise between law and custom' is necessary in managing natural resources, rather than a strict application of state environmental laws (Interview 16, 2011). Some Cambodian provincial government officials also regarded human rights, as reflected in domestic law, as part of their work (Interview 14, 2011). Some affected Cambodian villagers expressed confidence in the local authorities to resolve disputes (Interview 18, 2011).

Some Cambodian NGOs, however, noted that local authorities may be answerable to political parties, have limited capacity, or be shareholders in developer companies (Interview 17, 2011). Local authorities may also be wedded to the central government's development agenda (Interview 10, 2011), especially in the communist states of Vietnam and Laos. A senior provincial government official in Cambodia observed that public versus private interests are an issue in every country, and he has to 'educate' the local people about the public interest – which is, ultimately, also in their 'family interest' (Interview 16, 2011).

However, local authorities are not the decision-makers or project operators in hydropower projects of national significance that have large-scale social and environmental impacts. They may have little input into the national-level decisions about project design or impacts. A senior provincial government official in Cambodia noted, for instance, that he had not been contacted for any EIA studies for hydropower development on the 3S rivers (Interview 16, 2011). Consequently, the level of state authority with which communities are most familiar has little control over hydropower development projects. Local authorities are often then tasked with implementing national decisions, including administering compensation and resettlement, but have little autonomy to determine their parameters or moderate their adverse impacts. Local officials have even less authority to respond where community concerns relate to transboundary impacts.

Environmental impact assessment and Procedures for Notification, Prior Consultation and Agreement

Moving beyond local authorities, for larger hydropower projects, national EIA laws and procedures are a key site for civil society to contest development. As is apparent from Chapter 5, EIA processes have enabled civil society to be consulted on particular projects (although not to take part in making decisions). Opportunities for public participation both are stimulated by civil society demands and in turn play a role in formally constituting 'civil society' (by recognizing or not recognizing the right to participate – that is, the legal personality – of various actors). The limitations of EIA processes are, however, well known – particularly inadequate disclosure and consultation, official resistance and poor government responsiveness to EIA findings. Intimidation of civil society has also marred EIAs in Laos. For smaller hydropower projects, such as A Luoi dam in Vietnam, a full EIA may not be required, thus lessening opportunities for civil society participation.

In parallel to national EIAs, regional interstate processes have unexpectedly expanded civil society participation. Formally, the Procedures for Notification, Prior Consultation and Agreement (PNPCA) apply among states parties to the Mekong Agreement, as explained in Chapter 4. However, the first use of the Mekong River Commission's (MRC's) PNPCA process, for Laos's Xayaburi dam, involved public consultations in Cambodia, Thailand and Vietnam (Rieu-Clarke, 2014, pp. 10–11, 19–20), including community meetings at the project level and national consultations with wider stakeholders. The consultations identified environmental, livelihood and social impacts.

It is highly significant that even a soft law procedure conferring exclusively interstate rights evolved in its first practical application to include some civil society participation. A peak Cambodian NGO noted that this process enabled 'talk' with the Lao government (Interview 8, 2011). A Thai environmental NGO noted that the PNPCA process for Xayaburi provided a space around which 'people can organise' and claim 'accountability' (Interview 23, 2011). A review of this PNPCA process described it 'as an entry point and opportunity for dialogue' (Australian Government, 2014, p. 3). Civil society has, however, continued to call for improvements to consultation processes in subsequent PNPCAs, such as that conducted for the Don Sahong hydropower project in Laos (NGOs' Joint Statement, 2014).

The MRC's own research linked to the PNPCA has also provided for civil society participation. For example, the MRC commissioned an SEA in 2009 to review 12 proposed mainstream dams and received input from 40 civil society groups (ICEM, 2010, p. 9).

Other national authorities

Beyond EIA, civil society has sought to influence the plurality of governmental actors responsible for decision-making. Hydropower development in the Mekong 'is governed by multiple government agencies operating in a multi-level social/ legal field of more than one legal order' (Suhardiman *et al.*, 2011, p. 131). A sectoral rather than an integrated approach to planning tends to prevail, with a separation of hydropower, water and environmental planning. Some NGOs have criticized the lack of coordination as creating confusion and difficulty for advocacy (Interview 8, 2011).

However, the dispersal of legal authority is also an opportunity for civil society, because the pluralism of a fractured, non-unitary state (Rajagopal, 2005, p. 347) enables more entry points for intervention and pressure. One INGO observed that 'governments are not homogenous' and in Thailand, for example, some actors in government may be for and others against a project (Interview 20, 2011). Thus, some in civil society have engaged with energy utilities and ministries – 'which tend to have more sway in national planning than ministries and line agencies affiliated with the MRC' (Lee and Scurrah, 2009, p. 48). In Vietnam, too, the Ministry of Natural Resources and Environment is seen by NGOs as weak in influence, and less open to legal arguments than the Ministry of Foreign Affairs (Interview 35, 2011). In Cambodia, some NGOs seek to influence members of the ruling party and opposition parties (Interview 6, 2011). There are, however, hard legal limits to strategies of targeting different entry points; for example, in practice investment law in Cambodia often prevails over the water law (Interview 19, 2011), making interventions with water authorities less fruitful.

A particularly common form of deploying law in attempts to engage with administrative and political structures is through direct appeals (often publicized through the media) to senior government bodies or officials. Appeals are typically directed towards heads of government, the responsible ministers, the government at large, or government leaders in their capacity as members of the MRC Council or ASEAN. Less attention is directed towards members of parliament (except in Thailand), who are seen to lack influence (Interview 5, 2011), although even in Vietnam some NGOs and INGOs engage them (Interview 35, 2011; Interview 24, 2011).

As has occurred in other regions of the world, letter-writing to advocate for justice is a particularly prominent and low-cost strategy used by civil society – most commonly in Cambodia and Thailand or by organizations based there – including by local activist networks (Interview 10, 2011). Letters often describe community concerns about dam impacts, but also articulate legal arguments, often drafted by NGO experts. Letter campaigns have emphasized the importance of adequate EIA processes, resettlement and compensation measures, and other environmental and human rights safeguards. Issuing public statements directed at governments is also a common way of making often extensive legal arguments, particularly in Cambodia and Thailand, although less so in Vietnam and Laos.

There are important differences in the strategies used by different civil society actors to influence governments. As one foreign donor representative commented on Cambodia, letter-writing to ministers should only be done 'to say something nice, not to criticise' (Interview 2, 2011). On this view, 'overt confrontation gets you nowhere; better to talk to people quietly, off the record, providing information, a point of view, perspective, not necessarily critical' (Interview 2, 2011). Building and cultivating relationships and trust are vital, whereas government officials 'lose face if you resort to publicity' (Interview 2, 2011). Some government officials are also sensitive about being 'talked down' to: 'the government does know and understand the issues, they are smart people' (Interview 2, 2011). Or, as Cambodian Prime Minister Hun Sen reportedly said to a UN special rapporteur, 'you don't need to tell me it's raining when I'm walking in the rain' (Interview 4, 2011).

Mindful of such sensitivities, many local NGOs profess to work constructively 'with' not 'against' governments (Interview 8, 2011), including, in the words of the 3S Rivers Protection Network (3SPN), to 'ultimately reach a group consensus' (3SPN website). The private Samreth Law Group (now the Vishnu Law Group) in Cambodia seldom uses the media so as not to antagonize government (Interview 5, 2011). Policymakers are said to be more likely to listen to lawyers than to grassroots activists because the former enjoy higher social status (Interview 22, 2011).

Local actors too are conscious of these strategic issues. One villager in the 3S rivers region of Cambodia, Va Khamphai (Kon Mon), reportedly said at a 3SPN strategy workshop: 'To do advocacy work is just like to catch a cobra. If you catch it properly, it cannot bite you, otherwise you will be bitten' (3SPN, 2011, p. 13). Customary law in Mekong countries also emphasizes conciliation between the parties, maintaining social cohesion, and restoring harmony (IUCN, 2008, pp. 27–29; UNDP Cambodia, 2007, p. 3), rather than adversarial confrontation and the assertion of winner-takes-all rights.

Foreign actors are not so constrained in overtly criticizing governments. There is also more opportunity for communities affected downstream by transboundary dams to criticize upstream governments, and likewise for foreign NGOs or INGOs to publicly criticize. That does not, however, mean that such interventions are more effective in influencing those governments, only that they are outside the reach of adverse legal or other retribution. Some key INGOs, such as International Rivers and EarthRights International (ERI), note that they prefer to work in the background by supporting local communities and NGOs in deciding upon their own strategic actions. At the same time, however, these organizations have publicly criticized Mekong governments at various times, so their mode or style of intervention is usually a question of timing and context.

Appeals to high-level authorities often do not, however, produce negotiation, bargaining or the satisfactory resolution of disputes. There have been some impressive petitions presented to governments. Against Xayaburi, for example, there was a letter from 10,000 Thai villagers sent in April 2009 to the Laos Embassy and the Thai prime minister (Thabchumpon and Middleton, 2012); a transnational petition to MRC governments with 23,000 signatures in October 2009; and a letter in March 2011 from 263 NGOs to the prime ministers of Laos and Thailand. But complaint letters and petitions are seldom answered by Mekong governments – although the MRC secretariat says it 'always' tries to answer them, even if 'we might have a different interpretation on some rights' compared with civil society (Interview 48, 2012).

Civil society coalitions have also pleaded with UN procedures to engage with governments. For example, in January 2015, ERI, International Rivers, the Samreth Law Group and 3SPN made a detailed submission to the UN special rapporteur on human rights in Cambodia in relation to the Lower Sesan 2 and Stung Cheay hydropower projects, arguing state and corporate breaches of rights to housing, food, water, health, a healthy environment and cultural life; indigenous and minority rights; and rights to information, consultation and participation.

Unilaterally arguing the law in the public sphere, or participating in procedural opportunities to be consulted, is not the same as enforcing substantive legal rights in binding institutions. Law may provide one of many languages or discourses for political argumentation, and even carry a distinctive authority and influence of its own. Occasionally, it can even impel governments to justify their actions by mounting legal counternarratives. But none of these processes – monologic or dialogic – should be mistaken for the actual vindication of legal rights or the curtailment in practice of arbitrary state action.

The use of law in engaging private actors

Private sector actors in the region are often perceived as less readily accessible to civil society, and less receptive to being engaged, than governments, the MRC or regional or multilateral development banks. The diffusion of international environmental law and international human rights discourses has nonetheless provided civil society with tools for pressuring private actors, supplementing private law with stronger public law controls.

At the policy level, NGOs have attempted to persuade or pressure private actors to adopt 'best practice' standards, such as the voluntary Equator Principles for financial institutions (2003) (in turn based on International Finance Corporation (IFC)

Performance Standards on social and environmental sustainability) and the recommendations of the World Commission on Dams (Lee and Scurrah, 2009, p. 48). Some banks in the region have signed up to the Equator Principles (Interview 20, 2011). More recently, the UN's Business and Human Rights guidelines have also been utilized, including by the Mekong Legal Network, although some in the region do not believe that they adequately emphasize environmental matters (Interview 22, 2011). Bodies such as the UN Office of the High Commissioner for Human Rights have also promoted rights-based approaches to resource development, including state responsibility for human rights abuses by third parties (UNOHCHR, 2012, p. 11). UNDP has promoted the Global Compact Network in Vietnam.

In their dealings with private actors, the multilateral development banks (such as the World Bank and the Asian Development Bank) and Western donors have been increasingly sensitive to the environmental and social impacts of hydropower dams, including by adopting safeguards in the sensitive areas of information, consultation, resettlement and compensation. This was often in response to past development failures, including the lessons learned from the World Commission on Dams report in 2000. The Asian Development Bank was stung by a sustained NGO campaign against the impacts of the Theun-Hinboun dam in Laos (Soutar, 2007), completed in 1998. Partly as a result, much more stringent safeguards were included in Laos's Nam Theun 2 project, completed in 2005. Paradoxically, however, the more extensive safeguards and longer approval processes in that project encouraged Laos and regional investors to turn away from the development banks towards private financiers and regional governments, for which standards may be less strict (Middleton, 2007, pp. 12–13). There is also criticism that safeguard policies still advance a neo-liberal development agenda (Krever, 2011), which may be at odds with other visions and practices of organizing economic, social, environmental and cultural life.

In this respect, the concept of CSR is not well developed in Cambodia, Vietnam or Laos, although it is more advanced in Thailand and China (whose companies are involved in Mekong hydropower). Some state electricity entities, such as the Electricity Generating Authority of Thailand (EGAT), and most Thai commercial banks, have CSR policies (Thabchumpon and Middleton, 2012). Vietnamese state enterprises also have policies (Interview 31, 2011), as do Chinese state entities. The IFC has a regional programme to encourage banks investing in hydropower to sign up to the Equator Principles (Interview 51, 2012). The IFC seeks to help businesses to understand the financial and reputational risks of adverse social and environmental impacts (Interview 51, 2012).

Most Thai banks, however, have not signed up to the Equator Principles, and the Export-Import Bank of Thailand 'does not have an environmental policy' (Middleton, 2007, p. 12), although some Thai banks have basic CSR policies. While EGAT has a CSR policy, it takes the view that when it is merely the power purchaser, it is the project developer, owner and foreign government, not EGAT, which must follow foreign social, environmental laws (Thabchumpon and Middleton, 2012). Its subsidiaries take the same view: 'we rely on the Laos government' (Interview 28, 2011). Thai regulators also take the view that extraterritorial activities by Thai entities are outside their mandate where they do not have authority to issue licences in other countries.

Even where there is policy recognition of CSR by the state (as in Vietnam and China), it often does not influence corporate behaviour. The Export-Import Bank of China is not easily penetrated by civil society. The IFC, however, works with a range of Chinese banks on investment and advisory services (Interview 51, 2012). Consumer advocacy or shareholder boycotts are unlikely in a hydropower context (Interview 20, 2011), since companies are not necessarily sensitive to public criticism. One INGO expressed 'not much hope' for CSR in the region, particularly if premised on self-regulation (Interview 20, 2011). There is little public pressure for Thai banks to follow CSR, and banks are 'not entirely convinced to take it seriously' (Interview 51, 2012).

Few affected communities or NGOs reported having had meaningful contact with hydropower companies or private investors. NGO efforts to contact Chinese companies in Cambodia, for instance, did not work (Interview 6, 2011) and villagers have often been rebuffed. One senior Thai commercial banker involved in hydropower investment observed that 'we do not communicate directly with the public' and that 'it is not for us to try to govern others we have no authority over' (Interview 29, 2011). Companies themselves have generally not established grievance procedures in the hydropower area. They have relied on local authorities to negotiate compensation and resettlement, often under intimidating conditions (Subedi, 2012, p. 65), rather than dealing directly with communities. NGOs have, however, reported sporadic contact with some companies. One Vietnamese NGO, for instance, suggested that it is sometimes possible to engage with companies in Cambodia, but in Vietnam 'hydro companies do not talk to you' (Interview 35, 2011).

Local NGOs in particular may lack the capacity and resources to lobby companies (Interview 10, 2011). Because of the complexity of many large hydropower projects, there can be a diffusion of responsibility between different actors (state, company, investor), making it difficult to identify who is legally responsible for what and thus complicating advocacy efforts. The legal arrangements underpinning hydropower projects can also be impervious to civil society, such as concession agreements that are commercially confidential (Interview 46, 2012).

NGOs nonetheless persist in efforts to use law to pressure companies – usually by persuading, cajoling or criticizing them from the outside, not the inside. Thus, on Lower Sesan 2, BankTrack, a global network of civil society groups tracking the operations and investments of private banks, has continually called for national laws to be upheld and the World Commission on Dams guidelines to be followed (BankTrack, 2015a). Civil society has also sought to persuade governments to better regulate the activities of their corporations or state entities abroad, including Thai, Vietnamese and Chinese entities operating in the region.

Civil society has also made extensive legal representations through letter campaigns to companies. One prominent example is a letter from a civil society coalition, led by 3SPN, to the chair of Hydropower Lower Sesan 2 Co Ltd on 26 May 2014, and copied to the Chinese government, Electricity of Vietnam and

Royal Group Cambodia (3SPN and other NGOs, 2014). The letter makes extensive legal submissions and is signed by a coalition of INGOs and NGOs from Cambodia, Vietnam and Thailand. It requests the company to 'comply with legal and corporate social responsibilities with respect to the Lower Sesan 2 hydropower dam project … under national and international laws'. It then objects to the limited EIA and social impact assessments (SIAs) and the 'vastly inadequate' information disclosure and transparency. The letter assesses the EIA against specific Chinese environmental law and Cambodian law standards, as well as customary international law (including the International Court of Justice's *Pulp Mills case*, 2006) and international best practice.

Growing civil society demands for CSR in hydropower have stimulated an industry-wide response. In 2011, the Hydropower Sustainability Assessment Protocol (HSAP) was adopted by the International Hydropower Association (IHA) as a tool to assess project development and operation against global criteria. The tool is not intended to replace existing legal regimes, including EIA or SIA, but to assess the quality and extent of EIAs as well as project implementation and operation (HSAP, n.d.). The HSAP is not binding, in part because governments such as Laos feared that it would require them to do more, but there are 'soft law expectations' of it (Interview 54, 2010). The MRC plans to pilot it in the 3S region, and Australia's aid program supported its trial (AusAID, 2010, p. 10).

The HSAP relates to civil society in a number of ways. First, it was developed by the IHA, the peak hydropower industry body, in collaboration not only with states and financial sector actors, but also with environmental and social NGOs (including World Wide Fund for Nature, the Nature Conservancy, Transparency International and Oxfam). Second, the 'Chambers' of its governing council include 'environmental organisations' (IHA, 2010, p. 6). Third, the level and quality of stakeholder 'engagement' and 'support' (including civil society) are part of project assessment (IHA, 2010, p. 15). Finally, where a high level of transparency on a project is justified, an assessment could be observed by, or carried out in partnership with, an external party (such as an NGO or civil society) (IHA, 2010, p. 20).

Finally, civil society has used the OECD's Guidelines for Multinational Enterprises in an attempt to hold private actors accountable for hydropower project impacts. The OECD Guidelines, first adopted in 1976 and subsequently revised, are voluntary standards for responsible business conduct in areas including the environment and human rights. Under the guidelines, a 'specific instance' complaint can be made to the national contact points (NCPs) of the 46 adhering states parties. Specific instances are not legal cases and the NCPs are not judicial bodies that issue binding decisions. Rather, NCPs attempt to resolve alleged non-compliance by offering their 'good offices' or, with the agreement of the parties, facilitating access to consensual, non-adversarial procedures such as conciliation or mediation. NCPs are typically government authorities of the company's nationality. Complaints can be made by any party, including foreign NGOs or civil society, and complaints are normally resolved within one year.

While none of the Mekong countries has adhered to the OECD Guidelines, four specific instance complaints have been brought against foreign companies involved in Mekong hydropower, including a project operator (*Tractebel-Suez case*, 2004), a principal shareholder (*Électricité de France case*, 2004), a technical consultant (*Pyory case*, 2012) and an equipment supplier (*Andritz case*, 2014). All four cases concerned dams in Laos (Houay Ho, Nam Theun 2 and two cases concerning Xayaburi). Two of the complaints were brought by European NGOs against companies of the same European nationality. One of the Xayaburi cases was brought by 15 NGOs from the Mekong, Finland and Switzerland against Finnish companies. The other Xayaburi case was brought against a Swiss company by a coalition of NGOs from the Mekong region (CRC Thailand, Fisheries Action Coalition Team Cambodia, the Samreth Law Group Cambodia, LPSD Vietnam and CSRD Vietnam), Australia (ECA Watch Australia) and internationally (International Rivers and ERI, with the latter as the legal consultant).

In the three decided cases, the NCPs decided that there was no non-compliance by the companies. At the time of writing, one case is pending. The complaints have nevertheless served as lobbying tools for civil society to articulate legal concerns about corporate activities, to pressure companies to improve behaviour, and to encourage closer scrutiny by regulators. Most of the complaints made legal arguments extensively citing human rights and environmental standards.

While none of the complaints has so far been upheld, in two cases the NCPs recommended improvements to corporate practices. Thus, in 2005, the French NCP requested a French company to continue to cooperate with Laos to implement compensation in relation to Nam Theun 2, and to continue to evaluate and mitigate adverse impacts. It also encouraged multinational enterprises in countries with weak legal and regulatory controls to apply the same international good practices that they would apply at home. In 2012, the Finnish NCP observed that a Finnish company should have more clearly addressed ambiguities on the environment and human rights concerning Xayaburi dam. It also recommended that, in future, companies should assess the risks of major projects more carefully, act more transparently, give more consideration to stakeholders' views, and better communicate about risk prevention.

Legal education, 'empowerment' and applied legal research

As noted earlier, many people affected by hydropower development in the Mekong countries lack knowledge of relevant laws, procedures, rights and remedies. A core strategy of many local NGOs and INGOs, particularly in Cambodia and Thailand, is to provide legal education and training to affected groups. Such activities are often influenced by instrumental, ideological policy ideas and practices concerning 'legal empowerment' as promoted by foreign donors and NGOs. The Asian Development Bank (ADB, 2000, p. 7), for example, defines legal empowerment as 'the use of law to increase disadvantaged populations' control over their lives through a combination of education and action', including by promoting 'critical consciousness' by the poor about unequal power relationships and the ability to challenge and transform them.

As one Cambodian legal NGO noted, 'after people receive training about law, they stand up for their rights' (Interview 6, 2011). Another regional NGO argued that knowing the law allows people to challenge officials, buys them time, gives them status, empowers them, enables them to ally with sympathetic authorities and helps them to talk to the press (Interview 9, 2011). As the UN Office of the High Commissioner for Human Rights in Cambodia observed, legal education 'arm[s] ordinary people with law to use in arguments' (Interview 4, 2011), whether in interactions with authorities or in public campaigns. Training of officials can also sensitize them to people's rights.

At the regional level, ERI established the Mekong Legal Advocacy Institute (MLAI) in 2009. The MLAI brings together (in Chang Mai, Thailand) Mekong lawyers to share and learn about legal and advocacy strategies on environmental and social issues relating to development decision-making. The focus is on civil and criminal law, but not specifically environmental or public interest law (Interview 22, 2011). ERI also holds a regular 'EarthRights Law School', which now has about 60 alumni (Interview 22, 2011).

The MLAI's offshoot, the Mekong Legal Network (MLN), is a network of about 25 experienced legal professionals and civil society leaders that works to promote the rule of law in the Mekong and ASEAN regions, focusing on regional and transboundary development (ERI website). ERI aims to provide MLN with strategic information so that MLN members can decide their own strategy and action (Interview 22, 2011). ERI's philosophy is that it does not itself engage in negotiations with governments, but seeks to empower MLN members to do so; it works 'behind the scenes and not in the spotlight' (Interview 22, 2011). ERI also writes legal advices for MLN, then defers to it in determining the next steps. ERI's approach seeks to avoid the criticism from Mekong governments that, as a Western (US) INGO, it is interfering in sovereign development decisions in Southeast Asia. ERI tries 'to create legal arguments to be used in the political sphere' (Interview 22, 2011). A number of other INGOs have been involved in promoting legal information, training and education in relation to hydropower, such as Oxfam International and International Rivers.

Local NGOs have been particularly active in promoting legal awareness about development. In Cambodia, 3SPN has sought to strengthen the capacity of its grassroots community network by conducting EIA trainings for hundreds of people, organizing discussions on the draft EIA report on Lower Sesan 2 (3SPN, 2010, p. 6), and liaising with the expert NGO the Cambodian Human Rights and Development Association (ADHOC) to conduct human rights training (Interview 10, 2011). It has, however, been wary of training local people on international law and thereby raising expectations, because we 'could not use it − [we] do not know how for law enforcement, to force [change]' (Interview 10, 2011).

The Cambodian NGO Culture and Environment Preservation Association (CEPA) also conducts regular community radio programmes, which have included discussions of policies and guidelines, the Mekong Agreement, and national resource laws on fisheries, land, forestry, the obligations of owners, and the rights of communities (Interview 17, 2011). It has also addressed specific legal issues in meetings with villagers, such as clarifying for them the legal standard of compensation in national law − namely, what is the 'reasonable' or 'fair' market value of land (Interview 17, 2011). Radio is CEPA's preferred medium because villagers in Cambodia are sometimes afraid of being identified on television (Interview 17, 2011). Cambodia's Highlanders' Association, a local NGO in the 3S rivers region, has also used radio to provide information on indigenous rights (Interview 11, 2011). Posters and leaflets are also an NGO mode of disseminating legal information (Interview 6, 2011).

In Vietnam, the Vietnam Rivers Network is involved in enhancing the capacity of its members to research, monitor and participate in river protection and management. In Thailand, many NGOs are involved in providing training and education about river resource and environment issues, including legal frameworks. Many Cambodian NGOs – including the Cambodian Center for Human Rights, the Cambodian League for the Promotion and Defense of Human Rights, ADHOC and Equitable Cambodia – have provided general education about laws on land, natural resources and human rights, without focusing on hydropower. This can include skills training, such as the Community Legal Education Center's (CLEC) education on the judicial process, how to write a complaint, and the rights of the parties (Interview 6, 2011).

Some organizations have also provided training to government officials, such as CLEC's training for commune council officials on indigenous rights and customary law (Lui and Calzaroni, 2011, p. 26). Some local officials freely acknowledge that they have no understanding of international or regional norms (Interview 14, 2011; Interview 15, 2011).

Private actors have also been involved in raising awareness about the domestic, regional and international legal standards applicable to hydropower development. In Cambodia, for example, the former Samreth Law Group provided information to the Cambodian Ministry of Interior about EIA, as well as training to public authorities, the Bar Association of Cambodia and civil society (Interview 5, 2011).

A related strategy by which INGOs and local and national NGOS have attempted to empower communities is through conducting and disseminating research about law and hydropower development, including by commissioning experts. Various studies have examined national, regional and international laws, particularly environmental laws, but also international human rights laws (Lerner, 2003; NGO Forum on Cambodia, 2005) and CSR standards (BankTrack *et al.*, 2009; International Rivers, 2010, 2008b, pp. 14–15).

Research has also been strategically undertaken to respond to, or shadow, official studies produced by states, developers, EIA consultants or the MRC (see, for example, Baird, 2009; River Coalition in Cambodia, 2008). Often such research alleges particular legal violations and the findings become a platform for advocacy or campaigns. In one example, in response to official failures to conduct downstream impact assessments of Vietnam's Yali Falls dam, the Se San Working Group of Cambodian NGOs and INGOs commissioned expert rapid assessments of affected communities, triggering MRC investigations and meetings between Cambodia and Vietnam (Hirsch and Wyatt, 2004, pp. 56, 59). 'Grassroots' research has also been undertaken, as when the NGOs Mekong Watch and 3SPN assisted two Cambodian villages to prepare a map of and photo-document their local natural resources affected by the Lower Sesan 2 hydropower dam (Mekong Watch, 2014).

Civil society research is, however, most commonly directed towards securing better implementation and enforcement of existing laws in specific cases, or correcting and remedying bad decisions, rather than focusing on a wider agenda to transform laws governing the Mekong. This focus partly reflects a strategic pragmatism in the face of constrained political environments. It may also reflect comfort with existing legal frameworks, or a belief that the status quo is preferable to unpredictable reforms. A common sentiment among civil society is that the key problem is that national laws are 'good on paper' but not in practice (Interview 6, 2011; Interview 7, 2011; Interview 12, 2011; Subedi, 2012, p. 2).

Using the courts

Even where legal challenges are formally available, civil society has seldom challenged hydropower development in domestic courts. No cases have been brought in Cambodia, Laos or Vietnam, and only a few cases have been brought in Thailand. The dim prospects for adjudication contrast with the occasional (albeit mixed) successes against dams elsewhere, as against the Narmada dam in India (Rajagopal, 2005) or litigation in South America.

Adjudication is rare for a range of reasons. Litigation is not a common way to settle disputes generally, and especially not for politically 'hot' cases concerning resource development in places such as Cambodia (Interview 6, 2011). Affected communities may lack knowledge of legal rights, procedures and institutions. Lawyers are scarce, inaccessible and expensive. Public interest lawyering is uncommon and legal aid is generally unavailable.

Some communities, NGOs and lawyers fear retribution – including arrest or counterclaims – from the state or developers if they commence court action. Even in Thailand, hydropower developers have used the courts to discourage resistance. The EGAT filed charges of trespass and inciting chaos against protesters with regard to the Pak Mun dam, completed in 1994. While the Supreme Court of Justice dismissed the charges in April 2013 (*Bangkok Post*, 2014), the case took 20 years to be finalized, prolonging uncertainty and chilling dissent.

NGOs tend to avoid confronting the state through litigation, although some NGOs (such as CEPA and CLEC in Cambodia) encourage people to utilize the courts. The high risk of losing is an important strategic consideration. An adverse judgment can legitimate a hydropower project and strengthen the state's development agenda. INGOs too, such as ERI and International Rivers, are cautious about litigation (Interview 22, 2011; Interview 20, 2011), despite using it in other regions.

A lack of confidence in courts is a significant reason why civil society is reluctant to use them. In all four countries considered in this book, there are concerns about the independence and impartiality of the judiciary, including because of corruption and political interference. The courts are also often costly and difficult to access by affected communities, and proceedings are protracted. The quality of judicial decisions can be poor, producing unpredictable results or wrong decisions.

Above all, as one Thai NGO suggested, 'people do not believe in the rule of law to do justice' (Interview 23, 2011), often because the courts are seen to advance state or developer interests. Another Thai NGO believed that 'bad or unfair laws are implemented, but there is a lack of implementation of protective law' (Interview 9, 2011). In Cambodia, complaints filed by the government are vigorously prosecuted, whereas cases brought against the government are not (Peou, 2011, p. 132). Law depends on which group you belong to: 'rights have to be taken, they do not belong to anyone' (Interview 23, 2011).

The picture is, however, different in Thailand, where there is more potential for litigation (Interview 23, 2011; Interview 20, 2011). The Thai Administrative Court has suspended 76 state development projects, from logging to dams, including EGAT projects (Interview 23, 2011), and remitted them for reconsideration. The positivism of the Thai judiciary can be used advantageously by civil society: it is 'damaging if you can find even small noncompliance', and the 'courts use that if they agree with the result, to justify it' (Interview 23, 2011).

As discussed in Chapters 4 and 5, in August 2012, 37 Thai villagers filed a case in the Administrative Court against five Thai state entities (EGAT, Electricity Generating Public Company Ltd, CH. Karnchang Plc, PTT Public Company Ltd and the Thai Cabinet) to prevent EGAT from purchasing electricity from the Xayaburi dam in Laos (Kate, 2012). The villagers were assisted by NGOs including Living Rivers Siam and the Network of Thai People in Eight Mekong Provinces. The villagers stressed that they had exhausted all other avenues, including appeals to all of the parties (Radio Free Asia, 2012). The Administrative Court dismissed the case, but the villagers appealed.

In April 2014, the Thai Supreme Administrative Court found that while the PNPCA arose under a regional treaty (the Mekong Agreement) and was thus outside its jurisdiction, there had been incomplete information disclosure and public participation under the Thai Constitution and domestic regulations (*Niwat v Electricity Generating Authority of Thailand*, 2014). The court remitted the matter to trial. The court also found that the plaintiffs had no standing to contest the government's approval of the Xayaburi Power Purchase Agreement (PPA) or to seek to invalidate that contract.

Bringing a matter to court may mean that 'you have to stop other methods, for example public campaigns' (Interview 23, 2011), so strategic choices must be made. However, Thai court proceedings are still often used as part of wider campaigns. As one Thai NGO noted, 'protest campaigns can influence court decisions' and 'can stop some dam projects' (Interview 23, 2011). The filing of the above-mentioned Xayaburi case was accompanied by 100 villagers and fisherfolk from eight Mekong provinces 'carrying banners, mock fish and fishing gear', because the villagers 'wanted people to understand why they are asking for justice, and how they depend on the river for their food and livelihoods' (Deetes, 2012). The courts sometimes accept this performative dimension of legal action. Thus, in this case, court staff 'welcomed the villagers, allowing them to conduct a spiritual ceremony for the mighty Mekong River and demonstrate their traditional Mekong fishing methods to the journalists who were present' (Deetes, 2012).

In contrast to the possibilities for environmental litigation, human rights litigation in domestic courts is highly sensitive and scarcely conceivable in Vietnam or Laos; it is difficult, unlikely to succeed and risky in Cambodia. This is in spite of express constitutional protections for human rights in all of the Mekong countries. There are, however, possibilities for human rights claims in Thailand, where the National Human Rights Commission (NHRC) can consider complaints.

Thailand's NHRC has been successfully used to challenge the transnational impacts of Thai investment in hydropower projects in Thailand's Mekong neighbours. In May 2011, a complaint was filed against Thai involvement in Laos's Xayaburi dam project, pursuant to a PPA. In May 2012, the Thai NHRC found violations of Thai human rights law by the EGAT and the Thai government (Thai NHRC, 2012). First, the NHRC found that the EIA and SIA done in 2010 did not involve a transboundary EIA study and did not consider downstream impacts in Thailand and Vietnam on fisheries, agriculture, water consumption and communities.

Second, the NHRC found that a clause in the PPA that immunized it from legal challenges may contradict the NHRC's mandate and that of the Senate Committee on Anti-Corruption and Good Governance. Third, the NRHC found that EGAT and the Thai government did not publicly disclose the PPA or other relevant information before concluding the agreement, contrary to human rights protections in the Thai Constitution and resolutions of the Cabinet and the National Energy Policy Council (including relating to the Mekong Agreement and good governance). The NHRC proposed that the Prime Minister review the implementation of the dam's construction, comply with an MRC resolution of December 2011 to conduct a transboundary EIA, and suspend any actions under the PPA until further investigation is completed.

Law and 'non-legal' modes of resistance

In every society, law is not the only or necessarily the main frame of resistance. Direct political action, civil disobedience, social justice advocacy, mass mobilization, media campaigns and grassroots social movements are some other means of contesting statist development discourses. Whether a particular approach is effective or not can often drive the strategic choices of those seeking redress, including whether or not rights-based approaches are validated (Merry, 2006, p. 215). Particularly in Cambodia and Thailand, direct protest action has been a common method of contesting hydropower dams, while it has been rare in Vietnam and Laos.

At the same time, however, even seemingly non-legal resistance can be facilitated by, and creative of, law. A prominent example comes from Thailand. There, a decentralized, mass-based movement, the Assembly of the Poor (AOP) (*samatcha khon chon*), emerged in 1995 out of protests against the Pak Mun dam in northern Thailand. Mass mobilization and protest have been the defining tactic of the AOP (Missingham, 2002, p. 1651). The AOP has mounted mass road marches, major demonstrations around Government House in Bangkok, and a protest village and blockade of Pak Mun dam, often mobilizing tens of thousands of people. The aim of mass actions has been to force the government to negotiate on the AOP's demands, such as opposition to dams, reopening of dam gates, restoration of river ecology and fisheries, restoration of livelihoods, and compensation (Chalermsripinyorat, 2004, p. 547). The AOP had successes in negotiations with governments in 1997 and 2001–2002, later undermined by changes in government.

It may seem that the AOP's strategy of mass mobilization and direct action does not involve much law. As one AOP leader noted, 'if we fought using the law, we'd lose to them' (quoted in Missingham, 2002, p. 1653). The AOP's approach split it from villagers who, for instance, filed appeals with government agencies (Chalermsripinyorat, 2004, p. 548). The AOP has been described as drawing on 'non-legal' discourses of justice, instead emphasizing 'a traditional discourse of responsibility to the environment that predated the Constitution and was more meaningful than a legal discourse borrowed from international environmental rights advocates' (Munger, 2007, pp. 837–838).

The AOP's strategy is not, however, untouched by law, for a number of reasons. First, conventional legal claims and norms still permeate the AOP's demands in various ways. For instance, the Mun River Declaration of December 1995, which crystallized the AOP's agenda, refers to a variety of legal concepts in resource management, such as 'rights', 'equity', 'fairness', 'participation' and 'self-determination' (quoted in Missingham, 2003, p. 320). The references to law are not surprising, given that the AOP was advised by urban NGOs, academics and experts (Missingham, 2003, p. 326). The AOP 'deliberately draws on discourses of environmentalism and human rights' (Missingham, 2003, p. 330). Legal concepts are, accordingly, at least partly constitutive of the AOP's cause and identity, and a normative resource in forging and maintaining group solidarity.

Second, the AOP emerged when Thailand was democratizing and human rights were gaining prominence in political discourse. Accordingly, the AOP utilized the language of civil rights in defending its tactic of mass protests, as when an AOP member noted that '[i]t is our right' (*sitthi*) to access and use the public streets and footpaths surrounding Government House (quoted in Missingham, 2002, p. 1653). Law thus shapes and legitimates 'the political opportunity structures that movements have at specific moments' (Rajagopal, 2005, p. 384).

Third, the AOP's tactic of presenting grievance petitions (*koroni panha*) to the government, giving it a deadline within which to respond, entering into formal negotiations with it and reaching agreements with it are all law-like processes. In 1996, for instance, the AOP 'negotiated with government in the style of a peace treaty' (Baker, 2000, p. 22), presenting 47 grievances, negotiating with junior cabinet ministers, and agreeing an 18-page treaty-style document.

In the 1997 protests, the process became even more law-like. The AOP presented 125 grievances and took part in regular meetings with ministers and officials, across a formal adversarial negotiating table, with vast documentation (Baker, 2000, p. 22). Both sides took minutes; NGOs and academics translated 'between local dialect and central Thai' and 'the style of officialdom and the style of the village'; and the agreements 'were solemnly signed in duplicate copies', ratified by the cabinet and publicly announced (Baker, 2000, p. 22). Even a ceremonial photograph was taken, and before leaving Bangkok the AOP conducted a *bai sis u khwan* ceremony 'to settle the spirits and restore harmony' (Baker, 2000, pp. 22–23).

Accordingly, as Baker observes, 'the event established a precedent for a new form of politics – the peace treaty negotiation – involving groups which had been systematically denied political access in the past', and in so doing it broke 'some constraints of law and procedure which were biased against the poor' (Baker, 2000, p. 23). The AOP's mass protests thus illustrated consciousness of two senses of law: legal process (that is, adopting a legal style of negotiation and agreement) and law formation (the creation of new substantive law by circumventing the normal modes of law-making that failed to accommodate their interests). The new norms adopted as a result aligned more closely with international norms on environmental and human rights protection than the Thai state had thus far provided.

Fourth, the AOP's mass protests also revealed internal organizational law-making. The 'Village of the Poor' that surrounded Government House for three months in 1997 was so named to suggest social order, not mob rule, and the AOP deliberately created and maintained a community (Missingham, 2002, p. 1657). Basic rules were issued to ensure order and the good reputation of the protesters, including bans on alcohol and drugs. The village was arranged into self-governing units to organize activities, logistics and food, with a 'superstructure to take care of watch-and-ward, entertainment, dispute settlement, fund raising, and education' (Baker, 2000, p. 21). There was a welfare shop, and volunteers for security, health, well-being, cleanliness and sanitation (Missingham, 2002, p. 1658). Collective leadership was exercised through meetings of elected village representatives. During rallies there were 1,000 volunteer 'guards' for 'maintaining peace and order, ensuring that the rules (*kotrabiap*) were followed, and watching for outside agitators' (Missingham, 2002, p. 1658).

The AOP's strategy of mass mobilization accordingly should not be understood as excluding law. To be sure, the AOP chose not to pursue legal claims in the courts, and it constructed itself as a mass organization exerting principally political pressure. Yet, its agenda was framed by legal principles; it relied on legal protections of freedom of association, assembly and expression to operate; it negotiated through a law-like process that produced agreements containing new substantive norms; and it generated its own internal organizational law. Its politics thus mobilized through law and made law.

Mass mobilization of this kind is thus far unique, among the Mekong countries, to Thailand. Whereas the AOP can get 30,000 people to protest outside parliament and the government takes notice, according to one INGO that is not possible in Cambodia, Laos or Vietnam (Interview 22, 2011). Mass mobilization also has very definite limits. It is vulnerable to state and police repression. It is contingent on the goodwill of governments to agree to negotiate in the first place, and further to agree to real and binding concessions. Agreements reached can be reversed or degraded by subsequent governments. Such movements are susceptible to fragmentation and factionalization, as the state and developers seek to divide communities. The process of political bargaining can also undermine human and environmental rights that affected communities may be strictly entitled to as a matter of law. As in the case of Pak Mun, it can also take a very long time and may not produce long-term systemic change to development politics, policies and practices (World Commission on Dams, 2000b, p. vii).

Conclusion

In the Mekong River Basin, civil society is reluctant to use the courts, although it has occasionally done so in Thailand and through quasi-judicial bodies such as the National Human Rights Commission. Yet, civil society frequently utilizes law in a myriad of other ways: when engaging administrative and political authorities; in confronting private actors; and through legal education, empowerment and research. Even where legal strategies are not at the front and centre of resistance, as with the mass protest campaigns by the AOP, law still structures and enables resistance in key ways.

It remains to ask whether all of these uses of law have been productive or effective in achieving the goals of civil society actors that contest hydropower. There has been mixed success in using law in the Mekong. Outright victories are rare and dams are virtually never cancelled. Participatory processes often serve to moderate projects, but also legitimate them. Any positive change is usually incremental and sometimes reversed. To go back to the beginning, law is frequently used to further the prevailing political economy of development, even if the existence of law also constrains and shapes the 'free hand' of politics. Even INGOs that routinely utilize law in their advocacy query whether law can change the prevailing political economy of development (Interview 22, 2011).

Nonetheless, the appeal to law has had effects. Legal arguments exert normative pressure on national authorities to justify and reason their decisions, and sometimes lead to concessions or mitigate adverse impacts. Concepts of public consultation, mitigation, compensation and resettlement protections have become embedded in national practice across the region, and draw on international and regional norms. CSR is gaining traction, albeit slowly. Law has assisted and empowered grassroots movements in campaigns for justice.

The social processes of legalization are not linear or predictable. Civil society approaches to law in the Mekong oscillate between faith and despair, and engagement and disengagement. What is striking, however, is that law appears to maintain a potent appeal for civil society even in the face of episodes of its routine non-observance, abject failure or the endless deferral of justice – all in the face of fairly authoritarian political cultures. Law is curiously a repository of hope. There is a persistent belief in its unrealized potential to oppose and constrain power, to empower and emancipate, and to remedy injustice. It is even common to blame not the law itself but those who administer it, as if the rational instrument of law can be separated from its flawed human agents. The sentiment is encapsulated by those who, after exhausting other means, petition their grievances directly to the prime minister or the king: the feeling is that if only the letter gets into the right hands, *if I am heard*, justice will be done.

Law provides a mutually comprehensible language of claim, which can lend authority and credence to civil society demands, and clothe them in the norms and processes validated by the state's own legal system. Its structural conservatism can mask and facilitate radical demands, which might not be possible if expressed in more overtly political terms. It provides procedural and institutional openings for valid negotiation and dialogue with states and developers. Law enables state control, but its pluralism and internal contradictions also enable the state's control to be challenged and subverted. The dynamism of transnational legal influences and processes enables civil society to further prise open the state. Law is no elixir for the damage inflicted by rapacious development and the political economy that drives and sustains it. Law may be administering palliative care to a dying river and the ways of life that it 'traditionally' sustained prior to the arrival of 'modern' hydro-power development. But even this is not nothing.

8 Conclusion

This book opened with three vignettes. The first offered impressions of our trip to the visitor centre at the Nam Theun 2 hydropower project in Laos and meeting the project company's public relations manager. The second described our visit to one of the resettlement villages to which those formerly residing along the Theun River had been relocated in light of that project's development. Finally, we related an encounter with the three members of the Nam Theun 2 Panel of Experts charged with independent oversight of the project and public reporting on its environmental and social impacts.

This trio of meetings encapsulates the intersection of collective lives around a Mekong River Basin dam: a purpose-built corporation and its employees; a village and its inhabitants; an 'epistemic community' of international dam experts and its itinerant members (Haas, 1989); and a cross-disciplinary team of researchers studying hydropower development. The intersection also depicts the dam in question as a crucible of legal relations. The visitor centre represents a physical instantiation of the imperative of 'transparency' in cross-border investment and development; this imperative is sustained through the influence of a range of norms, some legally binding and some non-binding, as discussed in Chapter 6. The centre's operation is also a defensive installation in ongoing legal and political conflict surrounding the decision-making processes leading to hydropower development and its social and environmental impacts, as discussed in Chapter 7. These disputes have revolved especially around the plight of resettlement villages such as the one we visited, where the long-term human cost of displacement becomes apparent. In relation to both of these sites, the Panel of Experts embodies the surveillant presence of the 'international community', 'foreign experts' or 'civil society', as these are often perceived. As Chapter 5 made clear, those last are understood to temper some negative dimensions of hydropower development (in particular, by their recourse to international legal norms and comparative 'best practice' regarding environmental assessment). In so doing, however, they also help to sustain, renew and validate the justificatory or normative architecture on which hydropower development in the Mekong has long depended, successive iterations of which were described in Chapters 3 and 4.

This book has re-described that normative architecture surrounding Nam Theun 2 as well as other dams – Xayaburi and Don Sahong, also in Laos; A Luoi in Vietnam; Pak Mun in Thailand; and the dams on the Sesan, Srepok and Sekong

Rivers in Cambodia – through a socio-legal lens. It has tracked the shifting hybridizations of public and private law, 'hard' and 'soft' norms, and subnational, national, regional and international principles that comprise this architecture, highlighting the range of governmental, inter-governmental and non-governmental agents that work on, in, around and sometimes against it. In so doing, the book has shown how varied are law, legal ideas and legal practices in the Mekong River Basin and how vital they remain amid its contested development for hydropower. Law, we have seen, opens up both predictable and unexpected routes for contestation as much as it seeks to regulate and harmonize these, or sometimes close them down.

Nonetheless, as noted in Chapter 2, this book has aspired to do more than add greater legal colour to the prevailing picture of political life in the Mekong River Basin. It has sought, also, to do more than enrich 'law on the books' by recounting regionally specific stories of 'law in action' (Pound, 1910). From the outset, the book has had polemical goals. It has aimed to challenge some commonly held assumptions about law's actual and potential role in the Mekong River Basin. It has taken issue with the idea that political and economic life in the Mekong River Basin awaits greater juridification – whether to displace unrestrained sovereign politics, to countervail the unequal accumulation of capital, to 'civilize' development and better protect the environment or human rights, or otherwise to 'normalize' activities in the basin against international benchmarks. In this regard, we have also queried the reliability of existing development templates, especially with regard to any ideal relationship between law and development. Instead, we have maintained that the models of 'transparent', 'participatory' and 'science-based' legality often promoted in connection with hydropower development – including by those who advocate against such development – warrant renewed critical attention. These, we have argued, should be recognized for their political negotiability and navigated tactically on that basis.

In the interests of fostering such tactical navigation and sharpening this book's intervention in prevailing debates, let us recapitulate some of the 'common-sense' accounts of the workings of law in the Mekong River Basin surrounding hydropower development that this book calls into question.

The contributions and displacements of this socio–legal account: the common sense we question

Among the assumptions of many scholars working in and/or on the Mekong River Basin that are cast into some doubt by this book are the five claims set out below. These are assumptions shared, also, among many activists, donors, policymakers and experts concerned with water governance and hydropower development.

1. Sovereign prerogative prevails in the Mekong River Basin; law remains mostly exogenous to social and political life and distributive decision-making.

Accounts of a remarkable range of those who write about hydropower development in the Mekong River Basin coalesce around one theme: sovereign prerogative. This power – the apparent freedom of choice of national government leaders of riparian states – is cast as a perennial and intractable determinant of social and political life and distributive decision-making. National law is typically thought to facilitate the sovereign's will instrumentally, while international law tends to be cast in opposition to that. Relatively little account is taken of how both national and international law also move in other directions. For instance, this book has shown how national law can be utilized sometimes in unexpected ways to constrain sovereign power – even when the sovereign in question wrote the laws and anticipated that they would operate in ways favourable to it or its interests. Equally, little attention is paid to the role of international and regional law in affirming and informing the would-be determinacy of national sovereign authority and shaping perceptions of national interest. Instead, such law is commonly cast as an exogenous counterweight to 'the push and pull of national sovereignty' (Wouters, 2014, p. 68). To many, this consigns international institutions and basin-wide regulatory regimes more or less to 'irrelevance' (Hensengerth, 2008, p. 105). To the extent that they envisage any role for law, such scholars reserve for it a 'soft', procedural, mediating function: one of 'balancing' seemingly obdurate political formations and guiding decision-makers around associated 'bottlenecks' – for example, concerning the sharing of data and information (Li, 2012; Plengsaeng *et al.*, 2014).

Against this view, this book has cast a complex network of legal norms, practices and institutions as integral to, and partly constitutive of, contemporary understandings of sovereign prerogative in the Mekong River Basin. International and regional norms have not simply countered or been irrelevant to the exercise of national sovereignty. Instead, lawful cooperation fostered by the Mekong River Commission (MRC) and through multi-stakeholder financing, for example, has helped to encourage sovereignty's reimagining in a register of scientific and technological advancement and regulatory 'innovation', as Chapters 4 and 6 showed. It is this renewed understanding of sovereignty – authorized or motivated by the necessity of economic development – that underpins public statements such as the following, made by Laos's deputy energy and mines minister, Viraphonh Viravong, in March 2014:

> Laos, Viravong added, has over the past 50 years 'built an enviable record of achievement in developing environmentally and socially sustainable hydropower projects in accordance with our laws, decrees and globally accepted standards' … Viravong said Laos would continue with its hydropower vision out of 'duty' to overcome poverty and said it had the World Bank on its side. 'The World Bank has recently underscored the wisdom of using large-scale hydropower projects to create renewable energy, spur economic growth and social progress, and alleviate poverty in the least developed countries.'
>
> (Worrell, 2014)

2. Informal negotiations, relations and privileges tend to trump rules, standards and formal institutions in public sector decision-making concerning hydropower development in the Mekong River Basin.

Many representations of environmental governance in Southeast Asia convey a sense of the field as being characterized by informality and a preference for negotiating and resolving disputes outside the scope of formal review or accountability mechanisms (for example, Nurhidayah *et al.*, 2014). Some of our interviewees put forward this diagnosis. Sometimes, it is advanced using a trope familiar since the 1990s – that of the 'ASEAN way'. This is associated with a sense that informality correlates, necessarily, with manipulability, parochialism and non-transparency and the concept of 'external' non-interference in decision-making concerning Mekong River Basin development.

Assumptions along these lines frequently fold into a linear narrative about the relationship between 'soft' and 'hard' law, or between 'informal' and 'formal' norms. By way of illustration, the claims of one scholar writing about natural resource management under the Mekong Agreement are indicative:

> What needs to be firmed up are legal procedures, progressive scientific and management measures, and methods of financial and technological sharing in consonance with the ecological approach to international law.
>
> (Tolentino, 2014, p. 302)

This book casts these ideas into question in a number of respects. First, it depicts the negotiation of environmental and development issues in and between Lower Mekong countries being conducted, to a significant degree, by appeal to legal norms and institutions (in varying forms, and with differing degrees of success). Peoples of the Lower Mekong are shown in this book – in Chapter 7, for instance – to have a voracious appetite for, and considerable facility with, legal argument, even if this is often only one element of their strategies of social action and political contestation (with informal 'soft' approaches being especially emphasized where law and legal institutions are distrusted).

Second, it highlights how much the foregoing assumptions have downplayed or ignored the prevalence and significance of private law, such as contract law, in negotiations, structures and disputes surrounding Mekong River Basin development. Once the effect of private law is taken into account – the law of financing contracts, public company investors' conduct and reporting, and agreements with environmental and engineering contractors, for instance, referenced in Chapters 5 and 6 – then Mekong River Basin development can be seen to be densely regulated, and responsive to an array of institutional and normative influences outside the usual line-up of actors imagined as custodians of the 'ASEAN way'.

Third, as the operation of so-called soft or informal mechanisms and principles is canvassed in this book, rigidities, regularities and repetitions emerge. These confound standard stories about the predictable effects of, and relationship between, formal and informal, or between soft and hard norms. In different ways, Chapters 3, 4, 5 and 6 all suggest that opting out of, modifying or ignoring norms that take 'soft' form may, in fact, be far harder to accomplish than legislative reform at the national level. This includes global norms in areas such as good governance, transparency and public participation and other elements of so-called 'best practice' or 'standard market practice'. In this sense, the 'informal' may prove less flexible and more exacting than the 'formal'. In the reverse, some 'hard' norms enshrined in constitutional or statute law appear to have quite negligible effects on decision-making concerning hydropower, as Chapter 7 showed. Assumptions that the 'firming up' of legal norms will produce greater compliance with those norms may, accordingly, not be justified.

3. *Law serves powerful elites, consistently and reliably.*

Apparent in many of the interviews that we conducted for this book, and littered throughout socio-legal scholarship, is an expectation that law is a reliable and effective conduit for the interests of powerful elites. The first assumption outlined above presented one version of this expectation. Depending where one locates, or how one characterizes, directive power with regard to Mekong River Basin development, such power frequently seems to lie *behind* law. Many see law as a mask sported by some set of puppet-masters, whether the latter are imagined as political, economic or cultural elites or some combination of the same. This is believed to particularly be the case in the authoritarian political cultures of the Mekong, where strong, centralized state authorities are viewed by many to be writing the legal scripts necessary to realize their vision of development.

Contrary to this expectation, however, legal norms referenced in this book appear, often, too contested, unresolved or institutionally dispersed to prove instrumentally reliable. This book has shown that there is intense and ongoing conflict over the significance, scope and ramifications of legal norms relevant to Mekong River Basin development, especially around the practice of environmental assessment, demands for participation, and issues of transparency. Moreover, the conflict involves a broad array of lay and professional actors. We have not disavowed the idea that law can serve legitimating functions and other instrumental purposes. Nonetheless, the book highlights the extent to which legal mechanisms often fall short of delivering on the instrumental goals of elites, as well as the degree to which legal practices surpass or belie instrumental explanations of their operation. This accounts in part for the hope and promise that law and legal argumentation seem to hold for so many concerned with hydropower development in the Mekong River Basin. Law and policy in the Mekong River Basin have proven inconsistent and unreliable, both in their service of ruling elites and in their constraint of them. Accordingly, they may be amenable to re-inscription and re-mobilization, including among non-elite constituencies such as some of the civil society actors considered in Chapter 7.

4. *If better or more fully implemented in the Mekong River Basin, international law would reliably deliver more just outcomes for all.*

Related to preceding assumptions, and sometimes travelling alongside them in scholarly and popular writings, is a sense that deficiencies afflicting Mekong River Basin governance emanate primarily from a want of the implementation of international laws through national regulatory frameworks, policies and guidelines. In such diagnoses, international laws – especially the international law of watercourses, international human rights law and international environmental law – are deemed potentially curative of much that is ailing the region, especially with regard to decision-making surrounding, and adverse impacts of, hydropower development. If only national elites would follow through consistently on a pre-proven, internationally informed programme of law reform and legal compliance, it is claimed, much that is judged by commentators to be normatively, scientifically or politically substandard in the Mekong River Basin could be made to measure up to global standards (for example, Barrington *et al.*, 2012; Sparkes, 2014). Often implicit in such accounts is a sense that, for many who are currently aggrieved in the Mekong River Basin, the fuller implementation of international laws, in national law and institutions, would ensure justice, equity, fairness, reduced conflict and/or improved well-being.

Suggestions along these lines are not entirely absent from this book; we authors have harboured divergent intuitions about the extent to which the non-implementation of international laws is a critical problem with which we should be grappling. Nonetheless, we did ultimately concur that conflict surrounding hydropower development in the Mekong River Basin could not simply be disciplined out of existence by the fuller or more willing implementation of blueprints drawn from international law or global 'best practice'.

Against such ideas, this book shows just how much international law is already at work in the Mekong River Basin – via myriad processes and practices of claim, deal-making, consultation, participation, argument and justification – without that law having brought any definitive answers or substantive settlement to the questions with which people continue to grapple concerning hydropower development. Internationally mandated 'upgrading' of national or subnational practice has not been beneficial in all instances in the Mekong River Basin. At times, this may, arguably, have contributed to a 'downgrading' of pre-existing practice, or to the continuation and expansion of practices known to be flawed (see, generally, Richter *et al.*, 2010; Orr *et al.*, 2012; Ziegler *et al.*, 2013).

Consider, for instance, the effect that public participation and international monitoring appear to have had in relation to the Nam Theun 2 project, discussed in Chapter 6. Relocated villagers do not appear to have fared as well as expected in the long run, in material or political terms, despite their continual consultation and observation (World Bank, 2014, p. 10). It may nonetheless still be true that they would have been even worse off in the absence of international resettlement norms and the institutions that have sought to implement them. In the absence of a counterfactual comparator, we have no way of knowing – although the certitude that often surrounds these norms and institutions would suggest otherwise.

This book shows, further, how much international legal argument can serve the purposes of advancing hydropower development, as much as taming it. Some version or derivative of international human rights law, for example, usually stands – or can be made to stand – on both sides of most arguments about dam development. To say, as the World Commission on Dams has done, that dams must be built with due regard to 'international recognition of … the right to development and the right to a healthy environment' signals the *start* of a difficult and recurrent process of conflict and trade-off from which 'winners' and 'losers' continually emerge

(World Commission on Dams, 2000a, p. xxxiv). It does not signal an end to, or a way out of, that predicament. In some cases, economic development may be regarded as a legitimate public policy aim under international human rights norms justifying proportionate restriction on other rights of affected individuals and communities. Scope for argument concerning contending human rights standards and goals can produce uncertainty, confusion and disappointment as much as it can empower resistance from disadvantaged and marginalized groups.

It is also our combined view that to focus relentlessly on the 'implementation' of international norms and practices in national law and policy would be to miss much that is normatively distinctive and generative within the Lower Mekong. These nations are not merely recipients or would-be recipients of international legal norms; they are producing particular hybridizations and translations of those norms that are irreducible, in practice, to their formal legal sources or 'original' intent. Policy development and practice concerning freshwater waterways, basin planning, environmental protection, community organization and resettlement, administrative review and public participation are occurring within the Mekong River Basin in some distinctive ways described in this book. These processes may be imagined vertically (involving transfers between the international, regional and national planes), horizontally (involving transnational borrowing and movement between jurisdictions) or autochthonously (involving the embrace or enlivening of national and subnational normative currents, including customary law). In all three configurations, Mekong-specific developments are partially continuous with and inflected by legal and policy developments elsewhere, but not wholly commensurate with or explicable by reference to the latter. We might, accordingly, better understand Mekong River Basin governance as a field open to comparative inquiry within the terrain of global law and policy (that is, to comparativism among legal modes, styles, professional communities and subject areas), rather than a field external to international law awaiting its deployment (Koskenniemi, 2011; Kennedy, 1997).

5. *The natural world of the Mekong River Basin, and the society identified with it, pre-exist the workings of law; they comprise the 'ground' upon which law is a partial, derivative overlay.*

The last of the assumptions unsettled by this book concerns the relation of 'law' to 'nature' or 'society'. We have sought to work against a convention that has long been a mainstay of socio-legal analysis – namely, that law has no normative autonomy, properties or force that are independent of its social 'base'. Such a claim is restated by Bruno Latour, with some hyperbole, as follows:

> The claim now is … that the soft and superficial links provided by laws, culture, media, beliefs, religions, politics, economics are 'in reality' made of the harder stuff provided by the social frame of power relations. Such is the standard way for the social sciences and cultural studies [to which we might add: 'and for legal inquiry informed by those disciplines'] to explain why any thing holds: things do not stand upright because of the inner solidity of what

they claim to be built with, but because their superficial facades are propped up by the solid steelwork of society. Law for instance has no solidity of its own, it merely adds 'legitimacy' to the hidden strength of power: left to their own devices, laws are no more than a fragile layer of paint, a cover up for domination ... *Every* thing is made of one and the same *stuff*: the overarching, indisputable, always already there, all-powerful society.

(Latour, 2003, p. 29, emphasis in the original)

More or less the same claim is made, by legal scholars with a different set of interdisciplinary allegiances, in relation to 'nature'. That is, regardless of its particular content or character, law can always be explained ultimately by, and must in the end be answerable to, forces of nature and the vocabularies through which we have come to understand nature – primarily, today, those of natural science. By this account, the Mekong River itself, and the movements of sediment, water, fish, people and energy associated with it, will shape the ground upon which law can only ever rest lightly and passingly. To paraphrase Latour: *every* thing is made of one and the same *stuff*: the overarching, indisputable, always already there, all-powerful *natural world*.

Latour is among those who have recurrently highlighted the implausibility and inadequacy of such explanations of the 'making' of the world and the difficulty of their accounting for 'even the simplest task: baking a cake, weaving a basket, sewing a button – not to mention erecting sky-scrapers, discovering black holes or passing new bills' or carrying out the many variants of technical legal work (Latour, 2003, p. 33). Among the flaws of such accounts, Latour has demonstrated, is a tendency to vastly overstate the directive, creative power of the 'maker' and to consign all forms of the 'made' either to flaccid plasticity, or to mute and stubborn guardianship of the form in which they are found.

This book is by no means a direct application or specific elaboration of Latour's work. Nonetheless, it shares with that work a concern with mundane practices, composed of heterogeneous parts, which have not always been as they appear now and might have failed to take that form, but are nonetheless obligatory as they are: in our case, that concern revolves around legal practices. These demand to be understood on their own terms and are neither rigid, nor answerable to and manipulable by some maker, so that they might simply be made otherwise on a whim. To be sure, law does not have a life of its own, unhinged from those who made it or the society in which it circulates. Yet nor is it a passive mirror of society or a layer of fanciful ornament upon the natural world. As this book suggests, law's capacity to shape behaviour, conditions and experience is derived from practices of negotiation, specialist technique, contestation, postulation and negation involving a wide range of human and non-human agents.

Practices of legal work and argument are among the ways that the natural and societal phenomena of the Mekong River Basin are produced, articulated and experienced, but law is no more the 'stuff' of which they are decisively made than it is their mere protuberance. In engaging in socio-legal inquiry, this book refuses to prioritize or treat as commanding either the 'socio-' or the 'legal' (or, for that matter, the 'natural'). We have insisted, instead, on examining their inseparable combination in practice and the particular forms that emerge from this practice, endure as features of a common world or a world experienced in common.

What this might mean for those concerned with hydropower development in the Mekong River Basin

This book is no manual. It advances no programme for reform. Nevertheless, it is still possible – for the benefit of readers so oriented – to translate the five polemical points above into some practical insights for those engaged in law reform, policymaking, study and advocacy concerning hydropower development in the Mekong River Basin.

For environmental and human rights advocates, this book likely confirms strategic lessons many have learned through experience and practice in the region: the limits of litigation as a precipitant of change; the continued utility of pursuing quiet diplomacy with governments, as much as publicly pushing against them; the importance of building broad networks and coalitions, including sometimes counterintuitive combinations of actors; the need to situate legal interventions within wider campaigns and social movements; and the potential potency of creative thinking and strategic intervention in and around hubs of corporate activity. Activists are typically well aware of the limits of what they can get away with in the light of prevailing political, institutional and developmental conditions in each of the Mekong countries.

Yet, this book might also suggest looking beyond the most obvious laws and institutions on which activism in the Mekong River Basin has tended to focus. More attention might, perhaps, be directed to private law – equitable doctrines and allowance for the rights of third-party beneficiaries in contract law, for instance, or principles of employment law or stock exchange listing rules – among those seeking strategic leverage in relation to hydropower development projects. Greater account might also be taken of the force and significance of 'soft' norms and mechanisms – and their shifting interpretation and reach – alongside continuing activist preoccupation with judicial enforcement of 'hard' laws in the public interest. There is also value in better understanding the pluralism and diversity of actors within even the most seemingly authoritarian governments in the region, and the attendant openings (perhaps) for sympathetic hearings and novel alliances, as illustrated by some of our interviews.

Likewise, activists might well be encouraged by this book to be more critical of the international laws in which they sometimes invest hope. For instance, the championing of transparency and faith in information flow encouraged by international law and global policy (for example, in the 1998 Aarhus Convention, which straddles both the human rights and environmental fields) could warrant greater circumspection. Perhaps there may be good reasons to withhold information, boycott consultation or refrain from public participation in certain instances, from an activist standpoint. This might arguably be the case, for instance, when consultation is likely to legitimize the *fait accompli* of a development decision already made.

Insisting on international legal compliance might also counterproductively antagonize governments that are defensive about sovereignty; mobilizing norms coded in some other vernacular – in national interest terms, for instance – might sometimes work better.

For governments in the Mekong River Basin, it will come as no surprise that they cannot simply impose their sovereign will through law; that national law is permeated by and susceptible to myriad, multiscalar forms of influence; that legal argument sometimes yields unexpected or unintended outcomes; and that law can be utilized as a tool of resistance and to force accountability even in relatively authoritarian political cultures, as much as it may express and defend governmental agendas. Mekong River Basin governments should not read this book in order to better know their enemies or to work out new ways of promoting their preferred visions of development, nor should they expect to encounter unrelenting criticism in these pages.

The book shows, actually, that even governments that are politically closed – authoritarian, perhaps – are sometimes highly attuned to the environmental and social concerns of affected communities, civil society organizations and donors. This is not necessarily attributable to the permeation of international norms to that effect, although such norms do frequently inform national laws and procedures (such as those in relation to environmental assessment). Rather, this may be so because governments understand that attentiveness and responsiveness can erode resistance to development, and thus lower transaction costs in the longer term – for instance, by appeasing affected communities and NGOs, placating donors or fulfilling legal and political commitments concerning environmental or cultural protection and social impact mitigation. Sometimes this may entail a new instantiation of traditional forms of political rule and feedback that, in some cases, might be construed as a variant of what we referred to in Chapter 3 as (neo)liberal authoritarianism. These contingencies and divergences notwithstanding, this book may aid those in public service in the Mekong River Basin (whatever their capacity) to better understand how to navigate competing legal, social, economic and political interests at stake in their work. This is true also for those in government in the nations that are largely absent from this book – China and Myanmar – that face similar challenges.

For developers, financiers and investors involved in developing the hydropower resources of the Mekong, and the consultants and advisers who assist them, this book may provide a deeper understanding of the normative expectations of those affected by hydropower development and the attendant risks of inattentive or irresponsible corporate behaviour. In the Mekong River Basin, foreign corporations may to date have been legally shielded by governments from the full force of public dissatisfaction with hydropower projects. However, advocates, communities and activists are increasingly seeking to gain the ears of corporate investors, either directly or by appeal to their suppliers, customers and extant or would-be regulators around the world. Many of the corporations active in the region have tended to have limited engagement with contemporary frameworks on business and human rights or corporate social responsibility, particularly in the case of state power utilities, export-import banks and state-owned banks, but also among private banks.

As it may do for governments in the region, this book highlights for corporate readers that civil society is not uniformly opposed to development, nor averse to private sector gain. Rather, civil society actors often raise legitimate and timely questions about the definition, terms and aims of development – matters in which corporate decision-makers have a significant stake and on which they should, therefore, engage. Again, as for governments, sometimes responsiveness to those concerns can lessen the overall transaction costs of doing business, even if it may require greater initial outlay to mitigate adverse social or environmental impacts. Sometimes, failure to meet prevailing social and normative expectations may, indeed, be financially damaging and ethically compromising for the decision-makers concerned.

To donors, aid experts and others concerned with the promulgation and promotion of law reform templates in the Mekong River Basin, this book challenges prescriptive models of law reform and expectations of law as a linear and predictable route towards an end state of development. In particular, it calls into question assumptions that movement from soft to harder legal regulation is inevitable or uniformly desirable. It may well be that normative hardening in certain areas would improve hydropower governance. It may equally be so, however, that transitions along a soft-to-hard spectrum risk political blowback, the ratcheting down of aspirations, or the unravelling of delicate normative compromises among diverse protagonists.

To cleave closer to traditional expectations for a moment, it might be the case that maintaining flexible, relatively unstructured, 'soft' approaches to regulation in certain areas is the best way of both preserving existing normative settlements and incrementally nudging protagonists forward to improve upon those arrangements. The unexpected evolution of the Procedures for Notification, Prior Consultation and Agreement (PNPCA), discussed in this book, is a good example. It is highly unlikely that Lower Mekong governments would have prospectively agreed to a procedural rule providing for region-wide public consultations on mainstream dams such as Xayaburi and Don Sahong. The PNPCA procedure was only envisaged and agreed as an interstate process of notification and consultation. In practice, however – and for all its defects and carve-outs – it rapidly morphed, in its first deployment at least, into a process of public and stakeholder engagement beyond the ambit of state authorities alone. A similar point can be made about the MRC's strategic environmental assessment, also discussed in this book, whereby a body (the MRC) and procedure (the strategic environmental assessment) formally directed towards advising governments, engaged a wide spectrum of external stakeholders, including the public, in unanticipated ways. Once these precedents are entrenched through practice and experience, they become near-impossible to wind back, and can generate a normative expectation in subsequent cases as strong as, or stronger than, the existence of a formal rule to the same effect.

For similar reasons, there is as much cause to scrutinize soft law standards as 'traditional' constitutional and legislative rules. That which is presented in the form of a guideline or a market expectation can be just as non-negotiable and demanding – indeed, arguably, even more so – than laws as conventionally under-stood. Consider, for example, the market expectation of long-term privatization and region-wide competition in the power sector, long maintained by international financial institutions such as the Asian Development Bank. Elite coalitions in the Mekong River Basin (of business, union and government leaders in Thailand and Vietnam, for example) have resisted the expectation (as discussed in Chapter 3), and there are countervailing trends in hydropower innovation towards small-scale self-generated pico-hydropower and renewable energy sources – both of which require public support to be sustainable in the long term. This expectation none-theless remains undiminished as a measure against which Mekong River Basin governments are recurrently asked to line up (see, for instance, Chapter 3).

For the MRC, this book offers a combination of the counsel offered to govern-ments, investors, donors and aid experts. In addition, it advocates caution among MRC personnel in thinking of their work and their institution exclusively in 'non-regulatory' terms (with the disavowal of power and responsibility that such terms may imply). Much of the difficult work that might, in another era, have been conducted and negotiated overtly in the register of politics is now carried out – for good and for ill – in a register of science and technical expertise. As a scientifi-cally oriented organization, the MRC can no longer rely on historical mainstays of objectivity: evidence, fact, nature. As historians of science Lorraine Daston and Peter Galison wrote at the end of their celebrated study of the travails of scientific image- and knowledge-making from the eighteenth to the twenty-first centuries:

> In the era of truth-to-nature, images were inspired passages to an idealized world; later, they became very much of this world, their automaticity aiming to make them, in their vaunted objectivity, all nature and none of us ... Now, as images become part toolkit and part art [with the advent of 'nanofacturing', for instance], what are they? ... The scientific image begins to shed its rep-resentational aspect altogether as it takes on the power to build. Once again, images are in flux. Once again, so too is the scientific self.
>
> (Daston and Galison, 2010, p. 415)

That the basis for its authority is in flux is as true of the MRC as of any organization that banks mostly on laying claim to, and making claims about, science. Science no longer purports to represent, decode or reflect the world; it is explicitly engaged in the making and remaking of worlds. Similarly, the MRC cannot but enter the political and regulatory fray surrounding hydropower in the Mekong – indeed, it has already done so, as Chapters 4, 5 and 6 have shown. Perhaps, in the light of those chapters and this book as a whole, the MRC would be better to do so deliberately, announcing itself as a witting and authoritative agent in the politics of technical work in the Mekong River Basin. This might entail the MRC assuming a role and responsibility beyond that of 'mere' fact-provider and dialogue-facilitator, even if doing so may complicate relations with MRC governments.

For those engaged in research and study of hydropower development in the Mekong River Basin, this book urges that greater attention be paid to law. More precisely, it encourages less recourse to 'law' in the abstract and more attentiveness to the plurality of practices through which lawful authority is conveyed, articulated and comes to find resting or gathering points. As in most places, in the Mekong River Basin there can be no viable compartmentalization of law into international, regional, national and local frames. Instead, those who track the movement of lawful authority encounter constantly moving, hybridized terrain, with a plurality of laws manifesting in highly contingent combinations at particular sites.

Often these juridical combinations are particular to certain hydropower projects and cannot be generalized across even a single national jurisdiction, let alone the region. This may be the case, for instance, in deals struck through financing and insurance arrangements, contractual covenants, the interplay of development bank safeguards, and negotiation with affected communities – all of which can produce highly idiosyncratic governance regimes. Elements of project-specific regimes are certainly replicated across projects, drawing on common standards, best practices, 'lessons learned', legal and financial precedents, and the influence of repeat actors. However, law governing hydropower development in the Mekong River Basin remains highly variegated and resistant to reformist attempts to fashion a more universally coherent or standardized regime.

None of this should foster a fatalistic conclusion that law reform or improvements in governance are simply too hard to accomplish. It does suggest, nonetheless, that reformers take fuller account of the particularity, hybridity, diffusion, nuances and unanticipated effects of law as it currently operates and circulates in this field. There are lessons here not only for the future governance of the Mekong River Basin, but also for the many other regions where law is expected or mobilized to govern transboundary watercourses contested by a plethora of actors with often divergent interests.

Appendix 1: Selected national legislation, regulations, decisions and ministerial instructions

Category	Cambodia	Lao PDR	Thailand	Vietnam
Constitution	Constitution of the Kingdom of Cambodia, 21.9.1993	Constitution of the Lao PDR, No.25/NA, 6.5.2003	Constitution of the Kingdom of Thailand, 24.8.2007 (supended on 20.5.2014 under Martial Law, B.E. 2457, 1914). Constitution of the Kingdom of Thailand (Interim), 22.7.2014	Constitution of the Socialist Republic of Vietnam, 28.11.2013
Land	Land Law, 30.8.2001 Law on Concessions, 19.9.2007 Law on Expropriation, 26.02.2010 Law on Expropriation, 26.2.2010	Land Law, No.04/NA, 2003 Decree on Compensation and Resettlement of the Development Project, No.192/PM, 7.7.2005 Regulations for Implementing Decree 192/PM on Compensation and Resettlement, No.2432/STEA, 11.11.2005	Land Code, B.E. 2497, 1954 Expropriation Act, B.E. 2530, 1987 Land Division Act, B.E. 2543, 2000 Land Development Act, B.E. 2551, 2007 Land Structuring for Agriculture Act, B.E. 2558, 2015	Land Law, No.45/2013/QH13, 2013 Decree on Compensation, Support and Resettlement when Land is Recovered by the State, No. 197/2004/ND-CP, 2004 Decree and Circular on the Roles and Responsibilities for Implementation of Resettlement Projects, No.116/2004/TT-BTC, 2004 Decision Promulgating the Regulation on Irrigation and Hydropower Project Related Compensation, Support and Resettlement, No.34/2010/QD-TTg, 2010
Investment	Law on Investment, 4.8.1994	Law on the Promotion of Foreign Investment, No.11/NA, 2004 Law on Investment Promotion, No.02/NA, 2009 Prime Minister's Decree on Implementation of the Investment Promotion Law, No.119/PM, 2009 Law on Enterprises, No.11/NA, 2005	Foreign Business Act, B.E. 2542, 1999 Investment Promotion Act, B.E. 2520, 1977 amended by Act (No.3), B.E. 2544, 2001 The Act to Allow Private Entities to Invest in Public Business, B.E. 2556, 2013	Enterprise Law, No.60/2005/QH11, 2005 Investment Law, No.67/2014/QH13, 2014

Category	Cambodia	Lao PDR	Thailand	Vietnam
Water	Law on Water Resources Management, 29.6.2007 Law on Fisheries, 21.5.2006 Royal Decree on the Establishment of the Tonle Sap Basin Management Authority, 8.9.2007	Law on Water and Water Resources, No.02/96, 1996 (under revision) Law on Fisheries, No.03/NA, 2009 Decree on Establishment and Activities of River Basin Committee No.293/PM, 2010	Fisheries Act, B.E. 2490, 1947 The State Irrigation Act, B.E. 2485, 1942	Law on Water Resources, No.17/2012/QH13, 2012 Government Decree No.120/2008/ND-CP, 2008 on river basin management
Environment	Law on Environmental Protection and Natural Resource Management, 18.11.1996 Sub Decree No. 72 on Environmental Impact Assessment, 11.8.1999 Protected Areas Law, 15.2.2008 Forestry Law, 31.8.2002	Law on Environmental Protection, No.02-99/NA, 1999, revised by No.29/NA, 2013 Law on Wildlife and Aquatic Animals, No.07/NA, 2008 Law on Forestry, No.06/NA, 2007 Environmental Impact Assessment Decree, No.112/PM, 2010 Ministerial Instruction on the Process of Initial Environmental Examination of the Investment Projects and Activities No. 8029/MONRE 2013	Forestry Act, B.E. 2484, 1941 National Park Act, B.E. 2504, 1961 Enhancement and Conservation of National Environmental Quality Act, B.E. 2535, 1992 Wildlife Reservation and Protection Act, B.E. 2535, 1992 National Preserved Forestry Act, B.E. 2535, 1992 Plant Varieties Protection Act, B.E. 2542, 1999	Circular 05/2008/TT-BTMNT on strategic environmental assessment/environmental impact assessment, 2008 Environmental Protection Law, No.55/2014/QH13, 2014 Law on Environmental Protection Tax, No.57/2010/QH12, 2010 Government Decree No.29/2011/ND-CP, 2011 establishing regulations on strategic environmental assessment and environmental impact assessment Government Decree No.112/2008/ND-CP, 2008 on management, protection and integrated exploitation of resources and environment of hydropower and irrigation reservoirs. Decree No.18/2015/ND-CP on environmental protection, planning, strategic environmental assessment, environmental impact assessment and environmental protection plans, 2015

Category	Cambodia	Lao PDR	Thailand	Vietnam
Energy	Law on Electricity, 2.2.2001	Electricity Law, No.02/97/NA, 1997, revised by No.03/NA, 2011	Development and Promotion of Energy Act B.E. 2535, 1992 Energy Industry Act, B.E. 2550, 2007 Energy Conservation Promotion Act, B.E. 2535, 1992, amended by Act (No.2), B.E.2550, 2007	Law on Electricity, No.28/2004/QH11, 2004 Law on Economical and Efficient Use of Energy, No.50/2010/QH12, 2010
Dispute Resolution	Law on the Organization and Functioning of the Supreme Council of the Magistracy, 16.7.2014 Law on the Statute of Judges and Prosecutors, 16.7.2014 Law on the Organization of the Courts, 16.7.2014 Civil Procedure Code, 6.7.2006 Commercial Arbitration Law, 6.3.2006	Law on the Resolution of Economic Disputes, No.02/NA, 2005, revised by No.06/NA, 2010 Law on Judgment Enforcement, No.03/NA, 2004, revised by No.04/NA, 2008 Law on the People's Courts, No.05/PSA, 2003, revised by No.09/NA, 2009	Act on the Establishment of Administrative Courts and Administrative Court Procedure, B.E. 2542, 1999 Arbitration Act, B.E. 2545, 2002 Civil Procedure Code, B.E. 2477, 1934, amended by the Civil Procedure Code Amendment Act, B.E. 2551, 2008	Commercial Arbitration Law, No.54/2010/QH12, 2010 Law on Administrative Procedures, No. 64/2010/QH10, 2010 Law on Complaints, No.02/2011/QH13, 2011 Law on Denunciations, No.03/2011/QH13, 2011 Law on Handling Administrative Violations, No.15/2012/QH13, 2012

Appendix 2: Selected international, regional and bilateral treaties pertaining to Mekong states, and selected 'soft' law human rights, environmental and financial standards

Treaty	Cambodia (date in force or as otherwise noted)	Lao PDR (date in force or as otherwise noted)	Thailand (date in force or as otherwise noted)	Vietnam (date in force or as otherwise noted)
Charter of the United Nations and Statute of the International Court of Justice ('ICJ'), San Francisco, 26 June 1945 (entered into force 24 October 1945)	14.12.1955[a]	14.12.1955	16.12.1946	1.9.1978
Convention on the Recognition and Enforcement of Foreign Arbitral Awards, New York, 10 June 1958 (entered into force 7 June 1959)	4.4.1960	15.9.1998	20.3.1960	11.12.1995
International Convention on the Elimination of all Forms of Racial Discrimination, New York, 7 March 1966 (entered into force 4 January 1969)	28.11.1983	22.2.1974	28.1.2003	9.6.1982
International Covenant on Economic, Social and Cultural Rights, New York, 16 December 1966 (entered into force 3 January 1976)	26.5.1992	13.2.2007	5.9.1999	24.9.1982
International Covenant on Civil and Political Rights, New York, 16 December 1966 (entered into force 23 March 1976; 28 March 1979 for Article 41)	26.5.1992	25.9.2009	29.10.1996	24.9.1982

Treaty	Cambodia (date in force or as otherwise noted)	Lao PDR (date in force or as otherwise noted)	Thailand (date in force or as otherwise noted)	Vietnam (date in force or as otherwise noted)
Convention on Wetlands of International Importance, Ramsar, 2 February 1971 (entered into force 21 December 1975)	23.10.1999	28.9.2010	13.9.1998	20.1.1989
Convention on the Elimination of All Forms of Discrimination against Women, New York, 18 December 1979 (entered into force 3 September 1981)	15.10.1992	3.9.1981	9.8.1985	17.2.1982
Convention Concerning the Protection of the World's Cultural and Natural Heritage, Paris, 16 November 1972 (entered into force 17 September 1975)	28.11.1991	20.3.1987	17.9.1987	19.10.1987
Convention on the Rights of the Child, New York, 20 November 1989 (entered into force 2 September 1990)	15.10.1992	8.5.1991	27.3.1992	2.9.1990
Convention on Environmental Impact Assessment in a Transboundary Context, Espoo, 25 February 1991 (entered into force 10 September1997)	Not a party	Not a party	Not a party	Not a party
Convention on the Protection and Use of Transboundary Watercourses and International Lakes, Helsinki, 17 March 1992 (entered into force 6 Oct 1996)	Not a party	Not a party	Not a party	Not a party
United Nations Framework Convention on Climate Change, New York, 9 May 1992 (entered into force 21 March 1994)	18.12.1995	4.1.1995	28.12.1994	16.11.1994
Protocol on Strategic Environmental Assessment, Kyiv, 21 May 2003 (entered into force 10 July 2010)	Not a party	Not a party	Not a party	Not a party
Kyoto Protocol to the United Nations Framework Convention on Climate Change, Kyoto, 11 December 1997 (entered into force 16 February 2005)	16.2.2005	16.2.2005	16.2.2005	16.2.2005

Treaty	Cambodia (date in force or as otherwise noted)	Lao PDR (date in force or as otherwise noted)	Thailand (date in force or as otherwise noted)	Vietnam (date in force or as otherwise noted)
Convention on Biological Diversity, Rio de Janeiro, 5 June 1992 (entered into force 29 December 1993)	9.2.1995	20.9.1996	Party 31.10.2003	16.11.1994
Convention on the Law of the Non-Navigational Uses of International Watercourses, New York, 21 May 1997 (entered into force 17 August 2014)	Not a party	Not a party	Not a party	17.8.2014
Convention on Access to Information, Public Participation in Decision-Making and Access to Justice in Environmental Matters, Aarhus, 25 June 1998 (entered into force 30 October 2001)	Not a party	Not a party	Not a party	Not a party
International Convention for the Protection of all Persons from Enforced Disappearance, New York, 20 December 2006 (entered into force 23 December 2010)	27.6.2013	Signed 29.9.2008; not yet ratified	Signed 9.1.2012; not yet ratified	Not a party

Selected regional treaties among Mekong states

Treaty	Cambodia	Lao PDR	Thailand	Vietnam
Declaration establishing the Association of Southeast Asian Nations ('ASEAN Declaration'), Bangkok, 8 August 1967	Member as of 30.4.1999	Member as of 23.7.1997	Member as of 8.8.1967	Member as of 28.7.1995
Treaty of Amity and Cooperation in Southeast Asia ('ASEAN Treaty'), Indonesia, 24 February 1976 (entered into force 21 June 1976)	23.1.1995	29.6.1992	21.6.1976	16.6.1992
ASEAN Agreement on the Conservation of Nature and Natural Resources, Kuala Lumpur, 9 July 1985 (not yet in force)	30.4.1999	Not a party	16.11.1997	16.11.1997

Treaty	Cambodia	Lao PDR	Thailand	Vietnam
ASEAN Agreement on Energy Cooperation, Manila, 24 June 1986 (entered into force 2 April 1987); amended 1995 and 1997	30.4.1999	16.10.1997	2.4.1987	22.6.1996
Agreement on the Cooperation for the Sustainable Development of the Mekong River Basin ('Mekong Agreement'), and Protocol for the Establishment and Commencement of the Mekong River Commission, 5 April 1995 (entered into force 5 April 1995)	5.4.1995	5.4.1995	5.4.1995	5.4.1995
Intergovernmental Agreement on Regional Power Trade in the Greater Mekong Subregion, Phnom Penh, 3 November 2002 (entered into force November 2003)	Ratified; date of ratification unknown	Ratified; date of ratification unknown	Ratified; date of ratification unknown	Ratified; date of ratification unknown
ASEAN–China Framework Agreement on Economic Cooperation, Phnom Penh, 4 November 2002, amended 2003, 2006 and 2012 (entered into force 1 July 2003)	1.7.2003	1.7.2003	1.7.2003	1.7.2003
Memorandum of Understanding on the ASEAN Power Grid, Singapore, 23 August 2007 (entered into force 19 March 2009)	19.3.2009	19.3.2009	19.3.2009	19.3.2009
Charter of the Association of Southeast Asian Nations ('ASEAN Charter'), Singapore, 20 November 2007 (entered into force 15 December 2008)	15.12.2008	15.12.2008	15.12.2008	15.12.2008

Treaty	Cambodia	Lao PDR	Thailand	Vietnam
ASEAN Comprehensive Investment Agreement, Cha-am, 26 February 2009 (entered into force 29 March 2012)[b]	29.3.2012	29.3.2012	29.3.2012	29.3.2012
ASEAN-China Free Trade Agreement on Investment, Bangkok, 15 August 2009 (entered into force 15 February 2010)	15.2.2010	15.2.2010	15.2.2010	15.2.2010

Selected bilateral treaties among Mekong states

Bilateral Investment Treaty Thailand-China, 12 March 1985 (entered into force 13 December 1985)	N/A	N/A	13.12.1985	N/A
Bilateral Investment Treaty Lao PDR-Thailand, 22 August 1990 (entered into force 7 December 90)	N/A	7.12.1990	7.12.1990	N/A
Bilateral Investment Treaty Vietnam-China, 2 December 1992 (entered into force 1 September 1993)	N/A	N/A	N/A	1.9.1993
Bilateral Investment Treaty Lao PDR-China, 31 January 1993 (entered into force 1 June 1993)	N/A	1.6.1993	N/A	N/A
Bilateral Investment Treaty Lao PDR-Vietnam, 14 January 1996 (entered into force 23 June 1996)	N/A	23.6.1996	N/A	23.6.1996
Bilateral Investment Treaty Cambodia-China, 19 July 1996 (entered into force 1 February 2000)	1.2.2000	N/A	N/A	N/A
Bilateral Investment Treaty Cambodia-Lao PDR, 24 November 2008 (not yet in force)	Not yet in force	Not yet in force	N/A	N/A

Selected 'soft' law human rights, environmental and financial standards

Standards	Year
International human rights norms	
Universal Declaration of Human Rights, UN General Assembly resolution 217A(III)	1948
UN Declaration on the Right to Exploit Freely Natural Wealth and Resources, UN General Assembly resolution 626 (VII)	1952
UN Declaration on the Right to Development, UN General Assembly resolution 41/128	1986
Limburg Principles on the Implementation of the International Covenant on Economic, Social and Cultural Rights	1987
UN Committee on Economic, Social and Cultural Rights (CESCR), General Comment No. 4: The Right to Adequate Housing (article 11, ICESCR)	1991
UN Declaration on the Rights of Persons Belonging to National or Ethnic, Religious and Linguistic Minorities, UN General Assembly resolution 47/135	1992
UN CESCR, General Comment No. 7: The Right to Adequate Housing: Forced Evictions (article 11(1), ICESCR)	1997
UN Comprehensive Human Rights Guidelines on Development-based Displacement, UN Document E/CN.4/Sub.2/1997/7, Annex	1997
UN Committee on the Elimination of Racial Discrimination (CERD), General Recommendation No. 23: Rights of Indigenous Peoples	1997
Maastricht Guidelines on Violations of Economic, Social and Cultural Rights; republished in UN Document E/C.12/2000/13 (2000)	1997
UN Guiding Principles on Internal Displacement, UN Document E/CN.4/1998/53/Add.2	1998
UN CESCR, General Comment No. 12: The Right to Adequate Food (article 11, ICESCR)	1999
UN CESCR, General Comment No. 14: The Right to the Highest Attainable Standard of Health (article 12, ICESCR)	2000
UN CESCR, General Comment No. 15: The Right to Water (articles 11 and 12, ICESCR)	2002
UN Basic Principles and Guidelines on the Right to a Remedy and Reparation for Victims of Gross Violations of International Human Rights Law and Serious Violations of International Humanitarian Law, UN General Assembly resolution 60/147 (2006)	2006
UN Declaration on the Rights of Indigenous Peoples, UN General Assembly resolution 61/295 (2007)	2007
UN Basic Principles and Guidelines on Development-based Evictions and Displacement, in Report of the Special Rapporteur on Adequate Housing, UN Document A/HRC/4/18, Annex I	2007
UN CESCR, General Comment No. 21: The Right of Everyone to Take Part in Cultural Life (article 15, ICESCR)	2009
UN Guiding Principles on Business and Human Rights: Implementing the United Nations 'Protect, Respect and Remedy' Framework, endorsed by UN Human Rights Council resolution 17/4 (2011)	2011

Standards	*Year*
International environmental norms	
Rio Declaration on Environment and Development, UN Document A/CONF.151/26 (Vol. I)	1992
The Future We Want, UNGA Resolution A/RES/66/288	2012
Regional declarations	
Association of Southeast Asian Nations (ASEAN) Human Rights Declaration	2012
Financial institution policies	
Multilateral banks	
The World Bank, Operational Manual 4.12: Involuntary Resettlement OP 4.12	2001, revised 2013
The World Bank, Operational Manual 4.20: Gender and Development	2003, revised 2012
The World Bank, Operational Manual 4.10: Indigenous Peoples, OP 4.10	2005, revised 2013
The World Bank, Operational Manual 4.11: Physical Cultural Resources OP 4.11	2006, revised 2013
International Finance Corporation (IFC), Policy and Performance Standards on Social and Environmental Sustainability, and Policy on Disclosure of Information	2006; updated 2012
IFC, International Business Leaders Forum (IBLF), in association with UN Global Compact, Guide to Human Rights Impact Assessment and Management	2010
Asian Development Bank (ADB), Safeguard Policy Statement (superseding the ADB's Involuntary Resettlement Policy of 1995, Policy on Indigenous Peoples of 1998, and Environment Policy of 2002)	2009
The World Bank, Operational Manual 4.03: Performance Standards for Private Sector Activities (including Performance Standards 1 (Assessment and Management of Environmental and Social Risks and Impacts), 2 (Labor and Working Conditions), 3 (Resource Efficiency and Pollution Prevention), 4 (Community Health, Safety, and Security), 5 (Land Acquisition and Involuntary Resettlement), 6 (Biodiversity Conservation and Sustainable Management of Living Natural Resources), 7 (Indigenous Peoples) and 8 (Cultural Heritage)	2013
Private banks	
Equator Principles III (based on the IFC Performance Standards)	2013
National banks	
People's Republic of China Banking and Regulatory Commission, Green Credit Guidelines	2012
People's Republic of China (Ministry of Commerce and Ministry of Environmental Protection), Environmental Protection Guidelines for Foreign Investment and Cooperation	2013
Export credit agencies	
OECD, Common Approaches for Officially Supported Export Credits and Environmental and Social Due Diligence	2005
China Export and Import Bank, Guidelines for the Environmental and Social Impact Assessment of China EXIM Bank Loan Projects	2004; updated 2008
Export Import Bank of Thailand (EXIM), Corporate Social Responsibility Policy	2007

Standards	Year
Corporate social responsibility policies	
International Labour Organization, Principles concerning Multinational Enterprises and Social Policy (MNE Declaration)	1977
United Nations Global Compact	2000
Stock Exchange of Thailand, Principles of Corporate Governance of the Organization for Economic Cooperation and Development	2002; revised 2006
International Organization for Standardization, ISO26000: Social Responsibility	2010; reviewed 2014
Stock Exchange of Thailand, Approach to Social Responsibility Implementation for Corporations; and Sustainability Reporting Guidelines	2012
Multilateral guidelines, recommendations and auditing tools	
Organisation for Economic Cooperation and Development (OECD), Guidelines for Multinational Enterprises	1979; revised in 2011
International Commission on Large Dams (ICOLD), Position Paper on Dams and Environment	1997
World Commission on Dams, Guidelines and Recommendations	2000
Mekong River Commission, Preliminary Design Guidance for Proposed Mainstream Dams in the Lower Mekong Basin	2009
International Hydropower Association (IHA), Hydropower Sustainability Assessment Protocol (HSAP)	2011
World Wildlife Fund, Mekong River Commission and Asian Development Bank, Rapid Sustainability Assessment Tool (RSAT)	2010
ISEAL Credibility Principles	2013

^a Also accepted compulsory ICJ jurisdiction on reciprocal basis on 19 September 1957.
^b Upon its entry into force, this instrument terminated the ASEAN Agreement on Promotion and Protection of Investments, Manila, 15 December 1987.

Bibliography

3SPN (2010) *Living Rivers* (newsletter), vol. 3, May–October

3SPN (2011) *Annual Report 2010*, January

3SPN (2014) Letter to Cambodia's minister of industry, mines and energy and chief of Inter-Ministries Committee on Compensation and Resettlement concerning Lower Sesan 2 in Stung Treng, February, www.3spn.org/wp-content/uploads/2014/02/Final_Communities-letter-regarding-policies-and-mechanisms_English.pdf, accessed 7 May 2015

3SPN and other NGOs (2014) Letter to the Chinese Ambassador to Cambodia, 'Urgent request to reconsider China's investment in the Lower Sesan 2 Dam, Stung Treng Province, Cambodia', 26 May, www.internationalrivers.org/files/attached-files/140512_cso_follow_up_letter_to_chinese_ambassador_-_final_-_english.pdf, accessed 7 May 2015

Aarhus Convention (1998) United Nations Economic Commission for Europe (UNECE), Convention on Access to Information, Public Participation in Decision-making and Access to Justice in Environmental Matters, adopted 25 June 1998, entered into force 30 October 2001, 2161 UNTS 447, http://ec.europa.eu/environment/aarhus, accessed 7 May 2015

Abbott, K.W. and Snidal, D. (2000) 'Hard and soft law in international governance', *International Organization*, 54: 421–456

ACC/ISGWR (1992) The Dublin Statement on Water and Sustainable Development, International Conference on Water and the Environment: Development Issues for the 21st Century, Dublin, 26–31 January

ADB (2000) *Law and Policy Reform at the Asian Development Bank*

ADB (2003) 'Environmental Assessment Guidelines', www.adb.org/documents/adb-environmental-assessment-guidelines, accessed 7 May 2015

ADB (2006) *Sesan, Sre Pok and Sekong River Basins Development Study in Kingdom of Cambodia, Lao PDR, and Socialist Republic of Viet Nam*, ADB Technical Assistance Report, December

ADB (2009) *Building a Sustainable Energy Future: The Greater Mekong Subregion*, Manila

ADB (2010) 'Developing decision tools in the 3S', 3S Technical Briefing Sheet No 3, RETA 40082

ADB (2011a) *Civil Society Briefs: Cambodia*, November

ADB (2011b) *Civil Society Briefs: Lao Democratic Republic*, September

ADB (2011c) *Civil Society Briefs: Thailand*, November

ADB (2011d) *Civil Society Briefs: Viet Nam*, September

ADB (2012a) *The Greater Mekong Subregion at 20: Progress and Prospects*, Manila

ADB (2012b) *Greater Mekong Subregion Power Trade and Interconnection: 2 Decades of Cooperation*, Manila

ADB (2014) *Assessing Impact in the Greater Mekong Subregion: An Analysis of Regional Cooperation Projects*, Manila

Advocacy Forum, Nepal, and 64 others (2013) 'Open Letter to H. E. Thonsing Thammavong, Prime Minister, Lao People's Democratic Republic, Re: "Disappearance" of Mr. Sombath Somphone', www.forum-asia.org/uploads/statements/2013/january/Open-Letter-Sombath-2013-01-17-FINAL.pdf, accessed 7 May 2015

AECEN (2012) 'Cambodia: Few companies conduct environmental studies', 25 November, www.aecen.org/stories/cambodia-few-companies-conduct-environmental-studies, accessed 7 May 2015

Ahmed. M. and Hirsch, P. (eds.) (2000) *Common Property in the Mekong: Issues of Sustainability and Subsistence*, ICLARM Studies and Reviews, vol 26, Manila

Al Jazeera (2012) 'Laos to start building controversial dam', 6 November, www.aljazeera.com/news/asia-pacific/2012/11/201211611050564931.html, accessed 7 May 2015

Alston, P. (1988) 'Making space for new human rights: the case of the right to development', *Harvard Human Rights Year Book*, 1: 3–40

Althusser, L. (1977) *For Marx*, NLB, London

Alvarez, J. (2005) *International Organizations as Law-makers*, Oxford University Press, Oxford

Anghie, A. (2006) 'The evolution of international law: colonial and postcolonial realities', *Third World Quarterly*, 27(5): 739–753

Anghie, T. (2005) *Imperialism, Sovereignty and the Making of International Law*, Cambridge University Press, Cambridge

Antikainen, J., Gebert, R. and Møller, U. (2011) *Review of the Greater Mekong Sub-region Regional Power Trade: Final Report*, commissioned by Sida, Stockholm

Anton, D.K. and Shelton, D.L. (2011) *Environmental Protection and Human Rights*, Cambridge University Press, Cambridge

Antsis, S. (2012) 'Access to justice in Cambodia: The experience of grassroots networks in land rights issues', Legal Working Paper Series, Centre for International Sustainable Development Law, McGill University, Montreal, June

AusAID (2010) *Water: The Heart of Development in the Mekong – Australia's Mekong Water Resources Program*, April

Australian Government (2014) 'Brief summary of 2012 PNPCA Research', 12 April, www.internationalrivers.org/files/attached-files/responseausaid.pdf, accessed 7 May 2015

Bach, H., Glennie, P., Taylor, R., Clausen, T.J., Holzwarth, F., Jensen, K.M., Mejia, A. and Schmeier, S. (2014) *Cooperation for Water, Energy, and Food Security in Transboundary Basins under Changing Climate*, MRC, www.mrcmekong.org/assets/Publications/conference/MRC-intl-conf-publ-2014.pdf, accessed 7 May 2015

Baird, I.G. (1999) *Towards Sustainable Co-management of Mekong River Inland Aquatic Resources, including Fisheries, in Southern Lao PDR*, CESVI (Cooperazione e Sviluppo), Centre for Protected Areas and Watershed Management, Department of Forestry, Agriculture and Forestry Division, Champasak Province, Pakse, Lao PDR

Baird, I.G. (2009) *Best Practices in Compensation and Resettlement for Large Dams: The Case of the Planned Lower Sesan 2 Hydropower Project in Northeastern Cambodia*, Rivers Coalition in Cambodia, May

Baker, C. (2000) 'Thailand's Assembly of the Poor: background, drama, reaction', *South East Asia Research*, 8(1): 5–29

Bakker, K. (1999) 'The politics of hydropower: developing the Mekong', *Political Geography*, 18: 209–232

Bakker, K. (2010) 'The limits of "neoliberal natures": debating green neoliberalism', *Progress in Human Geography*, 34(6): 715–735

Bangkok Post (2014) 'The river of dreams', *Bangkok Post*, 31 March, www.bangkokpost.com/print/402566, accessed 7 May 2015

Bangkok Post (2015) 'Fish fix adds B10bn to Xayaburi cost', *Bangkok Post*, 3 February, www.bangkokpost.com/business/news/464681/fish-fix-adds-b10bn-to-xayaburi-cost, accessed 7 May 2015

BankTrack (2015a) 'Dodgy deal: Lower Sesan 2 dam, Cambodia', www.banktrack.org/show/dodgydeals/lower_sesan_2_dam, accessed 7 May 2015

BankTrack (2015b) 'Dodgy deal: Xayaburi dam, Laos', www.banktrack.org/manage/ajax/ems_dodgydeals/createPDF/xayaburi_dam, accessed 7 May 2015

BankTrack, FIVAS, IRN Les Amis de la Terre and Uniting Church in Australia (2009) *Expanding Failure: An Assessment of the Theun-Hinboun Hydropower Expansion Project's Compliance with Equator Principles and Lao Law*, October

Banpasirichote, C. (2004) 'Civil society discourse and the future of radical environmental movements in Thailand', in L.H. Guan (ed.), *Civil Society in Southeast Asia*, Institute of Southeast Asian Studies, Singapore, pp. 234–264

Bao Moi (2013) 'PM highlights public feedback on revised constitution', *Peoples' Army Newspaper*, 20 March, http://en.baomoi.com/Home/society/www.qdnd.vn/PM-highlights-public-feedback-on-revised-Constitution/343578.epi, accessed 7 May 2015

Baran, E. (2010) *Fisheries Sections of the Strategic Environmental Assessment of Hydropower on the Mekong Mainstream,* prepared for the MRC by ICEM, October, http://pubs.iclarm.net/resource_centre/WF_2736.pdf, accessed 7 May 2015

Barney, K. (2012) 'Locating "green neoliberalism", and other forms of environmental governance in Southeast Asia', Center for Southeast Asian Studies Kyoto University (CSEAS), *Newsletter*, 66: 25–28

Barrington, D.J., Dobbs, S. and Loden, D.I. (2012) 'Social and environmental justice for communities of the Mekong River', *International Journal of Engineering, Social Science, and Peace*, 1(1): 31–49

Bearden, B.L. (2010) 'The legal regime of the Mekong River: a look back and some proposals for the way ahead', *Water Policy*, 12(6): 798–821

Bello, W. and Guttal, S. (2006) 'The limits of reform: the Wolfensohn era at the World Bank', *Race & Class*, 47(3): 68–81

Bellver A. and Kaufmann, D. (2005) *Transparenting Transparency: Initial Empirics and Policy Applications*, World Bank Policy Research Working Paper, World Bank, Washington, DC

Belton, R.K. (2005) *Competing Definitions of the Rule of Law: Implications for Practitioners*, Carnegie Papers, Rule of Law Series No. 55, Carnegie Endowment for International Peace, Washington, DC, http://carnegieendowment.org/files/CP55.Belton.FINAL.pdf, accessed 7 May 2015

Bentham, J. (1843) 'Of publicity', in *The Works of Jeremy Bentham*, William Tait, Edinburgh, vol. 2, pp. 310–317

Bergling, P. (1999) *Legal Reform and Private Enterprise: The Vietnamese Experience*, Umeå Studies in Law, no. 1, Department of Law, Umeå University, Sweden

Bianchi, A. (2013) 'On power and illusion: the concept of transparency in international law', in A. Bianchi and A. Peters (eds.), *Transparency in International Law*, Cambridge University Press, Cambridge

Bianchi, A. and Peters, A. (eds.) (2013) *Transparency in International Law*, Cambridge University Press, Cambridge

Bilder, R. (2000) 'Beyond compliance: helping nations cooperate', in D. Shelton (ed.), *Commitment and Compliance: The Role of Non-binding Norms in the International Legal System*, Oxford University Press, Oxford, pp. 65–73

Bishop, P., Sanderson, D.C. and Stark, M.T. (2004) 'OSL and radiocarbon dating of a pre-Angkorian canal in the Mekong Delta, southern Cambodia', *Journal of Archaeological Science*, 31(3): 319–336

Biswas, A. (2004) 'Integrated water resources management: a re-assessment', *Water International*, 29(2): 248–256

Blyth, M. (2003) 'The political power of financial ideas: transparency, risk and distribution in global finance', in J. Kirshner (ed.), *Monetary Orders: Ambiguous Economics, Ubiquitous Politics*, Cornell University Press, Ithaca, NY, pp. 239–259

Bogdan, M. (1991) 'Legal aspects of the re-introduction of a market economy in Laos', *Review of Socialist Law*, 17(2): 101–123

Both ENDS and Gomukh (2005) *River Basin Management: A Negotiated Approach*, Amsterdam and Pune, India

Bourdieu, P. (1988) 'Vive la crise!', *Theory and Society*, 17(5): 773–787

Bourquain, K. (2008) *Freshwater Access from a Human Rights Perspective: A Challenge to International Water and Human Rights Law*, Martinus Nijhoff, Leiden

Boyle, A. and Chinkin, C. (2007) *The Making of International Law*, Oxford University Press, Oxford

Boyle, D. and Seangly, P. (2010) 'Villagers write to Hun Sen over dam', *Phnom Penh Post*, 15 March

Braithwaite, J. and Drahos, P. (2000) *Global Business Regulation*, Cambridge University Press, Cambridge

Bridges Across Borders (2012) *DERAILED: A Study on the Resettlement Impacts of the Rehabilitation of the Cambodia Railway*, February

Brocheux, P. and Hémery, P. (2009) *Indochina: An Ambiguous Colonization 1858–1954*, trans. L.L. Dill-Klein, University of California Press, Berkeley, CA

Browder, G. and Ortolano, L. (2000) 'The evolution of an international water resources management regime in the Mekong River Basin', *Natural Resources Journal*, 40: 499–531

Brunnée, J. and Hey, E. (2013) 'Transparency and international environmental institutions', in A. Bianchi and A. Peters (eds.), *Transparency in International Law*, Cambridge University Press, Cambridge, pp. 23–48

Bui, T. (2013) 'Vietnam's civil society', *East Asia Forum*, 5 September, www.eastasiaforum. org/2013/09/05/vietnams-civil-society-undergoing-vital-changes, accessed 7 May 2015

Burg, E.M. (1977) 'Law and development: a review of the literature and a critique of "Scholars in Self-estrangement"', *American Journal of Comparative Law*, 25(3): 492–530

Calavita, K. (2010) *Invitation to Law & Society: An Introduction to the Study of Real Law*, University of Chicago Press, Chicago, IL

Calvino, I. (1988) *Six Memos for the Next Millennium*, Harvard University Press, Cambridge, MA

Cambodia (2014) *Draft Law on Environmental Impact Assessment*, www.opendevelopmentcambodia.net/download/law/Revised_Draft_EIA_Law%20_Eng_06_05_2014.pdf, accessed 7 May 2015

Cambodia Daily (2013) 'SRP commune told to stop talking to media about dam', *Cambodia Daily*, 3 April

Can, Ö. and Leader, S. (2005) *Nam Theun 2 Hydroelectric Project: Memorandum of Legal Issues in Relation to the Concession Agreement – An Analysis by Human Rights Centre, Essex for Mekong*

Watch, London, 30 May, www.internationalrivers.org/files/attached-files/060118analysis.pdf, accessed 7 May 2015

Caney, S. (2005) *Justice beyond Borders: A Global Political Theory*, Oxford University Press, Oxford

Carew-Reid, J. (2014) 'Environment assessment tools for improving the sustainability of development in the Mekong region', policy brief, ICEM, http://icem.com.au/wp-content/uploads/2014/05/Environment-assessment-tools_Brief.pdf, accessed 7 May 2015

Carew-Reid, J. and Roop, J. (2013) 'Strategic environmental assessment as a tool to improve climate change adaptation in the Greater Mekong Subregion', policy brief, ICEM, http://icem.com.au/wp-content/uploads/2014/05/ICEM-SEA-CC-in-GMS-Brief.pdf, accessed 7 May 2015

Carothers, T. (1998) 'The rule of law revival', *Foreign Affairs*, 77(2): 95–106

Carothers, T. (2006) 'The rule of law revival', in T. Carothers (ed.), *Promoting the Rule of Law Abroad: In Search of Knowledge*, Carnegie Endowment for International Peace, Washington, DC

Central Region Hydropower Stock Company (2006) *A Luoi Hydropower Project – Huong River Basin – Thua Thien Hu Province Environmental Impact Assessment in Lower Basin in Lao PDR*, informal translation

Chachavalpongpun, P. (2011) 'Bounding the Mekong: the Asian Development Bank, China, and Thailand (review)', *ASEAN Economic Bulletin*, 28: 97–98

Chalermsripinyorat, R. (2004) 'Politics of representation: a case study of Thailand's Assembly of the Poor', *Critical Asian Studies*, 36(4): 541–566

Chen, H., Rieu-Clarke, A. and Wouters, P. (2013) 'Exploring China's transboundary water treaty practice through the prism of the UN Watercourses Convention', *Water International*, 38(2): 217–230

Chen, P.-H. (2014) 'The vulnerability of Thai democracy: coups d'état and political changes in modern Thailand', in P. Liamputtong (ed.), *Contemporary Socio-cultural and Political Perspectives in Thailand*, Springer, Dordrecht, pp. 185–207

Chinkin, C. (1989) 'The challenge of soft law: development and change in international law', *International and Comparative Law Quarterly*, 38: 850–866

CIVICUS (2005) Civil Society Index: Shortened Assessment Tool

Clarke, G. (1998) 'Non-governmental organizations (NGOs) and politics in the developing world', *Political Studies*, 46(1): 36–52

Clarkson, M. (1994) 'A risk based model of stakeholder theory', *Proceedings of the Second Toronto Conference on Stakeholder Theory*, Centre for Corporate Social Performance and Ethics, University of Toronto, Toronto

CLICK Laos (2014) http://clicklaos.org, accessed 7 May 2015

Cohen, A. (2008) 'Negotiation, meet new governance: interests, skills, and selves', *Law & Social Inquiry*, 33: 503–562

Cohen, A. (2010) 'Governance legalism: Hayek and Sabel on reason and rules, organization and law', *Wisconsin Law Review*, 2010(2): 357–387

Collier, S.J. (2009) 'Topologies of power: Foucault's analysis of political government beyond "Governmentality"', *Theory Culture Society*, 26(6): 78–108

Connors, M. (2011) 'Ambivalent about human rights: Thai democracy', in T. Davis and B. Galligan (eds.), *Human Rights in Asia,* Edward Elgar, Cheltenham, pp. 103–122

Cooper, R. (2012) 'The potential of MRC to pursue IWRM in the Mekong: trade-offs and public participation', in J. Öjendal, S. Hansson and S. Hellberg (eds.), *Politics and*

Development in a Transboundary Watershed: The Case of the Lower Mekong Basin, Springer, Dordrecht, pp. 61–82

Corless, R. (1989) *The Vision of Buddhism,* Paragon House, New York

Corsetti, G., Pesenti, P. and Roubini, N. (1998) *What Caused the Asian Currency and Financial Crisis? Part 1: A Macroeconomic Overview,* Working Paper 6833, National Bureau of Economic Research, Cambridge, December, www.nber.org/papers/w6833.pdf, accessed 7 May 2015

Costanza, R., Kubiszewski, I., Paquet, P., King, J., Halimi, S., Sanguanngoi, H., Bach, N.L., Frankel, R., Ganaseni, J., Intralawan, A. and Morell, D. (2011) *Planning Approaches for Water Resources Development in the Lower Mekong Basin,* Portland State University and Mae Fah Luang University, http://web.pdx.edu/~kub/publicfiles/Mekong/LMB_Report_FullReport.pdf, accessed 7 May 2015

Cotterrell, R. (2002) 'Subverting orthodoxy, making law central: a view of sociolegal studies', *Journal of Law and Society,* 29(4): 632–644

Crawford, P.W. (2014) 'Promoting enabling mechanisms for trans-boundary impact assessments', *International Journal of Hydropower and Dams,* 21(3): 82–86

Creighton, J.L. (2005) *The Public Participation Handbook: Making Better Decisions through Citizen Involvement,* John Wiley & Sons, San Francisco, CA

Cronin, R. and Weatherby, C. (2014) *Letters from the Mekong: Obstacles to Equitable Hydropower Development Planning in the Lower Mekong Basin,* Stimson Center, September, www.stimson.org/images/uploads/research-pdfs/Letters_from_the_Mekong_Sept_2014-WEB.pdf, accessed 7 May 2015

d'Aspremont, J. (2008) 'Softness in international law: a self-serving quest for new legal materials', *European Journal of International Law,* 19: 1075–1093

Daniel, R., Lebel, L. and Manorom, K. (eds.) (2013) *Governing the Mekong: Engaging in the Politics of Knowledge,* Strategic Information and Research Development Centre, Petaling Jaya, Malaysia

Dao, N. (2010) 'Dam development in Vietnam: the development of dam-induced resettlement policy', *Water Alternatives,* 3(2): 324–340

Darian-Smith, E. (2013) *Laws and Societies in Global Contexts,* Cambridge University Press, Cambridge

Daston, L. and Galison, P. (2010) *Objectivity,* Zone Books, Brooklyn, NY

Dean, M. (1999) *Governmentality: Power and Rule in Modern Society,* Sage, London

de Búrca, G., Keohane, R.O. and Sabel, C. (2014) 'Global experimentalist governance', *British Journal of Political Science,* 44(3), http://ssrn.com/abstract=2423810, accessed 7 May 2015de Chazournes, L.B. (2013) *Fresh Water in International Law,* Oxford University Press, Oxford

Deetes, P. (2012) 'Thai villagers file lawsuit on Xayaburi dam', *International Rivers,* 8 August

Deetes, P. (2014) 'Justice for the Mekong – Thai villagers back at court', *International Rivers,* 20 June, www.internationalrivers.org/blogs/259-0 (accessed 25 March 2015)

de Martens, G.F. (1888) *Nouveau Recueil Général de traits,* Leipzig, Dieterich Weicher

Department of Foreign Affairs and Trade, Australia (2015) *Water Resources Management in the Mekong Region*

Dewey, J. (1927) *The Public and Its Problems,* Henry Holt and Company, New York

Dezalay, Y. and Garth, B.G. (eds) (2002) *Global Prescriptions: The Production, Exportation, and Importation of a New Legal Orthodoxy,* Michigan University Press, Ann Arbor, MI

Dezalay, Y. and Garth, B.G. (2006) 'From the Cold War to Kosovo: the rise and renewal of the field of international human rights', *Annual Review of Law and Social Science,* 2: 231–255

DFDL Legal and Tax Services (2014a) *Deal List: Focus on Energy, Mining and Infrastructure,* December, www.dfdl.com/images/stories/Articles/Deal_Lists/DFDL_Deal_List_-_Regional_EMI_111214.pdf, accessed 7 May 2015

DFDL Legal and Tax Services (2014b) *Legal, Tax & Investment Guide: Lao PDR,* 4th edn, www.dfdl.com/resources/news/741-dfdl-investment-guides-2014-edition-cambodia-laos-myanmar-thailand-and-vietnam-available-online, accessed 7 May 2015

DFDL Legal and Tax Services (2014c) *Legal, Tax & Investment Guide: Thailand,* 4th edn, www.dfdl.com/resources/news/741-dfdl-investment-guides-2014-edition-cambodia-laos-myanmar-thailand-and-vietnam-available-online, accessed 7 May 2015

DFDL Legal and Tax Services (2014d) *Legal, Tax & Investment Guide: Vietnam,* 4th edn, www.dfdl.com/resources/news/741-dfdl-investment-guides-2014-edition-cambodia-laos-myanmar-thailand-and-vietnam-available-online, accessed 7 May 2015

Di Certo, B. and Titthara, M. (2012) 'Villagers bemoan scant information on dam project', *Phnom Penh Post,* 29 November, www.phnompenhpost.com/national/villagers-bemoan-scant-information-dam-project, accessed 7 May 2015

Dinar, S. (2008) *International Water Treaties: Negotiation and Cooperation along Transboundary Rivers,* Routledge, London

Donaldson, M. and Kingsbury, B. (2013) 'The adoption of transparency policies in global governance institutions: Justifications, effects, and implications', *Annual Review of Law and Social Science,* 9: 119–147

Doran, D.D. and Christensen, M. (2014) 'Cross-border hydro projects in Asia: legal issues, hurdles and solutions', *International Journal on Hydropower & Dams,* 21: 66–72

Dore, J. and Lazarus, K. (2009) 'Demarginalizing the Mekong River Commission', in F. Molle, T. Foran and M. Käkönen (eds.), *Contested Waterscapes in the Mekong Region: Hydropower, Livelihoods and Governance,* Earthscan, London, pp. 357–381

Dore, J. and Lebel, L. (2010) 'Gaining public acceptance: a critical strategic priority of the World Commission on Dams', *Water Alternatives,* 3: 124–141

Dore, J. and Xiaogang, Y. (2004) *Yunnan Hydropower Expansion: Update on China's Energy Industry Reforms and the Nu, Lancang and Jinsha Hydropower Dams,* Chiang Mai University's Unit for Social and Environmental Research and Green Watershed, Kunming, PR of China (Working Paper), www.sea-user.org/download_pubdoc.php?doc=2586, accessed 7 May 2015

Dore, J., Molle, F., Lebel, L., Foran, T. and Lazarus, K. (2010) *CPWF Project Report: Improving Mekong Water Resources Investment and Allocation Choices,* Project No 67, CGIAR Challenge Programme on Water and Food, www.mpowernetwork.org/Knowledge_Bank/Key_Reports/PDF/Research_Reports/PN67_Final_Report.pdf, accessed 7 May 2015

Duong, M. N. (2004) *Grassroots Democracy in Vietnamese Communes,* research paper, Centre for Democratic Institutions, Australian National University, Canberra

Dupuy, P.-M. (1991) 'Soft law and the international law of the environment', *Michigan Journal of International Law,* 12, 420–435

Dupuy, P.-M. and Vierucci, L. (eds.) (2008) *NGOs in International Law: Efficiency in Flexibility?,* Edward Elgar, Cheltenham

Ear, S. (2007) 'The political economy of aid and governance in Cambodia', *Asian Journal of Political Science,* 15(1): 68–96

Ebbesson, J. (1997) 'The notion of public participation in international environmental law', *Yearbook of International Environmental Law,* 8: 51–97

Ebbesson, J. (2013) 'Global or European only? International law on transparency in environmental matters for members of the public', in A. Bianchi and A. Peters (eds.), *Transparency in International Law,* Cambridge University Press, Cambridge

ECAFE (1957) *Development of Water Resources in the Lower Mekong Basin, Economic Commission for Asia and the Far East*, United Nations, Bangkok

Economist (2014) 'Why China is creating a new World Bank for Asia', *Economist*, 11 November, www.economist.com/blogs/economist-explains/2014/11/economist-explains-6, accessed 7 May 2015

Economist Intelligence Unit (2014) *Creative Productivity Index in Asia: Analysing Creativity and Innovation in Asia*, www.adb.org/sites/default/files/pub/2014/creative-productivity-index. pdf, accessed 7 May 2015

Environmental Law Institute (2009) *Establishing a Transboundary Environmental Impact Assessment Framework for the Mekong River Basin*, prepared on behalf of ECO-Asia and USAID for the MRC, www.eli.org/research-report/establishing-transboundary-environmental-impact-assessment-framework-mekong-river-ba, accessed 7 May 2015

Epstein, G.A. (2005) *Financialization and the World Economy*, Edward Elgar, Cheltenham

ERI (n.d.a) *I Want to Eat Fish. I Cannot Eat Electricity: Public Participation in Mekong Basin Development*

ERI (n.d.b) *Mekong Legal Advocacy Institute*, www.earthrights.org/training/mekong-legal-advocacy-institute, accessed 7 May 2015

ERI (n.d.c) *Mekong Legal Network*, www.earthrights.org/legal/mekong-legal-network, accessed 7 May 2015

Espoo Convention (1991) Convention on Environmental Impact Assessment in a Transboundary Context, entered into force 10 September 1997, www.unece.org/env/eia/eia.html, accessed 7 May 2015

Etzioni, A. (2010) 'Is transparency the best disinfectant?', *Journal of Political Philosophy*, 18: 389–404

FAO (1998) *Sources of International Water Law*, Food and Agriculture Organization of the United Nations, Rome

Fenster, M. (2006) 'The opacity of transparency', *Iowa Law Review*, 91: 885–950

Finger, M., Tamiotti, L. and Allouche, J. (eds.) (2006) *The Multi-governance of Water: Four Case Studies*, SUNY Press, Albany, NY

Fischer, T.B. (2003) 'Strategic environmental assessment in post-modern times', *Environmental Impact Assessment Review*, 23, 155–170

Fischer-Lescano, A. and Teubner, G. (2004) 'Reply to Andreas L. Paulus: consensus as fiction of global law', *Michigan Journal of International Law*, 25: 1059–1073

Fisher, P. (2013) 'Why environmental impact assessments fail to protect rivers', *World Rivers Review*, 28(1), www.internationalrivers.org/files/attached-files/wrr_march13.pdf, accessed 7 May 2015

Fitzpatrick, D. and McWilliam, A. (2013) 'Bright-line fever: simple legal rules and complex property customs among the Fataluku of East Timor', *Law and Society Review*, 47(2): 311–343

Fitzpatrick, P. (1992) *The Mythology of Modern Law*, Routledge, London and New York

Fong, J. (2009) 'Sacred nationalism: the Thai monarchy and primordial nation construction', *Journal of Contemporary Asia*, 39(4): 673–696

Foran, T. and Manorom, K. (2009) 'Pak Mun dam: perpetually contested?', in F. Molle, T. Foran and M. Käkönen (eds.), *Contested Waterscapes in the Mekong Region: Hydropower, Livelihoods and Governance*, Earthscan, London, pp. 55–80

Foucault, M. (1979) *The History of Sexuality, Volume I: An Introduction*, trans. R. Hurley, Allen Lane, London

Foucault, M. (1982) 'The subject and the power', in H.L. Dreyfus and P. Rabinow (eds.), *Michel Foucault: Beyond Structuralism and Hermeneutics*, Chicago University Press, Chicago, pp. 208–226

Foucault, M. (1991) 'Governmentality', in G. Burchell, C. Gordon and P. Miller (eds), *The Foucault Effect: Studies in Governmentality*, University of Chicago Press, Chicago, pp. 87–104

Foucault, M. (2007) *Security, Territory, Population: Lectures at the Collège de France 1977–1978*, trans. G. Burchell, Palgrave Macmillan, Basingstoke

Foucault, M. (2008) *The Birth of Biopolitics: Lectures at the Collège de France 1978–1979*, ed. M. Senellart, trans. G. Burchell, Palgrave Macmillan, New York

Freeman, R.E. (1984) *Strategic Management: A Stakeholder Approach*, Pitman Publishing, Boston, MA

Freeman, R.E. and McVea, J. (2001) *A Stakeholder Approach to Strategic Management*, Darden Business School Working Paper No 01-02, http://papers.ssrn.com/sol3/papers. cfm?abstract_id=263511, accessed 7 May 2015

Friend, R., Arthur, R. and Keskinen, M. (2009) 'Songs of the doomed: the continuing neglect of capture fisheries in hydropower development in the Meong', in F. Molle, T. Foran and M. Käkönen (eds.), *Contested Waterscapes in the Mekong Region: Hydropower, Livelihoods and Governance*, Earthscan, London, pp. 307–332

Gao, Qi (2014) *A Procedural Framework for Transboundary Water Management in the Mekong River Basin: Shared Mekong for a Common Future*, Martinus Nijhoff, Leiden

Gathii, J.T. (2008) 'Third World approaches to international economic governance', in R. Falk, B. Rajagopal and J. Stevens (eds.), *International Law and the Third World: Reshaping Justice*, Routledge, London

Geheb, K., West, N. and Matthews, N. (2015) 'The invisible dam: hydropower and its narration in the Lao People's Democratic Republic', in N. Matthews and K. Geheb (eds.), *Hydropower Development in the Mekong Region: Political, Socio-economic and Environmental Perspectives*, Routledge, London, pp. 101–126

Gereffi, G. (1996) 'The elusive last lap in the quest for developed-country status', in J.H. Mittelman (ed.), *Globalization: Critical Reflections*, Lynne Rienner, Boulder, CO, pp. 53–81

Gillespie, J. (1994) 'Private commercial rights in Vietnam: a comparative analysis', *Stanford Journal of International Law*, 30(2): 325–377

Gillespie, J. (2004) 'Concepts of law in Vietnam: transforming statist socialism', in R. Peerenboom (ed.), *Asian Discourses of Rule of Law: Theories and Implementation of Rule of Law in Twelve Asian Countries, France and the US*, Routledge, London and New York

Gillespie, J. (2005) 'Changing concepts of socialist law in Vietnam', in J. Gillespie and P. Nicholson (eds.), *Asian Socialism and Legal Change: The Dynamics of Vietnamese and Chinese Reform*, ANU E Press and Asia Pacific Press, Canberra, pp. 45–75

Ginsburg, T. (2000) 'Does law matter for economic development? Evidence from East Asia', *Law & Society Review*, 34(3): 829–856

Ginsburg, T. (2009) 'Constitutional afterlife: the continuing impact of Thailand's post-political constitution', *International Journal of Constitutional Law*, 7: 83–105

Global Water Partnership (2000) *Integrated Water Resources Management*, Global Water Partnership, Stockholm

Goldman, M. (2001) 'Constructing an environmental state: eco-governmentality and other transnational practices of a "green" World Bank', *Social Problems*, 48(4): 499–523

Goldsmith, E. (1992) 'Development as enclosure', *Ecologist*, 22(4): 132–149

Gooch, G. and Stalnacke, P. (eds.) (2010) *Science, Policy and Stakeholders in Water Management: An Integrated Approach to River Basin Management*, Earthscan, London

Gracean, C.S. and Gracean, C. (2004) 'Thailand's electricity reforms: privatization of benefits and socialization of costs and risks', *Pacific Affairs*, 77(3): 517–541

Grigorescu, A. (2007) 'Transparency of intergovernmental organizations: the roles of member states, international bureaucracies and nongovernmental organizations', *International Studies Quarterly*, 51(3): 625–648

Grimsditch, M. (2012) *3S Rivers under Threat: Understanding New Threats and Challenges from Hydropower Development to Biodiversity and Community Rights in the 3S River Basin*, 3SPN and International Rivers, April

Grumbine, R.E. and Xu, J. (2011) 'Mekong hydropower development', *Science*, 8 April

Gustafsson, S. (2005) 'Laos, law and education', *International Journal for Education Law and Policy*, 1(1–2): 28–32

Guttal, S. and Shoemaker, B. (2004) 'Manipulating consent: the World Bank and public consultation in the Nam Theun 2 hydroelectric project', *Watershed*, 10(2): 18–25

Haas, P.M. (1989) 'Do regimes matter? Epistemic communities and Mediterranean pollution control', *International Organization*, 43: 377–403

Hajer, M. (1997) *The Politics of Environmental Discourse: Ecological Modernization and the Policy Process*, Clarendon Press, London

Hall, S. (1986) 'The problem of ideology – Marxism without guarantees', *Journal of Communication Inquiry*, 10: 28–44

Harding, A. (2012) 'The politics of law and development in Thailand: seeking Rousseau, finding Hobbes', in G.P. McAlinn and C. Pejovic (eds.), *Law and Development in Asia*, Routledge, London and New York

Hardy, A. (1998) 'The economics of French rule in Indochina: a biography of Paul Bernard (1892–1960)', *Modern Asian Studies*, 32: 807–848

Hayek, F. (1944) *The Road to Serfdom*, Routledge, Abingdon

Hayek, F. (2011) *The Constitution of Liberty: The Definitive Edition*, Routledge, Abingdon

Heinrich Böll Stiftung, WWF and IISD (2008) *Rethinking Investments in Natural Resources: China's Emerging Role in the Mekong Region*, policy brief, http://d2ouvy59p0dg6k. cloudfront.net/downloads/china_study_executive_summary.pdf, accessed 7 May 2015

Hensengerth, O. (2008) 'Vietnam's security objectives in Mekong Basin governance', *Journal of Vietnamese Studies*, 3: 101–127

Herbertson, K. (2012) *The Xayaburi Dam: Threatening Food Security in the Mekong*, International Rivers, Berkeley, CA, www.internationalrivers.org/resources/the-xayaburi-dam-threatening-food-security-in-the-mekong-7675, accessed 6 July 2015

Herbertson, K. (2013) *Xayaburi Dam: How Laos Violated the 1995 Mekong Agreement*, International Rivers, Berkeley, CA, www.internationalrivers.org/blogs/267/xayaburi-dam-how-laos-violated-the-1995-mekong-agreement, accessed 6 July 2015

Hiang-Khng Heng, R. (2004) 'Civil society effectiveness and the Vietnamese state – despite or because of the lack of autonomy', in L.H. Guan (ed.), *Civil Society in Southeast Asia*, Institute of Southeast Asian Studies, Singapore, pp. 144–166

Higgins, R. (2003) *Problems and Process: International Law and How to Use it*, Clarendon Press, Oxford

Higgs, S. (2011) *Analysis of International Environmental Laws Implicated by Decision to Approve Construction of the Xayaburi Dam*, prepared by Perkins Coie LLP for International Rivers and Environmental Defenders Law Center, 12 October, www.internationalrivers.org/files/attached-files/xayaburi_legal_analysis_en.pdf, accessed 7 May 2015

Hill, H. and Menon, H. (2013) 'Cambodia: rapid growth with weak institutions', *Asian Economic Policy Review*, 8: 46–65

Hillgenberg, H. (1999) 'A fresh look at soft law', *European Journal of International Law*, 10(3): 499–515

Hirsch, P. (1991) 'Development dilemmas facing Laos', *Nation*, 24 April, p. C1

Hirsch, P. (1994) 'Where are the roots of Thai environmentalism?', *Thai Environmental Institute Quarterly Environmental Journal*, 2(2): 5–15

Hirsch, P. (1997) *Seeing Forests for Trees: Environment and Environmentalism in Thailand*, Silkworm Books, Chiang Mai, Thailand

Hirsch, P. (2000) 'Managing the Mekong commons: local, national and regional issues', in M. Ahmed and P. Hirsch (eds.), *Common Property in the Mekong: Issues of Sustainability and Subsistence*, World Fish Centre and Australian Mekong Resource Centre, Manila, pp. 19–26

Hirsch, P. (2001) 'Globalisation, regionalisation and local voices: the Asian Development Bank and rescaled politics of environment in the Mekong region', *Singapore Journal of Tropical Geography*, 22(3): 237–251

Hirsch, P. (2002) 'Global norms, local compliance and the human rights–environment nexus: a case study of the Nam Theun II dam in Laos', in L. Zarsky (ed.), *Human Rights and the Environment: Conflicts and Norms in a Globalizing World*, Earthscan, London, pp. 147–171

Hirsch, P. (2006a) 'The politics of fisheries knowledge in the Mekong River Basin', in R.L. Welcomme and T. Petr (eds.), *Fisheries Bethesda*, pp. 91–102, www.fao.org/docrep/007/ad526e/ad526e0a.htm, accessed 7 May 2015

Hirsch, P. (2006b) 'Water governance reform and catchment management in the Mekong region', *Journal of Environment & Development*, 15(2): 184–201

Hirsch, P. (2007a) 'Advocacy, civil society and the state in the Mekong region', in B. Rugendyke (ed.), *NGOs as Advocates for Development in a Globalizing World*, Routledge, London and New York, pp. 185–199

Hirsch, P. (2007b) 'Civil society and interdependencies: towards a regional political ecology of Mekong development', in J. Connell and E. Waddell (eds.), *Environment, Development and Change in Rural Asia-Pacific*, Routledge, London, pp. 226–246

Hirsch, P. (2008) '13 years of bad luck? A reflection on MRC and civil society in the Mekong', *Watershed*, 12(3): 38–43

Hirsch, P. (2011) 'Cascade effect', *China Dialogue*, www.chinadialogue.net/article/4093-Cascade-effect, accessed 7 May 2015

Hirsch, P. (2012) 'IWRM as a participatory governance framework for the Mekong River Basin?', in J. Öjendal, S. Hansson and S. Hellberg (eds.), *Politics and Development in a Transboundary Watershed: The Case of the Lower Mekong Basin*, Springer Verlag, London, pp. 155–170

Hirsch, P. (2015) 'The shifting geopolitics of the Mekong: a multi-dimensional approach', *Political Geography*, forthcoming

Hirsch, P. and Lohmann, L. (1989) 'Contemporary politics of environment in Thailand', *Asian Survey*, 29(4): 439–451

Hirsch, P. and Wyatt, A. (2004) 'Negotiating local livelihoods: scales of conflict in the Se San River Basin', *Asia Pacific Viewpoint*, 45(1): 51–68

Hirsch, P. and Jensen, K.M. with Boer, B., Carrard, N., FitzGerald, S. and Lyster, R. (2006) *National Interests and Transboundary Water Governance in the Mekong*, Australian Mekong Resource Centre, School of Geosciences, University of Sydney in collaboration with Danish International Development Assistance, http://sydney.edu.au/mekong/documents/mekwatgov_mainreport.pdf, accessed 28 April 2015

Hoang, D.H. (2012) 'Conflict between resettlement policy and customary regulations of indigenous people in natural resources management: Case study in A Luoi hydropower resettlement area, Thua Thien, Hue province, Vietnam', www.slideshare.net/CPWFMekong/presentation-hn-1-15291401, accessed 7 May 2015

Hooker, M.B. (1975) *Legal Pluralism: An Introduction to Colonial and Neo-colonial Laws*, Clarendon Press, Oxford

Hooper, B. (2005) *Integrated River Basin Governance: Learning from International Experiences*, IWA Publishing, London

Hortle, K. (2007) *Consumption and the Yield of Fish and Other Aquatic Animals from the Lower Mekong Basin*, MRC Technical Paper No 16, Vientiane, October, www.mrcmekong.org/assets/Publications/technical/tech-No16-consumption-n-yield-of-fish.pdf, accessed 7 May 2015

HSAP (n.d.) www.hydrosustainability.org/Protocol.aspx, accessed 7 May 2015

HSAP (2011) General Comment No 34 – Article 19: Freedoms of opinion and expression, CCPR/C/GC/34, 12 September, www2.ohchr.org/english/bodies/hrc/docs/gc34.pdf, accessed 7 May 2015

Hurwitz, J. (2014) *Dam Standards: A Rights-based Approach. A Guidebook for Civil Society*, International Rivers, Berkeley, CA, www.internationalrivers.org/files/attached-files/intlrivers_dam_standards_final.pdf, accessed 7 May 2015

Huxley, A. (1991) 'The Draft Constitution of the Laotian People's Democratic Republic', *Review of Socialist Law*, 17(1): 75–78

ICEM (2010) *Strategic Environmental Assessment of Hydropower on the Mekong Mainstream: Final Report*, prepared for the MRC, October, www.mrcmekong.org/assets/Publications/Consultations/SEA-Hydropower/SEA-Main-Final-Report.pdf, accessed 7 May 2015

ICEM (2013) *Impact Assessment Report: Ensuring Sustainability of GMS Regional Power Development*, prepared for the ADB

IFC (2012) 'Are Thai bankers ready for the Equator Principles? An interview with the secretary-general of the Thai Bankers Association', 4 April, www.ifc.org/wps/wcm/connect/region__ext_content/regions/east+asia+and+the+pacific/news/are+thai+banks+ready+for+the+equator+principles, accessed 7 May 2015

IFC (2015) 'IFC promotes sustainability of the hydropower sector in Lao PRD', www.ifc.org, accessed 16 April 2015

IFReDI (2013) *Food and Nutrition Security Vulnerability to Mainstream Hydropower Dam Development in Cambodia*, Inland Fisheries Research and Development Institute, Fisheries Administration, Ministry of Agriculture, Forestry and Fisheries, Cambodia, supported by Danida, WWF and Oxfam, June, www.oxfam.org.au/wp-content/uploads/2014/02/pdf_food-and-nutrition-for-print-2.pdf, accessed 7 May 2015

IHA (2010) *Hydropower Sustainability Assessment Protocol: Background Document,* November

IMF (2015) 'Uneven growth: short- and long-term factors, world economic outlook, April 2015', www.imf.org/external/pubs/ft/weo/2015/01, accessed 7 May 2015

International Rivers (2008a) 'Nam Theun 2 hydropower project: risky business in Laos', June

International Rivers (2008b) 'Power surge: the impacts of rapid dam development in Laos', September

International Rivers (2010) 'Nam Theun 2 hydropower project: the real cost of a controversial dam', December

International Rivers (2011) 'The Xayaburi dam: a looming threat to the Mekong River', www.internationalrivers.org/files/attached-files/the_xayaburi_dam_eng.pdf, accessed 7 May 2015

International Rivers (2013a) 'Lancang River dams: threatening the flow of the Lower Mekong', www.internationalrivers.org/files/attached-files/ir_lacang_dams_2013_5.pdf, accessed 7 May 2015

International Rivers (2013b) 'Xayaburi dam: timeline of events'

International Rivers (2014) 'The World Commission on Dams', www.internationalrivers.org/campaigns/the-world-commission-on-dams, accessed 7 May 2015

International Rivers Network (2002) 'Damming the Sesan River: impacts in Cambodia and Vietnam', Berkeley, CA, www.internationalrivers.org/files/attached-files/04.sesan.pdf, accessed 7 May 2015

International Service for Human Rights (2014) *The Situation of Human Rights Defenders in Lao PDR*, Universal Periodic Review Briefing Paper

International Waterpower and Dam Construction (2009) 'Developing A Luoi', *Waterpower Magazine*, 19 June, www.waterpowermagazine.com/features/featuredeveloping-a-luoi, accessed 7 May 2015

Islam, R. (2003) *Do More Transparent Governments Govern Better?*, World Bank Institute Policy Research Working Paper No 3077, World Bank, Washington, DC

IUCN (2008) *Statutory and Customary Forest Rights and Their Governance Implications: The Case of Vietnam*, Hanoi, 8 July

Ivarsson, S. (2008) *Creating Laos: The Making of a Lao Space between Indochina and Siam, 1862–1945*, NIAS Press, Copenhagen

Jacobs, H.M. (2013) 'Private property and human rights: a mismatch in the 21st century?', *International Journal of Social Welfare*, 22(S1): S85–S101

Jacobs, W.J. (2002) 'The Mekong River Commission: transboundary water resources planning and regional security', *Geographical Journal*, 168(4): 354–364

Jasanoff, S. and Kim, S.-H. (2013) 'Sociotechnical imaginaries and national energy policies', *Science as Culture*, 22(2): 189–196

Jayasankar, S. and Porter, I. (2011) *Doing a Dam Better: The Lao People's Democratic Republic and the Story of Nam Theun 2 (NT2)*, World Bank, https://openknowledge.worldbank.org/handle/10986/2540, accessed 7 May 2015

Jayasuriya, K. (ed.) (1999) *Law, Capitalism, and Power in Asia: The Rule of Law and Legal Institutions*, Routledge, London and New York

Jessup, P.C. (1956) *Transnational Law*, Yale University Press, New Haven, CT

Johns, F.E. (2015) 'On failing forward: neoliberal legality in the Mekong River Basin', *Cornell International Law Journal*, 48

Jones, K.A. (2010) *The Doha Blues: Institutional Crisis and Reform in the WTO*, Oxford University Press, Oxford

Käkönen, M. and Hirsch, P. (2009) 'The antipolitics of Mekong knowledge production', in F. Molle, T. Foran and M. Käkönen (eds.), *Contested Waterscapes in the Mekong Region: Hydropower, Livelihoods and Governance*, Earthscan, London, pp. 333–365

Kaosa-ard, M. and Dore, J. (eds) (2003) *Social Challenges for the Mekong Region*, Chiang Mai University, Thailand

Karki, S.K., Mann, M.D. and Salehfar, H. (2005) 'Energy and environment in the ASEAN: challenges and opportunities', *Energy Policy*, 33(4): 499–509

Kasian, T. (2006) 'Toppling Thaksin', *New Left Review*, 39: 5–37

Kate, D.T. (2012) 'Thai lawsuit threatens to derail Laos plans for Mekong River dam', *Bloomberg News*, 7 August

Kaufmann, D., Kraay, A. and Mastruzzi, M. (2003) *Governance Matters III: Governance Indicators for 1996–2002*, World Bank Policy Research Working Paper 3106, World Bank, Washington, DC

Ke, J. and Gao, Q. (2013) 'Only one Mekong: developing transboundary EIA procedures of Mekong River Basin', *Pace Environmental Law Review*, 30(3): 950–1278

Kelly, P.J. (2001) 'Classical utilitarianism and the concept of freedom: a response to the republican critique', *Journal of Political Ideologies*, 6(1): 13–31

Kennedy, D. (1997) 'New approaches to comparative law: comparativism and international governance', *Utah Law Review*, 2: 545–637

Kennedy, D. (2003) 'The methods and the politics of comparative law', in P. Legrand and R. Munday (eds.), *Comparative Legal Studies: Traditions and Transitions*, Cambridge University Press, Cambridge, pp. 345–433

Kennedy, D. (2004) *The Dark Sides of Virtue: Reassessing International Humanitarianism*, Princeton University Press, Princeton, NJ

Kennedy, D. (2006a) 'The "rule of law", political choices and development common sense', in D.M. Trubek and A. Santos (eds.), *The New Law and Economic Development*, Cambridge University Press, Cambridge, pp. 95–173

Kennedy, D. (2006b) 'Three globalizations of law and legal thought: 1850–2000', in D. Trubek and A. Santos (eds.), *The New Law and Economic Development: A Critical Appraisal*, Cambridge University Press, Cambridge, pp. 19–73

Kepa (2011) *Reflections on Vietnamese Civil Society*, Kepa

Kerkvliet, B., Nguyen Quang, A. and Sinh, B.T. (2008) *Forms of Engagement between State Agencies & Civil Society in Vietnam: Study Report*, Hanoi, December

Keskinen, M., Chinvanno, S., Kummu, M., Nuorteva, P., Snidvongs, A., Varis, O. and Västilä, K. (2010) 'Climate change and water resources in the Lower Mekong River Basin: putting adaptation into the context', *Journal of Water and Climate Change*, 1: 103–117

Keskinen, M., Kummu, M., Käkönen, M. and Varis, O. (2012) 'Mekong at the crossroads: Next steps for impact assessment of large dams', *AMBIO*, 41, 319–324

King, D. (2011) 'Xayaburi dam poses test for Mekong regional cooperation', blog posting, 7 October, www.earthrights.org/blog/xayaburi-dam-poses-test-mekong-regional-cooperation, accessed 7 May 2015

Kirton, J. and Trebilcock, M.J. (eds.) (2004) *Hard Choices, Soft Law: Voluntary Standards in Global Trade, Environment and Social Governance*, Ashgate, Burlington, VT

Kititasnasorchai, V. and Tasneeyanond, P. (2000) 'Thai environmental law', *Singapore Journal of International and Comparative Law*, 4: 1–35

Kitthananan, A. (2008) 'Developmental states and global neoliberalism', in P. Kennett (ed.), *Governance, Globalization and Public Policy*, Edward Elgar, Northampton, MA, pp. 77–106

Klaaren, J. (2013) 'The human right to information and transparency', in A. Bianchi and A. Peters (eds.), *Transparency in International Law*, Cambridge University Press, Cambridge, pp. 223–238

Klabbers, J. (2001) 'Institutional ambivalence by design: soft organisations in international law', *Nordic Journal of International Law*, 70: 403–421

Klabbers, J. (2006) 'Reflections on soft law in a privatized world', *Lakimies*, 104: 1191–1205

Knox, J. (2002) 'The myth and reality of transboundary environmental impact assessment', *American Journal of International Law*, 96(2): 291–319

Koh, H.H. (1996) 'Transnational legal process', *Nebraska Law Review*, 75: 181–207

Kopits, G. and Craig, J. (1998) *Transparency in Government Operations*, IMF Occasional Paper No 158, IMF, Washington, DC

Koskenniemi, M. (2001) *The Gentle Civilizer of Nations: The Rise and Fall of International Law 1870–1960*, Cambridge University Press, Cambridge

Koskenniemi, M. (2011) 'The case for comparative international law', *Finnish Yearbook of International Law*, 20: 1–8

Krever, T. (2011) 'The legal turn in late development theory: the rule of law and the World Bank's development model', *Harvard International Law Journal*, 52(1): 288–318

Krever, T. (2013) 'Quantifying law: legal indicator projects and the reproduction of neoliberal common sense', *Third World Quarterly*, 34(1): 131–150

Krygier, M. (2014) 'The rule of law after the short twentieth century: launching a global career', in R. Nobles and D. Schiff (eds.), *Law, Society and Community: Socio-Legal Essays in Honour of Roger Cotterrell*, Ashgate, Farnham, pp. 327–346

Kummu, M. and Sarkkula, J. (2008) 'Impact of the Mekong River flow alteration on the Tonle Sap flood pulse', *Ambio*, 37(3): 185–192, http://dx.doi.org/10.1579/0044-7447(2008)37[185:IOTMRF]2.0.CO;2, accessed 7 May 2015

Laking, R. (2008) *The Nam Theun 2 Hydro Project: A Better Kind of Dam?*, paper for the Annual Conference of the Aotearoa New Zealand International Development Studies Network, Victoria University of Wellington, New Zealand, 3–5 December, www.devnet.org.nz, accessed 7 May 2015

Lamberts, D. (2008) 'Little impact, much damage: the consequences of Meong River flow alterations for the Tonle Sap ecosystem', in M. Kummu, M. Keskinen and O. Varis (eds.), *Modern Myths of the Mekong*, Water & Development Publications, Helsinki University of Technology, Helsinki, pp. 3–18

Lao PDR (2000) Regulation on Environment Assessment in the Lao PDR (No. 1770/STEA, 10, 2000)

Lao PDR (2010) Decree on Environmental Impact Assessment, No. 112/PM, 16 February

Lao PDR (2012) *Environmental Impact Assessment Guidelines*, October, http://202.123.178.235/emsp/images/doc/eia_guidelines20121001.pdf, accessed 7 May 2015

Lao PDR Ministry of Justice and UNDP (2012) *Report on Customary Law Survey and Practice in Laos*

Lao Research Team MK8 (2013) *Research Report on Decision-making on Hydropower Development in Lao PDR*, National University of Laos, Third Mekong Forum on Water, Food & Energy

Laos Constitution (2003) Constitution of the Lao PDR, No. 25/NA, 6 May 2003, www.na.gov.la/index.php?option=com_content&view=category&id=35%3Aconstitution-of-lao-pdr&Itemid=186&layout=default&lang=en, accessed 7 May 2015

Latour, B. (1990) 'Technology is society made durable', *Sociological Review*, 38: 103–131

Latour, B. (2003) 'The promises of constructivism', in D. Ihde (ed.), *Chasing Technology: Matrix of Materiality*, Indiana University Press, Bloomington, IN, pp. 27–46

Lawrence, D.P. (2013) *Impact Assessment: Practical Solutions to Recurrent Problems and Contemporary Challenges*, 2nd edn, Wiley, Hoboken, NJ

Lawrence, M. and Logevall, F. (2007) *The First Indochina War: Colonial Conflict and Cold War Crisis*, Harvard University Press, Cambridge, MA

Lazarus, K., Badenoch, N., Dao, N. and Resurreccion, B. (eds.) (2011) *Water Rights and Social Justice in the Mekong Region*, Routledge, London

Lebel, L., Dore, J., Daniel, R. and Koma, Y.S. (2007) *Democratizing Water Governance in the Mekong Region*, USER and Mekong Press, Chiang Mai, Thailand

Lee, G. and Scurrah, N. (2009) *Power and Responsibility: The Mekong River Commission and Lower Mekong Mainstream Dams*, joint report of the Australian Mekong Resource Centre and Oxfam Australia, Sydney, http://sydney.edu.au/mekong/documents/power_and_responsibility_fullreport_2009.pdf, accessed 7 May 2015

Lerner, M. (2003) *Dangerous Waters: Violations of International Law and Hydropower Development along the Sesan River*, Oxfam America, Phnom Penh

Letter from Sesan communities to Suy Sem, Minister of Industry, Mines and Energy and Chief of Inter-Ministries Committee, 13 February 2014

Li, T.M. (2007) *The Will to Improve: Governmentality, Development, and the Practice of Politics*, Duke University Press, Durham, NC

Li, X. (2012) 'Hydropower in the Mekong River Basin: a balancing test', *Environmental Claims Journal*, 24: 51–69

Lobel, O. (2004) 'The renew deal: the fall of regulation and the rise of governance in contemporary legal thought', *Minnesota Law Review*, 89: 342–470

Lockhart, B.M. (2003) 'Narrating 1945 in Lao historiography', in C. Goscha and S. Ivarsson (eds), *Contesting Visions of the Lao Past: Lao Historiography at the Crossroads*, NIAS Press, Copenhagen, pp. 129–164

Lohmann, F. (1990) 'Remaking the Mekong', *Ecologist*, 20(2): 61–66

Long, N.V. (1973) *Before the Revolution: The Vietnamese Peasants under the French*, MIT Press, Cambridge, MA

Lui, G. and Calzaroni, C. (2011) 'Access to justice: Community Legal Education Center', in Hurights Osaka (ed.), *Human Rights Education in Asia-Pacific – Volume 2*, Asia-Pacific Human Rights Information Center, Osaka, pp. 23–42

Luong, H.V. (1996) 'The Mekong Delta: ecology, economy, and revolution, 1860–1960', *Pacific Affairs*, 69(2): 283

Mahiou, A. (2013) 'Development, international law of', in *Max Planck Encyclopedia of Public International Law*, http://opil.ouplaw.com/view/10.1093/law:epil/9780199231690/law-9780199231690-e1428?rskey=iJn7JD&result=1&prd=EPIL, accessed 7 May 2015

Maierbrugger, A. (2014) 'Laos exports up 6% in 2013', *Investvine*, http://investvine.com/laos-exports-up-6-in-2013, accessed 7 May 2015

Makim, A. (1997) *The Politics of Regional Cooperation on the Mekong 1957–1995*, PhD Thesis, Griffith University, Queensland, Australia

Makim, A. (2002) *The Changing Face of Mekong Resource Politics in the Post-Cold War Era: Re-negotiating Arrangements for Water Resource Management in the Lower Mekong River Basin (1991–1995)*, Working Paper No 6, Australian Mekong Resource Centre, University of Sydney

Manorom, K. (2007) 'People's EIA: a mechanism for grassroots participation in environmental decision-making', *Watershed*, 12(1): 26–30

Marks, J. (2001) 'Jean-Jacques Rousseau, Michael Sandel and the politics of transparency', *Polity*, 33(4): 619–642

Marong, A. (2010) 'Development, right to, international protection', in *Max Planck Encyclopedia of Public International Law*, http://opil.ouplaw.com/view/10.1093/law:epil/9780199231690/law-9780199231690-e931?rskey=tZ00ZG&result=2&prd=E PIL, accessed 7 May 2015

Marx, K. (1843) 'On the Jewish question', in R.C. Tucker (ed.), *The Marx–Engels Reader*, 2nd edn, W.W. Norton & Co, New York and London, pp. 26–52

Marx, K. (1875) 'Critique of the Gotha program', in R.C. Tucker (ed.), *The Marx–Engels Reader*, 2nd edn, W.W. Norton & Co, New York and London, pp. 525–541

Marx, K. (1926) 'After the revolution: Marx debates Bakunin', in R.C. Tucker (ed.), *The Marx–Engels Reader*, 2nd edn, W.W. Norton & Co, New York and London, pp. 542–549

Matthews, N. (2012) 'Water grabbing in the Mekong Basin – an analysis of the winners and losers of Thailand's hydropower development in Lao PDR', *Water Alternatives*, 5(2): 392–411

Matthews, N. and Geheb, K. (eds.) (2015) *Hydropower Development in the Mekong Region: Political, Socio-economic and Environmental Perspectives*, Routledge, London

McAlinn, P. and Pejovic, C. (eds.) (2012) *Law and Development in Asia*, Routledge, London

McCaffrey, S. (2007) *The Law of International Watercourses*, Oxford University Press, 2nd edn, Oxford University Press, Oxford

McCargo, D. (2005) 'Network monarchy and legitimacy crises in Thailand', *The Pacific Review*, 18(4): 499–519

McCormack, G. (2001) 'Water margins: competing paradigms in China', *Critical Asian Studies*, 33(1): 5–31

McGillivray, M., Carpenter, D. and Norup, S. (2012) *Evaluation Study of Long-term Development Co-operation between Laos and Sweden*, Sida

McPhail, K. and C. Callieri (1998) *Use of Public Consultation in the Nam Theun 2 Hydroelectric Project*, World Bank, Washington, DC

Mean, M. (2010) 'Damming the Lower Sesan River: a case study of the potential effects on a local community', in ERI, *I Want to Eat Fish. I Cannot Eat Electricity: Public Participation in Mekong Basin Development*, pp. 148–167

Mekong Agreement (1995) *Agreement on the Cooperation for the Sustainable Development of the Mekong River Basin*, 5 April

Mekong Secretariat (1970) *Indicative Basin Plan: Committee for Coordination and Investigation of the Lower Mekong Basin*, Bangkok

Mekong Secretariat (1988) *Perspectives for Mekong Development: Revised Indicative Plan (1987) for the Development of Land, Water and Related Resources of the Lower Mekong Basin – Summary Report*

Mekong Secretariat (1994) *Mekong Mainstream Run-of-river Hydropower*, study conducted by Mekong Secretariat in collaboration with Compagnie Nationale du Rhône and Acres International Ltd, Mekong Secretariat, Bangkok

Mekong Watch (2014) *Making a Community Resources Map in Kbal Romeas and Sraekor Villages in Stung Treng Province, Cambodia*, 16 October, www.mekongwatch.org/english/projects/MM_LS2.html, accessed 7 May 2015

Menon, J. and Melendez, A.C. (2011) *Trade and Investment in the Greater Mekong Subregion: Remaining Challenges and the Unfinished Policy Agenda*, ADB Working Paper Series on Regional Economic Integration No 78, ADB, www.nottingham.ac.uk/gep/documents/conferences/2011/malyasia-conf-january/jay-menon.pdf, accessed 7 May 2015

Mercer, C. (2002) 'NGOs, civil society and democratization: A critical review of the literature', *Progress in Development Studies*, 2(1): 5–22

Merme, V., Ahlers, R. and Gupta, J. (2014) 'Private equity, public affair: hydropower financing in the Mekong Basin', *Global Environmental Change*, 24: 20–29, http://dx.doi.org/10.1016/j.gloenvcha.2013.11.007, accessed 7 May 2015

Merry, S.E. (1992) 'Culture, power and the discourse of law', *New York Law School Law Review*, 37: 209–225

Merry, S.E. (2006) *Human Rights and Gender Violence: Translating International Law into Local Justice*, University of Chicago Press, Chicago

Merry, S.E. (2007) 'International law and sociolegal scholarship: toward a spatial global legal pluralism', *Studies in Law, Politics, and Society*, 41: 149–168

Merry, S.E. ((2013) 'McGill Convocation Address: legal pluralism in practice', *McGill Law Journal/Revue de droit de McGill*, 59: 1–8

Meyer, T. (2009) 'Soft law as delegation', *Fordham International Law Journal*, 32(3): 888–942

Middleton, C. (2007) 'New dam builders raid the Mekong', *World Rivers Review*, 22(2): 12–13

Middleton, C. (2008) *Cambodia's Hydropower Development and China's Involvement*, International Rivers and Rivers Coalition in Cambodia, Berkeley, CA and Phnom Penh, www.internationalrivers.org/files/attached-files/cambodia_hydropower_and_chinese_involvement_jan_2008.pdf, accessed 7 May 2015

Middleton, C. (2012) *Contestation, Cooperation and the Transborder Commons: The Hydropolitics of Mainstream Dams on the Mekong River*, International Conference on Sustainability & Rural Reconstruction, http://commons.ln.edu.hk/southsouthforum/2012/s4/3/, accessed 7 May 2015

Middleton, C. (2014) 'The politics of uncertainty: knowledge production, power and politics on the Mekong River', in *Proceedings of the International Conference on Development and Cooperation of the Mekong Region*, Seoul National University, 4–5 December, pp. 1–19

Middleton, C., Garcia, J. and Foran, T. (2009) 'Old and new hydropower players in the Mekong region: agendas and strategies', in F. Molle, T. Foran and M. Käkönen (eds.), *Contested Waterscapes in the Mekong Region: Hydropower, Livelihoods and Governance*, Earthscan, London, pp. 23–54

Middleton, C., Grundy-Warr, C. and Yong, M.L. (2013) 'Neoliberalizing hydropower in the Mekong Basin: The political economy of partial enclosure', *Social Science Journal*, 43: 299–334

Middleton, C., Matthews, N. and Mirumachi, N. (2015) 'Whose risky business? Public–private partnerships, build–operate–transfer and large hydropower dams in the Mekong region', in N. Matthews and K. Geheb (eds.), *Hydropower Development in the Mekong Region: Political, Socio-economic and Environmental Perspectives*, Routledge, London

Miles, K. (2013) *The Origins of International Investment Law: Empire, Environment and the Safeguarding of Capital*, Cambridge University Press, Cambridge

Miller, F. and Hirsch, P. (2002) 'Reflections on international experience of transferring river basin development models', in *River Symposium*, Brisbane, www.riverfestival.com.au/2002/content/papers2002Index.htm, accessed 7 May 2015

Miller, F. and Hirsch, P. (2003) *Civil Society and Internationalised River Basin Management*, Working Paper No 7, Australian Mekong Resource Centre, University of Sydney, June

Ministry of Energy and Mines, Lao PDR (n.d.) 'Xayabury hydropower project', http://poweringprogress.org/new/10-projects-under-construction/30-xayaboury-mekong-1285mw, accessed 27 March 2015

Ministry of Energy and Mines, Lao PDR (2014) 'Power sector', www.poweringprogress.org/new/power-sector, accessed 27 March 2015

Ministry of Industry and Trade, Vietnam (2007) *Result of Hydropower Dam Construction Survey Nationwide*, MOIT, Ha Noi

Ministry of Industry and Trade, Vietnam (2013) *Result of Hydropower Dam Construction Survey Nationwide*, MOIT, Ha Noi

Ministry of Industry and Trade, Vietnam (2014) 'Promoting cooperation in hydro power between Vietnam and Norway', 19 March, www.moit.gov.vn/en/News/570/promoting-cooperation-in-hydro-power-between-vietnam-and-norway.aspx, accessed 7 May 2015

Mirumachi, N. and Torriti, J. (2012) 'The use of public participation and economic appraisal for public involvement in large-scale hydropower projects: case study of the Nam Theun 2 hydropower project', *Energy Policy*, 47: 125–132

Missingham, B. (2002) 'The Village of the Poor confronts the state: a geography of protest in the Assembly of the Poor', *Urban Studies*, 39(9): 1647–1663

Missingham, B. (2003) 'Forging solidarity and identity in the Assembly of the Poor: from local struggles to a national social movement in Thailand', *Asian Studies Review*, 27(3): 317–340

Missingham, B. (2005) 'The Assembly of the Poor in Thailand: from local struggles to national protest movement', *Journal of Asian Studies*, 64(3): 800–802

Mitchell, A. and Souche, A. (2014) 'Exporting megawatts', *Bangkok Post*, 16 March

Mitchell, R.B. (2011) 'Transparency for governance: the mechanisms and effectiveness of disclosure-based and education-based transparency policies', *Ecological Economics*, 70: 1882–1890

Mitchell, R.K., Agle, B.R. and Wood, D.J. (1997) 'Toward a theory of stakeholder identification and salience: defining the principle of who and what really counts', *Academy of Management Review*, 22(4): 853–886

Mock, W.B.T. (1999–2000) 'An interdisciplinary introduction to legal transparency: a tool for rational development', *Dickinson Journal of International Law*, 18: 293–304

Molle, F. (2008) 'Nirvana concepts, storylines and policy models: insights from the water sector', *Water Alternatives*, 1(1): 131–156

Molle, F. and Floch, P. (2007) *Water, Poverty and the Governance of Megaprojects: The Thai 'Water Grid'*, M-POWER Working Paper, Unit for Social and Environmental Research, Chiang Mai University, Chiang Mai

Molle, F., Foran, T. and Käkönen, M. (2009) *Contested Waterscapes in the Mekong Region: Hydropower, Livelihoods and Governance*, Earthscan, London

Molle, F., Wester, P. and Hirsch, P. (2010) 'River basin closure: processes, implications and responses', *Agricultural Water Management*, 97(4): 569–577

Montini, M. (2013) 'Towards a new instrument for promoting sustainability beyond the EIA and the SEA: the holistic impact assessment', in C.Voigt (ed.), *Rule of Law for Nature: New Dimensions and Ideas in Environmental Law*, Cambridge University Press, Cambridge, pp. 243–258

Morris, R.C. (2000) 'Modernism's media and the end of mediumship? On the aesthetic economy of transparency in Thailand', *Public Culture*, 12(2): 457–475

Mostert, E. (2003) 'The challenge of public participation', *Water Policy*, 5: 179–197

MRC (n.d.) 'Basin Development Plan Programme', www.mrcmekong.org/about-mrc/programmes/basin-development-plan-programme, accessed 7 May 2015

MRC (2001) *Procedures for Data and Information Exchange and Sharing*, MRC, Phnom Penh, www.mrcmekong.org/assets/Publications/policies/Procedures-Data-Info-Exchange-n-Sharing.pdf, accessed 7 May 2015

MRC (2003a) *Procedures for Notification, Prior Consultation and Agreement under the Mekong Agreement*, MRC, Phnom Penh, www.mrcmekong.org/assets/Publications/policies/Procedures-Notification-Prior-Consultation-Agreement.pdf, accessed 7 May 2015

MRC (2003b) *Procedures for Water Use Monitoring*, MRC, Phnom Penh, www.mrcmekong.org/assets/Publications/policies/Procedures-Water-Use-Monitoring.pdf, accessed 7 May 2015

MRC (2005) *Overview of the Hydrology of the Mekong Basin*, MRC, Phnom Penh

MRC (2006) *Procedures for Maintenance of Flows on the Mainstream*, MRC, Vientiane, www.mrcmekong.org/assets/Publications/policies/Procedures-Maintenance-Flows.pdf, accessed 7 May 2015

MRC (2009) *The Flow of the Mekong*, MRC Management Information Booklet Series No 2,Vientiane

MRC (2011a) *Procedures for Water Quality*, MRC, Vientiane, www.mrcmekong.org/assets/Publications/policies/Procedures-for-Water-Quality-council-approved260111.pdf, accessed 7 May 2015

MRC (2011b) *IWRM-based Basin Development Strategy for the Lower Mekong Basin*, MRC, Vientiane

MRC (2011c) *Strategic Plan 2011–2015*, www.mrcmekong.org/assets/Publications/strategies-workprog/Stratigic-Plan-2011-2015-council-approved25012011-final-.pdf, accessed 30 October 2014

MRC (2012a) *Draft Technical Guidance for Conducting and Considering Transboundary Environmental Impact Assessment Process for Proposed Development Projects/Activities in Connection with the National EIA Process*

MRC (2012b) 'Mekong2Rio', www.mrcmekong.org/news-and-events/events/mekong2rio/?url=/mekong2rio, accessed 7 May 2015

MRC (2013) *Mekong Basin Planning: The Basin Development Plan Story*, www.mrcmekong.org/assets/Publications/basin-reports/BDP-Story-2013-small.pdf, accessed 7 May 2015

MRC (2014a) 'Decision support framework', MRC, Vientiane, http://archive-org.com/page/3481774/2014-01-06/http://ns1.mrcmekong.org/programmes/wup/DSF/DSF_introduction.htm, accessed 7 May 2015

MRC (2014b) 'Watershed management', www.mrcmekong.org/about-the-mrc/programmes/basin-development-plan-programme/watershed-management, accessed 7 May 2015

MRC (2015a) 'Don Sahong hydropower project, Vientiane, Lao PDR, 15th Oct 2014 – 31st Jan 2015', www.mrcmekong.org/news-and-events/consultations/don-sahong-hydropower-project, accessed 7 May 2015

MRC (2015b) 'Mekong integrated water resources management project', www.mrcmekong.org/about-mrc/programmes/mekong-integrated-water-resources-management-project, accessed 26 March 2015

MRC Council (2013) *Joint Development Partner Statement: 19th MRC Council Meeting*, 17 January, www.mrcmekong.org/news-and-events/speeches/joint-development-partner-statement-19th-mrc-council-meeting-17-january-2013/, accessed 30 October 2014

MRC Secretariat (2013) *The BDP Story – Mekong Basin Planning: The Story behind the Basin Development Plan*, www.mrcmekong.org/assets/Publications/basin-reports/BDP-Story-2013-small.pdf, accessed 7 May 2015

Mukhtarov, F. and Cherp, A. (2014) 'The hegemony of integrated water resources management as a global water discourse', in V.R. Squires, H.M. Milner and K.A. Daniell (eds.), *River Basin Management in the Twenty-first Century: Understanding People and Place*, CRC Press, Boca Raton, FL, pp. 3–21

Munger, F. (2007) 'Culture, power and law: thinking about the anthropology of rights in Thailand in an era of globalization', *New York Law School Law Review*, 51: 817–838

Myint, T. (2012) *Governing International Rivers: Polycentric Politics in the Mekong and the Rhine*, Edward Elgar, Cheltenham

Myrdal, G. (1974) 'What is development?', *Journal of Economic Issues*, 8(4): 729–736

Nam Theun 2 Power Company (2005) *Social Development Plan*, www.namtheun2.com/documents/project-documents.html, accessed 7 May 2015

NARBO (2013a) *NARBO Member Organizations List*, www.narbo.jp/whats/materials/memberlist_2013.5.pdf, accessed 27 January 2015

NARBO (2013b) *Network of Asian River Basin Associations Charter (NARBO): Charter (Revised)*, www.narbo.jp/whats/materials/charter2013.05.pdf, accessed 27 January 2015

National Consulting Company (2014) *Environmental Impact Assessment: Don Sahong Hydropower Project*, Vientiane, www.mrcmekong.org/assets/Other-Documents/Don-Sahong/DSHPP-EIA-FINAL.pdf, accessed 7 May 2015

Neal, A.N. (2009) 'Rethinking Foucault in international relations: promiscuity and unfaithfulness', *Global Society*, 23(4): 539–543

NeJaime, D. (2009) 'When new governance fails', *Ohio State Law Journal*, 70: 323–401

Ngan Anh (2014) Vietnam scraps a dozen planned hydropower projects, *Than Nien News*, 13 August, www.thanhniennews.com/business/vietnam-scraps-a-dozen-planned-hydropower-projects-29891.html, accessed 7 May 2015

NGO Forum on Cambodia (2005) *Down River: The Consequences of Vietnam's Se San River Dams on Life in Cambodia and Their Meaning in International Law*, December

NGOs' Joint Statement (2014) 'NGOs' Joint Statement on Concerns about *Procedures for Notification, Prior Consultation and Agreement (PNPCA)* for Don Sahong hydropower project', 12 November 2014, www.mrcmekong.org/assets/Other-Documents/stakeholder-submissions/Joint-Statement-NGOs-on-DSD-Consultation-Eng-fin al-141114.pdf, accessed 7 May 2015

Nguyen, A.T. (2012) *A Case Study on Power Sector Restructuring in Vietnam*, Pacific Energy Summit Papers, www.nbr.org, accessed 7 May 2015

Nguyen, T.D. (1999) *The Mekong River and the Struggle for Indochina: Water, War, and Peace*, Praeger, New York

Nguyen, T.L. (2013) *Statement by Deputy Minister of Natural Resources and Environment of Viet Nam to the 19th Meeting of the MRC Council*, 17 January, www.mrcmekong.org/news-and-events/speeches/statement-by-h-e-dr-nguyen-thai-lai-19th-meeting-of-the-mrc-council, accessed 30 October 2014

Nicholson, P. and Low, S. (2013) 'Local accounts of rule of law aid: implications for donors', *Hague Journal on the Rule of Law*, 5(1): 1–43

Nørlund, I. (2007a) 'Civil society in Vietnam: social organisations and approaches to new concepts', *ASIEN*, 105: 68–90

Nørlund, I. (2007b) *Filling the Gap: The Emerging Civil Society in Viet Nam*, UNDP/SVN, Ha Noi

North, D. C. (1991) 'Institutions', *Journal of Economic Perspectives*, 5(1): 97–112

Nou, L. (2008) 'The making of political mavericks and globalization: a quest for symbolic participatory democracy in Cambodia', *Santa Clara Journal of International Law*, 6(1): 161–204

Nurhidayah, L., Lipman, Z. and Alam, S. (2014) 'Regional environmental governance: an evaluation of the ASEAN legal framework for addressing transboundary haze pollution', *Australian Journal of Asian Law*, 15(1): 1–17, http://ssrn.com/abstract=2457727, accessed 7 May 2015

OED Online (2014) *Oxford English Dictionary*, Oxford University Press, Oxford, www.oed.com/view/Entry/246856?redirectedFrom=stakeholder, accessed 7 May 2015

Office of Natural Resources and Environmental Policy and Planning (2012) *Environmental Impact Assessment in Thailand*

Öjendal, J., Hansson, S. and Hellberg, S. (eds.) (2011) *Politics and Development in a Transboundary Watershed: The Case of the Lower Mekong Basin*, Springer, London

Open Development, 2015, Cambodia, Environmental Law and Environmental Impact Assessments, http://www.opendevelopmentcambodia.net/briefing/eia/

Orr, S., Pittock, J., Chapagain, A. and Dumaresq, D. (2012) 'Dams on the Mekong River: lost fish protein and the implications for land and water resources', *Global Environmental Change*, 22: 925–932

Osborne, M. (1975) *River Road to China: The Mekong River Expedition 1866–1873*, Liveright, New York

Osborne, M. ([1975] 1997) *River Road to China: The Search for the Source of the Mekong 1866–73*, Allen & Unwin, Sydney

Osborne, M. (2000) *The Mekong: Turbulent Past, Uncertain Future*, Allen & Unwin, Sydney

Osborne, M. (2004) *River at Risk: The Mekong and the Water Politics of China and Southeast Asia*, Lowy Institute Paper 02, Sydney

Osborne, M. (2009) *The Mekong: River under Threat*, Lowy Institute, Sydney

Otomo, Y. and Eslava, L. (2010) *Air-conditioned Nation: The Magic of International Law or, the Difficulty of Living in a Cyrogenic Age*, South of International Law Workshop, Melbourne Law School, Melbourne, 8–9 July,

Ounstead, M. (2010) 'Taking a river basin approach: the Oxfam Mekong Initiative', in L. Roper (ed.), *Change Not Charity: Essays on Oxfam America's First 40 Years*, Oxfam America, pp. 393–406

Oxfam Australia (2001) *Breaking the Banks: The Impact of the Asian Development Bank and Australia's Role in the Mekong Region*, Melbourne

Oxfam International (2012) 'Lao communities' land and natural resources are not for sale', www.rightsandresources.org/news/oxfam-lao-communities-land-and-natural-resourc es-are-not-for-sale, accessed 7 May 2015

Palacio, A. (2006) *The Way Forward: Human Rights and the World Bank*, World Bank Institute, http://go.worldbank.org/RR8FOU4RG0, accessed 7 May 2015

PanNature (2010) *Small and Medium Hydropower: Great Impacts on Environment but Less Supervision of Environmental Impact Assessment*

Parker, C. (2008) 'The pluralization of regulation', *Theoretical Inquiries in Law*, 9: 349–369

Pathmanand, U. (2008) 'A different coup d'état?', *Journal of Contemporary Asia*, 38(1): 124–142

Pearce, F. (2012) *The Land Grabbers: The New Fight over Who Owns the Earth*, Beacon Press, Boston, MA

Peck, J. (2010) *Constructions of Neoliberal Reason*, Oxford University Press, New York

Peck, J. (2013) 'Explaining (with) neoliberalism', *Territory, Politics, Governance*, 1(2): 132–157

Peou, S. (2011) 'The challenge for human rights in Cambodia', in T. Davis and B. Galligan (eds.), *Human Rights in Asia*, Edward Elgar, Cheltenham, pp. 123–143

Pettit, P. (1997) *Republicanism: A Theory of Freedom and Government*, Oxford University Press, Oxford

Pham, H.T. (2015) *Dilemmas of Hydropower Development in Vietnam: Between Dam-induced Displacement and Sustainable Development*, Eburon, Delft

Pham, H.T., Van Westen, A.C.M. and Zoomers, A. (2013) 'Compensation and resettlement policies after compulsory land acquisition for hydropower development in Vietnam: policy and practice', *Land*, 2: 678–704

Phongpaichit, P. and Baker, C. (2009) *History of Thailand*, 2nd edn, Cambridge University Press, Cambridge

Phuoc Bu (2013) 'Locals duped by power plant promises', *Vietnam News*, 13 October, http://vietnamnews.vn/economy/245755/locals-duped-by-power-plant-promises.html, accessed 7 May 2015

Piman, T., Cochrane, T.A., Arias, M.E., Green, A. and Dat, N.D. (2013) 'Assessment of flow changes from hydropower development and operations in Sekong, Sesan, and Srepok Rivers of the Mekong Basin', *Journal of Water Resources Planning and Management*, 139: 723–732

Pintoptaeng, P. (1998) *Kan muang bon thong thanon: 99 wan samatcha khon chon*, Krirk University, Bangkok

Pirie, F. (2013) *The Anthropology of Law*, Oxford University Press, Oxford

Plengsaeng, B., Wehn, U. and van der Zaag, P. (2014) 'Data-sharing bottlenecks in transboundary integrated water resources management: a case study of the Mekong River Commission's procedures for data sharing in the Thai context', *Water International*, 39: 933–951

Poch, K. and Tuy, S. (2012) 'Cambodia's electricity sector in the context of regional electricity market integration', in Y. Wu, X. Shi and F. Kimura (eds.), *Energy Market Integration in East Asia: Theories, Electricity Sector and Subsidies*, ERIA Research Project Report 2011–17,

ERIA, Jakarta, pp. 141–172, www.eria.org/Chapter%207-Cambodia's%20Electricity%20 Sector%20in%20the%20Context%20of%20Regional%20Electricity%20Market%20 Integration.pdf, accessed 7 May 2015

POE (2007) *Name Theun 2 Multipurpose Project: Twelfth Report*, International Social and Environmental Panel of Experts, www.namtheun2.com/images/stories/poe/poe12.pdf, accessed 7 May 2015

Pogge, T.W. (2002) *World Poverty and Human Rights: Cosmopolitan Responsibilities and Reforms*, Polity, Cambridge

Polkuamdee, N. (2015) 'Fish fix adds B10bn to Xayaburi cost', *Bangkok Post*, 3 February, www.bangkokpost.com/business/news/464681/fish-fix-adds-b10bn-to-xayaburi-cost, accessed 7 May 2015

Popkin, S.L. (1976) 'Corporatism and colonialism: the political economy of rural change in Vietnam', *Comparative Politics*, 8(3): 431–464

Porter, I.C. and Shivakumar, J. (eds.) (2011) *Doing a Dam Better: The Lao People's Democratic Republic and the Story of Nam Theun 2*, World Bank, Washington, DC

Pound, R. (1910) 'Law in books and law in action', *American Law Review*, 44: 12–36

Power-technology.com (n.d.) 'A Luoi Hydropower Plant, Thua Thien Hue, Vietnam', www. power-technology.com/projects/luoihydropower, accessed 27 March 2015

Pronto, A.N. (2008) 'Some thoughts on the making of international law', *European Journal of International Law*, 19(3): 601–616

Radio Free Asia (2012) 'Thai villagers sue over dam', 7 August

Radio Free Asia (2014) 'Cambodia's parliament passes laws "threatening judicial independence"', 23 May, www.rfa.org/english/news/cambodia/laws-05232014201001. html, accessed 7 May 2015

Rajagopal, B. (2003) *International Law from Below: Development, Social Movements and Third World Resistance*, Cambridge University Press, New York

Rajagopal, B. (2005) 'The role of law in counter-hegemonic globalization and global legal pluralism: lessons from the Narmada Valley struggle in India', *Leiden Journal of International Law*, 18: 345–387

Rajesh, D., Lebel, L. and Manorom, K. (eds.) (2013) *Governing the Mekong: Engaging in the Politics of Knowledge*, Strategic Information and Research Development Centre, Petaling Jaya, Malaysia

Razzaque, J. (2013) *Environmental Governance in Europe and Asia: A Comparative Study of Institutional and Legislative Frameworks*, Routledge, London

Relly, J.E. and Sabharwal, M. (2009) 'Perceptions of transparency of government policymaking: a cross-national study', *Government Information Quarterly*, 26(1): 148–157

Reynolds, C.J. and Lysa, H. (1983) 'Marxism in Thai historical studies', *Journal of Asian Studies*, 43(1): 77–104

Richter, B.D., Postel, S., Revenga, C., Scudder, T., Lehner, B., Churchill, A. and Chow, M. (2010) 'Lost in development's shadow: the downstream human consequences of dams', *Water Alternatives*, 3(2): 14–42

Rieu-Clarke, A. (2010) 'Good governance and IWRM – a legal perspective', *Irrigation & Drainage Systems*, 24(3–4): 239–248

Rieu-Clarke, A. (2014) 'Notification and consultation procedures under the Mekong Agreement: insights from the Xayaburi controversy', *Asian Journal of International Law*, 1–33

Rieu-Clarke, A. and Gooch, G. (2010) 'Governing the tributaries of the Mekong: the contribution of international law and institutions to enhancing equitable cooperation over the Sesan', *Pacific McGeorge Global Business & Development Law Journal*, 22(2): 193–224

Rigg, J. (1991) 'Thailand's Nam Choan project: a case study in the "greening" of South-East Asia', *Global Ecology and Biogeography Letters*, 1(2): 42–54

Riska, G. (1999a) *NGOS in the GMS: Involvement Related to Poverty Alleviation and Watershed Management – Thailand*

Riska, G. (1999b) *NGOS in the GMS: Involvement Related to Poverty Alleviation and Watershed Management – Viet Nam*

Riska, G. (2011) *NGOs in the GMS: Involvement Related to Poverty Alleviation and Watershed Management – Cambodia*

Rittich, K. (2006) 'The future of law and development: second generation reforms and the incorporation of the social', in D.M. Trubek and A. Santos (eds.), *The New Law and Economic Development: A Critical Appraisal*, Cambridge University Press, Cambridge, pp. 203–252

Rivers Coalition in Cambodia (2008) *Alleviating Dam Impacts along the Transboundary Se San River in Northeastern Cambodia: A Review of the Rapid Environmental Impact Assessment on the Cambodian Part of the Se San River Due to Hydropower Development in Vietnam (July 2007 version)*, February

Robichau, R.W. (2011) 'The mosaic of governance: creating a picture with definitions, theories, and debates', *Policy Studies Journal*, 39: 113–131

Robison, R. (2009) 'Strange bedfellows: political alliances in the making of neo-liberal governance', in W. Hunt and R. Robison (eds.), *Governance and the Depoliticisation of Development*, Routledge, London, pp. 15–28

Robinson, M. (2005) 'What rights can add to good development practice', in P. Alston and M. Robinson (eds.), *Human Rights and Development: Towards Mutual Reinforcement*, Oxford University Press, New York, pp. 25–41

Rose, C.V. (1998) 'The "new" law and development movement in the post-Cold War era: a Vietnam case study, *Law & Society Review*, 32(1): 93–140

Rose, N. (1996) 'The death of the social? Re-figuring the territory of government', *Economy and Society*, 25: 327–356

Rousseau, J. (1997) *Rousseau: 'The Discourses' and Other Early Political Writings*, trans. V. Gourevitch, Cambridge University Press, Cambridge

Salman, S.M.A. (2007) 'The Helsinki Rules, the UN Watercourses Convention and the Berlin Rules: Perspectives on international water law', *International Journal of Water Resources Development*, 23(4): 625–640

Salvioni, D.M. (2002) 'Transparency culture and financial communication', *Symphony A: Emerging Issues in Management*, 2, www.unimib.it/upload/gestioneFiles/Symphonya/f2002issue2/salvionieng22002.pdf, accessed 7 May 2015

Samudavanija, C. (1989) 'Thailand: a stable semi-democracy', in L. Diamond, J.J. Linz and S.M. Lipset (eds.), *Democracy in Developing Countries*, Vol. 3, Adamantine Press, London

Sarat, A. (2014) 'From movement to mentality, from paradigm to perspective, from action to performance: law and society at mid-life', *Law and Social Inquiry*, 39: 217–225

Sarfaty, G. A. (2005) 'The World Bank and the internalization of indigenous rights norms', *Yale Law Journal*, 114: 1791–1818

Sarkkula, J., Keskinen, M., Koponen, J., Kummu, M., Nikula, J., Varis, O. and Virtanen, M. (2007) 'Mathematical modeling in integrated management of water resources: Magical tool, mathematical toy or something in between?', in L. Lebel, J. Dore, R. Daniel and Y.S. Koma (eds.), *Democratizing Water Governance in the Mekong Region*, Mekong Press, Chiang Mai, pp. 127–156

Sarkkula, J., Koponen, J., Lauri, H. and Virtanen, M. (2010) *Origin, Fate and Impacts of the Mekong Sediments*, Vientiane

Save the Mekong (2014a) Letter to the heads of government of Cambodia, Lao PDR, Vietnam, Thailand and China, 'Urgent request to reconsider the Lower Sesan 2 dam in Cambodia', 11 February, www.savethemekong.org, accessed 7 May 2015

Save the Mekong (2014b) 'Mekong mainstream dams are a major transboundary threat to the region's food security and people: civil society calls upon prime ministers to cancel mainstream dams', 25 June, www.savethemekong.org, accessed 7 May 2015

Schaaf, C.H. and Fifield, R.H. (1963) *The Lower Mekong: Challenge to Cooperation in Southeast Asia*, D.Van Nostrand, Princeton, NJ

Sen, A. (1999) *Development as Freedom*, Alfred A. Knopf, New York

Simon, J. (2000) 'Law after society', *Law and Social Inquiry*, 24: 143–194

Simon, W.H. (2004–2005) 'Solving problems vs. claiming rights: the pragmatist challenge to legal liberalism', *William and Mary Law Review*, 46: 127–212

Simorn, M. (2008) *Socio-economic Report for Lower Se San 2 Hydropower Plant*, April, Annex and Appendix of EIA Report for Lower Sesan 2 Hydropower Project

Sin, N. (2013) 'Statement by H.E. Mr. Sin Niny, Permanent Vice-Chairman of Cambodia National Mekong Committee, Alternate Member of the MRC Council for Kingdom of Cambodia, Head of Delegation of Cambodia' to Joint 19th Council and 17th Donor Consultative Group Meeting, MRC, 17 January, www.mrcmekong.org/news-and-events/speeches/statement-by-h-e-mr-sin-niny-19th-meeting-of-the-mrc-council, accessed 30 October 2014

Singh, S. (2009) 'World Bank-directed development? Negotiating participation in the Nam Theun 2 hydropower project in Laos', *Development and Change*, 40: 487–507

Siphana, S. (2005) *Lessons from Cambodia's Entry into the World Trade Organization*, ADB Institute, Tokyo

Smith, H. (2008) 'Governing water: the semicommons of fluid property rights', *Arizona Law Review*, 50: 445–478

Smits, M. and Bush, S.R. (2010) 'A light left in the dark: the practice and politics of pico-hydropower in the Lao PDR', *Energy Policy*, 38: 116–127

Sneddon, C. (2012) 'The "sinew of development": Cold War geopolitics, technical expertise, and water resource development in Southeast Asia 1954–1975', *Social Studies of Science*, 42(4): 564–590

Sneddon, C. and Fox, C. (2006) 'Rethinking transboundary waters: a critical hydropolitics of the Mekong Basin', *Political Geography*, 25(2): 181–202

Sneddon, C. and Fox, C. (2007) 'Power, development, and institutional change: participatory governance in the Lower Mekong Basin', *World Development*, 35(12): 2161–2181

Sneddon, C. and Fox, C. (2011) 'The Cold War, the US Bureau of Reclamation, and the technopolitics of river basin development, 1950–1970', *Political Geography*, 30(8): 450–460

Sneddon, C. and Fox, C. (2012) 'Water, geopolitics, and economic development in the conceptualization of a region', *Eurasian Geography and Economics*, 53(1) 143–160

Sokheng, V. and Kunmakara, M (2012) 'Hun Sen bans industry fishing at Tonle Sap lake permanently', *Phnom Penh Post*, 29 February

Sombath.org (n.d.) 'Sombath Somphone', www.sombath.org, accessed 27 March 2015

Somek, A. (2015) 'Authoritarian liberalism', *European Law Journal*, http://dx.doi.org/10.1111/eulj.12132, accessed 7 May 2015

Souksavath, B. and Nakayama, M. (2013) 'Reconstruction of the livelihood of resettlers from the Nam Theun 2 hydropower project in Laos', *International Journal of Water Resources Development*, vol 29, no 1, pp. 71–86

Soutar, L. (2007) 'Asian Development Bank: NGO encounters and the Theun-Hinboun dam, Laos', in B. Rugendyke (ed.), *NGOs as Advocates for Development in a Globalizing World*, Routledge, London and New York, pp. 200–221

Sparkes, S. (2014) 'Corporate social responsibility: benefits for youth in hydropower development in Laos', *International Review of Education*, 60: 261–277

Springer, S. (2011) 'Articulated neoliberalism: the specificity of patronage, kleptocracy, and violence in Cambodia's neoliberalization', *Environment and Planning A*, 43(11): 2254–2570

Starobinski, J. (1988) *Jean-Jacques Rousseau: Transparency and Obstruction*, trans. A. Goldhammer, University of Chicago Press, Chicago, IL

Stiglitz, J. (1999) 'More instruments and broader goals: moving toward the post-Washington Consensus', *Revista de Economia Política*, 19(1): 94–120

Stiglitz, J. (2002) 'Transparency in government', in R. Islam (ed.), *The Right to Tell: The Role of Mass Media in Economic Development*, World Bank, Washington, DC, pp. 27–44

St John, O.B.C., Lovett, B., Euan, C.B. and Goldsmid, F.J. (1876) *Eastern Persia, an Account of the Journeys of the Persian Boundary Commission, 1870–71–72*, Vol. 1, Macmillan and Co, London

Stone, R. (2011) 'Mayhem on the Mekong', *Science*, 333: 814–818

Stuart-Fox, M. (2003) 'Historiography, power and identity: history and political legitimization in Laos', in C. Goscha and S. Ivarsson (eds.), *Contesting Visions of the Lao Past: Lao Historiography at the Crossroads*, NIAS Press, Copenhagen, pp. 73–95

Subedi, S.P. (2010) *Report of the Special Rapporteur on the Situation of Human Rights in Cambodia*, 16 September, UN General Assembly Doc. A/HRC/15/46

Subedi, S.P. (2012) *Report of the Special Rapporteur on the Situation of Human Rights in Cambodia: A Human Rights Analysis of Economic and Other Land Concessions in Cambodia*, 24 September, UN General Assembly Doc. A/HRC/21/63/Add.1

Subedi, S.P. (2013) *Report of the Special Rapporteur on the Situation of Human Rights in Cambodia*, 5 August, UN General Assembly Doc. A/HRC/24/36, http://ap.ohchr.org/documents/dpage_e.aspx?si=A/HRC/24/36, accessed 7 May 2015

Suhardiman, D. and Giordano, M. (2012) 'Process-focused analysis in transboundary water governance research', *International Environmental Agreements*, 12: 299–308

Suhardiman, D. and Giordano, M. (2014) 'Legal plurality: an analysis of power interplay in Mekong hydropower', *Annals of the Association of American Geographers*, 104(5): 973–988

Suhardiman, D., de Silva, S. and Carew-Reid, J. (2011) *Policy Review and Institutional Analysis of the Hydropower Sector in Lao PDR, Cambodia, and Vietnam: Final Report*, IWMI/ICEM/CIGIAR Challenge Programme on Water and Food, May

Suhardiman, D., Giordano, M. and Molle, F. (2012) 'Scalar disconnect: the logic of transboundary water governance in the Mekong', *Society & Natural Resources*, 25(6): 572–586

Suhardiman, D., Giordano, M. and Molle, F. (2015) 'Between interests and worldviews: the narrow path of the Mekong River Commission', *Environment and Planning C: Government and Policy*, 33(1): 199–217

Tai Baan Research Team (2004) *The Return of Fish, River Ecology and Local Livelihoods of the Mun River: A Thai Baan (Villagers') Research*, Southeast Asia Rivers Network, www.livingriversiam.org/3river-thai/pm/tb_research/pm-fish-book-eng.pdf, accessed 7 May 2015

Tamanaha, B.Z. (1993) 'The folly of the "social scientific" concept of legal pluralism', *Journal of Law and Society*, 20(2): 192–217

Tangwisutijit, N. (2006) 'Royal decrees revoked: court ends privatisation of Egat', *Nation*, 29 March

Taylor, W., Nguyen, T.H., Pham, Q.T. and Huynh, T.N.T. (2012) *Civil Society in Vietnam: A Comparative Study of Civil Society Organisations in Hanoi and Ho Chi Minh City*, Asia Foundation, Hanoi, October

TEAM Consulting Engineering and Management (2010) *Environmental Impact Assessment: Xayaburi Hydroelectric Power Project, Lao PDR – Final Report*, Bangkok, August, www.mrcmekong.org/assets/Consultations/2010-Xayaburi/Xayaburi-EIA-August-2010.pdf, accessed 7 May 2015

Teclaff, L.A. (1996) 'Evolution of the river basin concept in national and international water law', *Natural Resources Journal*, 36: 359–392

TERRA (1995) 'Mekong politics: "new era", same old plans', *Watershed*, 1: 24–29

TERRA (1998) 'From the Mekong to the Chao Phraya: the Kok-Ing-Nan water diversion project', *Watershed: Water – a Source of Life or a Resource for Development?*, 4(2): 10–24

Teubner, G. (1997) 'Global Bukowina: legal pluralism in the world society', in G. Teubner (ed.), *Global Law without a State*, Dartmouth, Brookfield, VT, pp. 3–28

Thabchumpon, N. and Middleton, C. (2012) 'Thai foreign direct investment and human security implications: a case study of the Xayaburi dam in Lao PDR', *Asian Review*, 25: 91–117

Thai NHRC (2012) *Opinion on Its Human Rights Investigation into the Xayaburi Project*, 4 May

Thai Office of Natural Resources and Environmental Policy and Planning (2012) *Environmental Impact Assessment in Thailand*, May, www.onep.go.th/EIA/images/7handbook/Environmental_Impact_Assessment_in_Thailand.pdf, accessed 7 May 2015

Thomas, C. (1999) 'Does the good governance policy of the international financial institutions privilege markets at the expense of democracy', *Connecticut Journal of International Law*, 14: 551–562

Thomas, C. (2011) 'Law and neoclassical economic development in theory and practice: toward an institutionalist critique of institutionalism', *Cornell Law Review*, 96: 967–1024

Tolentino, A.S. (2014) 'Sovereignty over natural resources – change of concept or change of perception?', *Environmental Policy and Law*, 44: 300–306

Toope, S. J. (2007) 'Formality and informality', in D. Bodansky, J. Brunnée and E. Hay (eds.), *The Oxford Handbook of International Environmental Law*, Oxford University Press, Oxford, pp. 107–124

Trainer, T. (2000) 'What does development mean? A rejection of the unidimensional conception', *International Journal of Sociology and Social Policy*, 20(5/6): 95–113

Tran, V.H. (2011) 'Local people's participation in involuntary resettlement in Vietnam: a case study of the Son La hydropower project', in K. Lazarus, N. Badenock, N. Dao and B.P. Resurreccion (eds.), *Water Rights and Social Justice in the Mekong Region*, Earthscan, London

Trandem, A. (2008) 'A Vietnamese/Cambodian transboundary dialogue: impacts of dams on the Se San River', *Development*, 51(1): 108–113

Trandem, A. (2011) 'Fatally flawed Xayaburi EIA fails to uphold international standards: A preliminary review of the environmental impact assessment (EIA) report for the Xayaburi hydropower dam on the Mekong River mainstream in Northern Lao PDR', www.banktrack.org/manage/ems_files/download/fatally_flawed_xayaburi_eia_fails_to_uphold_international_standards/110707_preliminary_review_of_xayaburi_eia_14_03_11_final.pdf, accessed 7 May 2015

Trebilcock, M.J. and Prado, M.M. (2011) *What Makes Poor Countries Poor? Institutional Determinants of Development*, Edward Elgar, Cheltenham

Triantafillou, P. and Nielsen, M.R. (2001) 'Policing empowerment: the making of capable subjects', *History of the Human Sciences*, 14(2): 63–86

Trubek, D.M. (1984) 'Where the action is: critical legal studies and empiricism', *Stanford Law Review*, 36: 575–622

Trubek, D.M. (2001) 'Law and development', in N.J. Smelser and P.B. Baltes (eds.), *International Encyclopedia of the Social and Behavioral Sciences*, Pergamon, Oxford

Trubek, D.M. (2006) 'The "rule of law" in development assistance: past, present, and future', in D.M. Trubek and A. Santos (eds.), *The New Law and Economic Development: A Critical Appraisal*, Cambridge University Press, Cambridge, pp. 74–94

Trubek, D.M. and Galanter, M. (1974) 'Scholars in self-estrangement: some reflections on the crisis in law and development studies in the United States', *Wisconsin Law Review*, 4: 1062–1102

Trubek, D.M. and Santos, A. (eds.) (2006) *The New Law and Economic Development: A Critical Appraisal*, Cambridge University Press, Cambridge

Trubek, D.M. and Trubek, L.G. (2005) 'Hard and soft law in the construction of social Europe: the role of the open method of co-ordination', *European Law Journal*, 11: 343–364

Un, K. (2006) 'State, society and democratic consolidation: the case of Cambodia', *Pacific Affairs*, 79(2): 225–245

UN Working Group on Enforced or Involuntary Disappearances (2013) 'A year on, the enforced disappearance of Sombath Somphone continues with impunity in Lao PDR', 16 December, www.ohchr.org, accessed 7 May 2015

UNDP (2013) *Analysis of Environmental and Social Costs and Risks of Hydropower Dams, with a Case Study of Song Tranh 2 Hydropower Plant*, UNDP, Vietnam www.vn.undp.org/content/dam/vietnam/docs/Publications/Hydropower_Research_final%20Oct%202013.pdf, accessed 7 May 2015

UNDP Cambodia (2007) *A Case Study of Indigenous Traditional Legal Systems and Conflict Resolution in Rattanakiri and Mondulkiri Provinces, Cambodia: Executive Summary*, Ministry of Justice, Ministry of Interior and UNDP Cambodia, Phnom Penh

UNESCAP (2009) 'What is good governance?', www.unescap.org/resources/what-good-governance, accessed 7 May 2015

UNESCAP (2012) 'Transparent and accountable public–private partnerships enable inclusive and sustainable growth, ESCAP tells Regional Ministerial Conference', www.unescap.org/news/transparent-and-accountable-public-private-partnerships-enable-inclusive-and-sustainable-growth, accessed 7 May 2015

United Nations (2002) *Johannesburg Declaration on Sustainable Development*, www.un-documents.net/jburgdec.htm, accessed 7 May 2015

United Nations (2008) *Draft Articles on the Law of Transboundary Aquifers*, Official Records of the General Assembly, Sixty-third Session, Supplement No 10 (A/63/10), New York

United Nations (2012) *The Future We Want*, resolution adopted by the General Assembly on 27 July 2012, A/RES/66/288, http://daccess-dds-ny.un.org/doc/UNDOC/GEN/N11/476/10/PDF/N1147610.pdf?OpenElement, accessed 7 May 2015

United Nations Department of Economic and Social Affairs (2005) *Lao People's Democratic Republic Public Administration Country Profile*, http://unpan1.un.org/intradoc/groups/public/documents/un/unpan023237.pdf, accessed 7 May 2015

United States Bureau of Reclamation (1956) *Reconnaissance Report: Lower Mekong Basin, Prepared for the International Cooperation Administration*

UNOHCHR (2010) 'Cambodia: "Tremendous challenges in delivering justice for all" says UN human rights expert', www.ohchr.org/EN/NewsEvents/Pages/DisplayNews.aspx?NewsID=10162&LangID=E, accessed 7 May 2015

UNOHCHR (2012) 'The role and achievements of the Office of the United Nations High Commissioner for Human Rights in assisting the Government and people of Cambodia in the promotion and protection of human rights', Report of the Secretary General to the UN Human Rights Council, A/HRC/21/35, 20 September

Urban, F., Nordensvärd, J., Khatri, D. and Wang, Yu (2013) 'An analysis of China's investment in the hydropower sector in the Greater Mekong sub-region', *Environment, Development and Sustainability*, 15(2): 301–324

Vaidyanathan, G. (2011) 'Dam controversy: remaking the Mekong', *Nature*, 19 October

Van Duyen, N. (2001) 'The inadequacy of environmental protection mechanisms in the Mekong River Basin Agreement', *Asia Pacific Journal of Environmental Law*, 6(3–4): 349–376

Vanclay, F. (2003) 'International principles for social impact assessment', *Impact Assessment and Project Appraisal*, 21(1): 5–11

Varis, O., Tortajada, C. and Biswas, A. (eds.) (2008) *Management of Transboundary Rivers and Lakes*, Springer, Berlin

Vietnam (2010) Decision No. 34/2010/QD-TTg dated April 08, 2010 of the Prime Minister promulgating the Regulation on irrigation and hydropower project-related compensation, support and resettlement, http://faolex.fao.org/docs/pdf/vie98703.pdf, accessed 7 May 2015

Vietnam (2015) Decree on Environmental Protection Planning, Strategic Environmental Assessment, Environmental Impact Assessment and Environmental Protection Plans, 14 February, http://binhdinh.eregulations.org/media/18_2015_ND-CP_268489.pdf, accessed 7 May 2015

Vietnam Constitution (2013) Constitution of the Socialist Republic of Vietnam, 28 November 2013, http://vietnamnews.vn/politics-laws/250222/the-constitution-of-the-socialist-republic-of-viet-nam.html, accessed 7 May 2015

Viñuales, J. (ed.) (2015) *The Rio Declaration on Environment and Development. A Commentary*, Oxford University Press, Oxford

Viravong Viraphonh (2013) 'Statement to 19th MRC Council', www.mrcmekong.org/news-and-events/speeches/statement-by-h-e-mr-viraphonh-viravong-19th-meeting-of-the-mrc-council/, accessed 7 May 2015

Vishnu Law Group (2015) 'Cambodia environmental impact law', seventh draft, www.vishnulawgroup.com/index.php/what-we-do/project-to-create-a-new-eia-law.html, accessed 28 April 2015

von Mehren, P. and Sawers, T. (1992) 'Revitalizing the law and development movement: a case study of title in Thailand', *Harvard International Law Journal*, 33: 67–102

Watercourses Convention (1997) *Convention on the Law of the Non-navigational Uses of International Watercourses*, opened for signature 21 May 1997, 36 ILM 700 (entered into force 17 August 2014)

Watershed (2008) 'Coming full circle: new trends, same old plans', *Watershed*, 12(3)

Wayakone, S. and Makoto, I. (2012) 'Evaluation of the environmental impacts assessment (EIA) system in Lao PDR', *Journal of Environmental Protection*, 3: 1655–1670

White, S. (2011) 'The republican critique of capitalism', *Critical Review of International Social and Political Philosophy*, 14: 561–579

Whitehead, I. (2011) *Closed Channels or Open Waters? Mekong Dams, Transboundary Law and the Xayaburi Prior Consultation Process*, Honours in Geography thesis, University of Sydney

Wilkinson, M. A. (2013) 'The specter of authoritarian liberalism: reflections on the constitutional crisis of the European Union', *German Law Journal*, 14(5): 527–560

Williamson, J. (2004) 'The strange history of the Washington Consensus', *Journal of Post Keynesian Economics*, 27(2): 195–206

Winichakul, T. (2008) 'Toppling democracy', *Journal of Contemporary Asia*, 38: 11–27

Wisuttisak, P. (2012) 'Regulation and competition issues in Thai electricity sector', *Energy Policy*, 44: 185–198

Wolf, A.T., Natharius, J.A., Danielson, J.J., Ward, B.S. and Pender, J.K. (1999) 'International river basins of the world', *International Journal of Water Resources Development*, 15(4), www. transboundarywaters.orst.edu/publications/register/, accessed 7 May 2015

Wolfensohn, J. D. (1996) *People and Development*, World Bank Annual Meeting Address, 1 October, web.worldbank.org, accessed 7 May 2015

World Bank (1989) 'Operational Directive 4.00, Annex A: Environmental Assessment', reproduced in *Environmental Assessment Sourcebook, Vol I: Policies, Procedures, and Cross-Sectoral Issues*, World Bank Technical Paper No. 139, http://elibrary.worldbank.org/doi/pdf/10.1596/0-8213-1843-8 pp. 32-44, accessed 7 May 2015

World Bank (1994) *Governance: The World Bank's Experience*, Washington, DC

World Bank (1999) 'Operational Directive 4.01: policy on environmental assessment', updated March 2007, http://go.worldbank.org/K7F3DCUDD0, accessed 7 May 2015

World Bank (2002) 'Decision framework for processing the proposed NT2 project', http://siteresources.worldbank.org/INTLAOPRD/Resources/NT2_Decision_Framework.pdf, accessed 7 May 2015

World Bank (2005) *Economic Growth in the 1990s: Learning from a Decade of Reform*, www1. worldbank.org/prem/lessons1990s, accessed 7 May 2015

World Bank (2009) *Implementation Completion and Results Report (TF-23406)*, Report No ICR00001005, 30 June, http://documents.worldbank.org/curated/en/2009/06/10956556/cambodia-lao-pdr-thailand-viet-nam-mekong-water-utilization-project, accessed 7 May 2015

World Bank (2013) 'Defining civil society', http://go.worldbank.org/4CE7W046K0, accessed 7 May 2015

World Bank (2014) *Laos – Nam Theun 2 Multipurpose Development Project: Twenty-second Report of the Environmental and Social Panel of Experts*, World Bank Group, Washington, DC

World Commission on Dams (2000a) *Dams and Development: A New Framework for Decision-making – Report of the World Commission on Dams*, Earthscan, London

World Commission on Dams (2000b) 'The Pak Mun dam in Mekong River Basin, Thailand', WCD Case Study, September

World Commission on Dams (2001) 'The Report of the World Commission on Dams: Executive Summary', *American University International Law Review*, vol 16, no 6, pp. 1435–1452

World Commission on Environment and Development (1987) *Report of the World Commission on Environment and Development: Our Common Future* (Brundtland Report), www.un-documents.net/our-common-future.pdf, accessed 7 May 2015

Worrell, S. (2014) 'Xayaburi dam 30% finished, says Laos', *Phnom Penh Post*, 25 March

Wouters, P. (2014) 'The yin and yang of international water law: China's transboundary water practice and the changing contours of state sovereignty', *Review of European Community & International Environmental Law*, 23: 67–75

WWC (2000) *Ministerial Declaration of The Hague on Water Security in the 21st Century*, www. worldwatercouncil.org/fileadmin/world_water_council/documents/world_water_forum_2/The_Hague_Declaration.pdf, accessed 7 May 2015

Wyatt, A. and Baird, I.G. (2007) 'Transboundary impact assessment in the Sesan River Basin: the case of Yali Falls dam', *International Journal of Water Resources Development*, 23(3): 427–442

Yu, X. (2003) 'Regional cooperation and energy development in the Greater Mekong sub-region', *Energy Policy*, 31(12): 1221–1234

Zeitoun, M. and Warner, J. (2006) 'Hydro-hegemony – a framework for analysis of trans-boundary water conflicts', *Water Policy*, 8(5): 435–460

Ziegler, A.D., Petney, T.N., Grundy-Warr, C., Andrews, R.H., Baird, I.G., Wasson, R.J. and Sithothaworn, P. (2013) 'Dams and disease triggers on the Lower Mekong River', *PLOS Neglected Tropical Diseases*, 7(6): e2166

Zinoman, P. (2001) *The Colonial Bastille: A History of Imprisonment in Vietnam, 1862–1940*, University of California Press, Berkeley and Los Angeles, CA

Ziv, G., Baran, E., Nam, S., Rodriguez-Iturbe, I. and Levin, S. (2012) 'Trading-off fish biodiversity, food security and hydropower in the Mekong River Basin', *Proceedings of the National Academy of Sciences*, 109(15): 5609–5614

Zoellner, C. (2005–2006) 'Transparency: an analysis of an evolving fundamental principle in international economic law', *Michigan Journal of International Law*, 27: 579–628

Zumbansen, P. (2008) 'Transnational law', *Comparative Research in Law & Political Economy*, Research Paper No 9/2008, http://digitalcommons.osgoode.yorku.ca/clpe/181, accessed 7 May 2015

OECD cases

Specific instance notified to the Austrian National Contact Point regarding the hydroelectric activities of a multinational enterprise (ANDRITZ AG) operating in Laos (Xayaburi), 9 April 2014, http://mneguidelines.oecd.org/database/instances/at0005.htm, accessed 7 May 2015

Specific instance notified to the Belgian National Contact Point by the NGO Proyecto Gato regarding the hydroelectric activities of Tractebel-Suez operating in Laos, 15 April 2004, http://mneguidelines.oecd.org/database/instances/be0003.htm, accessed 7 May 2015

Specific instance notified to the Finnish National Contact Point by an NGO regarding the hydroelectric activities of a multinational enterprise (Pyory) operating in Laos, 11 June 2012, http://mneguidelines.oecd.org/database/instances/fi0003.htm, accessed 7 May 2015

Specific instance notified to the French National Contact Point by the NGO Friends of the Earth regarding the hydroelectric activities of EDF in Laos, 26 September 2004, http://mneguidelines.oecd.org/database/instances/fr0008.htm, accessed 7 May 2015

Other cases

Niwat et al v Electricity Generation Authority of Thailand et al, Decision of 17 April 2014, Thai Supreme Administrative Court (Unofficial Conclusion and Translation in Brief of Decision to Grant the Appeal Submitted by the 8 Mekong Provinces in order to Challenge the Xayaburi Dam Power Purchase Agreement)

Pulp Mills on the River Uruguay (Argentina v Uruguay), Provisional Measures, Order of 13 July 2006, (2010) ICJ Reports 113

List of interviews

Cambodia

Interview 1, 27 June 2011
Interview 2, 28 June 2011
Interview 3, 27 June 2011
Interview 4, 28 June 2011
Interview 5, 28 June 2011

Interview 6, 27 June 2011
Interview 7, 27 June 2011
Interview 8, 28 June 2011
Interview 9, 28 June 2011
Interview 10, 23 June 2011
Interview 11, 24 June 2011
Interview 12, 23 June 2011
Interview 13, 23 June 2011
Interview 14, 23 June 2011
Interview 15, 23 June 2011
Interview 16, 24 June 2011
Interview 17, 25 June 2011
Interview 18, 25 June 2011
Interview 19, 27 June 2011

Thailand

Interview 20, 29 June 2011
Interview 21, 29 June 2011
Interview 22, 30 June 2011
Interview 23, 30 June 2011
Interview 24, 30 June 2011
Interview 25, 1 July 2011
Interview 26, 17 September 2011
Interview 27, 22 September 2011
Interview 28, 26 November 2012
Interview 29, 26 November 2012

Vietnam

Interview 30, 28 November 2011
Interview 31, 28 November 2011
Interview 32, 28 November 2011
Interview 33, 29 November 2011
Interview 34, 28 November 2011
Interview 35, 29 November 2011
Interview 36, 29 November 2011
Interview 37, 29 November 2011
Interview 38, 29 November 2011
Interview 39, 30 November 2011
Interview 40, 2 December 2011
Interview 41, 1 December 2011
Interview 42, 1 December 2011
Interview 43, 1 December 2011

Laos

Interview 44, 19 November 2012
Interview 45, 20 November 2012

Interview 46, 20 November 2012
Interview 47, 21 November 2012
Interview 48, 22 November 2012
Interview 49, 21 November 2012
Interview 50, 21 November 2012
Interview 51, 21 November 2012
Interview 52, 22 November 2012
Interview 53, 22 November 2012

Australia

Interview 54, 28 October 2010

Index

Page numbers in *italics* are figures; with 't' are tables.

3S dams *27*, 28
3SPN (3S Rivers Protection Network) 19, 28, 172, 173, 175–6, 178, 179–80

Aarhus Convention 40, 195
access to information *see* information, access to
accountability 72, 90, 139, 189–90, 196; and Lao PDR 83, 85; and Thailand 76, 171; and transparency 155; UN Human Rights Committee on 139–40; and Vietnam 127
ADB (Asian Development Bank) 16, 17–18, 31, 77, 78–9, 81, 82, 198; on EA 151; GMS Economic Cooperation Programme 15; and legal empowerment 177; and RBOs 90–1; and Theun Hinboun dam 82, 174
ADHOC (Cambodian Human Rights and Development Association) 178, 179
AFD (Agence Française de Développement *or* French Agency for Development) 81
Agreement on the Cooperation for the Sustainable Development of the Mekong River Basin (1995) *see* Mekong Agreement
Althusser, Louis 142
A Luoi dam (Vietnam) 29–30, *30*, 37; and EA 127–30, 132, 136
Alvarez, José 149
Anghie, A. 56
AOP (Assembly of the Poor) 167, 182–4, 185
ASEAN (Association of Southeast Asian Nations) 15–16, 18, 77, 90, 94, 190
'ASEAN way' 90, 190
Australian Mekong Resource Centre 144

Baker, C. 183–4
banks 17–18; *see also* ADB; World Bank
Basin Development Strategy 10, 102, 107
BDP (Basin Development Plan) 101–3, 107, 110
Bentham, Jeremy 140
best practice 86, 176, 187, 190; and Lao PDR 83; and the private sector 173
Bianchi, A. 153
Bird, Jeremy 151
'black-boxing' effect 150
BOOT (build-own-operate-transfer) 17, 82
BOT (build-own-transfer) 17, 18
'bright line' property rights 59
Buddhism, and transparency 142–3
build-own-transfer (BOT) 17

Cambodia 7, 8t, 13–14, 94; ADHOC 178, 179; and ASEAN 94; and China 17; and civil society 20, 165, 166; and contestation of projects 167, 168–70, 171, 181; development common sense 146; and EA 115, 116, 117t, 118–19, 125, 135, 154–5; and governance 52; and law and development 55; laws/policies and decision-making 48, 49, 50, 203–7; legislation/regulations/decisions/ministerial instructions 200–2; and the Mekong 6; and the Mekong Agreement 8, 15; and the Mekong Committee 68, 70; and neoliberal law 74–6; NMC (National Mekong Committee) 126; and private-sector actors 17, 18; RCC 19, 126; socialist-style law 65, 66–7; TbEIAs (transboundary environmental impact assessments) 117t, 118, 119; Tonle Sap

Lake 9; and transparency 147, 154–5, 163; UNTAC 74–5; and Vietnam 94–5; and the Xayaburi dam 109, 133–4; *see also* Lower Sesan 2 dam
Cambodian Human Rights and Development Association (ADHOC) 178, 179
canals, Indochina 63
CEPA (Culture and Environment Preservation Agency) (Cambodia) 178–9, 180
China 7, 8t, 196; and ASEAN 95; and Asian regionalism 16–17; dams 10; and engagement with private actors 175; and investment in the Mekong 17, 31, 33; mainstream dams 24, *25*, 26; and the Mekong Committee 68; and the MRC 16
CIA (cumulative impact assessment) 114, 116, 117t, 122
civil society 11, 19–21, 36, 38, 137, 197; and EA 124–6; Tai Baan programme 101; Thailand 76–7; *see also* legal strategies for contestation; NGOs
Clean Development Mechanism 30
CLEC (Community Legal Education Center) (Cambodia) 179, 180
CLICK 19, 156
CNMC (Cambodian National Mekong Committee) 126
Cogels, Olivier 151
Cohen, Amy 52
Collier, Stephen 53
colonialism 36, 56; in French Indochina 63–5
commonality-for-development 138, 158, 164, 166
Common fund for Communities 57
common sense 141, 144, 188–95; and rule of law 145–77
communism 55, 61, 65–8, 70, 85; Lao PDR 20, 48, 66, 73–4, 94, 161, 170; Thailand 12; Vietnam 20, 48, 52, 65, 73–4, 94, 148, 170
conflict, geopolitical 10
consultants 85, 102, 119, 121, 129
contestation by civil society 38; *see also* legal strategies for contestation
Convention on Environmental Impact Assessment in a Transboundary Context 40
Convention on the Law of the Non-Navigational Uses of International Watercourses 8, 89, 97–8, 205

corporate social responsibility (CSR) 167, 174–5, 179
critical hydropolitics 93
cultural meaning, and rivers 5–6
cumulative impact assessment (CIA) 114, 116, 117t, 122
customary law 54, 61, 67, 169, 172, 193, 197; Cambodia 179; Vietnam 64

dams 21–35
Dao, N. 127
Daston, Lorraine 198
de Búrca, G. 52
decisions 200–2
Decision Support Framework (DSF) 144; MRC (Mekong River Commission) 144
Declaration on the Right to Development 57
democratization 55, 72
Denmark 100
development 5; and governance 53; law 36, 54–9, 65–7, 70–7, 85; and Mekong River Basin 37–8
development-oriented transparency 157
development-versus-conservation 5
Dewey, John 140–1
Donor Consultative Group 98, 100, 111–12
donors: and contestation or projects 174, 177; and EA 132; and law and development 48, 54–5, 59, 60; and legal reform 37, 61, 62–3, 67–9, 72–7, 78, 79, 83, 85, 197; and the MRC 15, 90, 98–100, 104, 105, 111–12, 150; and transparency 141, 146, 150; *see also* private sector actors
Don Sahong dam (Lao PDR) 37, 105–6, 107, 111, 171; and EA 130, 131, 132, 134; and private investors 50
Dore, J. 98–9
DOS (development opportunity space) 102
droughts 9
DSF (Decision Support Framework) 144, 150
Dublin Principles (1992) 91, 154
Dung, Nguyen Tan 148

EA *see* environmental assessment (EA) processes
EarthRights International (ERI) 19, 180
ECAFE (Economic Commission for Asia and the Far East) 68
ECA Watch (Export Credit Agencies Watch) 177

Economic Commission for Asia and the Far
    East (ECAFE) (UN) 68
economic development 54, 76, 78, 85, 94,
    129, 161, 189, 193; and colonialism 37,
    64; and transparency 59; and the World
    Bank 38
EDF International 31
EGAT (Electricity Generating Authority
    of Thailand) 12, 18, 80–1, 83, 160, 174,
    180, 182; *Niwat v Electricity Generating
    Authority of Thailand* 89, 124, 132, 181
EGCO (Electricity Generating Public
    Company Ltd) 17, 18, 22, 31, 81
EIAs (environmental impact assessments) 37,
    49, 121–4; and contestation of projects
    170–1; and environmental assessment 115;
    Thailand 119; and transparency 151–2,
    154–5; Xayaburi dam (Laos) 24, 108, 109,
    131–4; Yali Falls 132
elites 78, 191, 198; and colonialism 64–5
empowerment 177–80
Enhancement and Conservation of National
    Quality Act (1992) (Thailand) 45
environment; and China 16
environmental assessment (EA) processes
    114–15, 134–6; and the A Luoi dam
    127–30; decision-makers of 122; and
    impacts 122–4; legal requirements 115–20,
    117t; and public participation 124–6; and
    technical politics 114, 120–1, 127; and the
    Xayaburi/Dan Sahong dams 130–4
environmental impact assessments *see* EIAs
environmental law 5, 73, 167, 169, 179, 191;
    international 19, 40, 153, 173, 174, 176
environmental management, and
    governance 51–3
Environment Programme 100–1
EPF (Electricity Power Forum) 79
Equator Principles 45, 173–4, 174
ERI (EarthRights International) 19, 135,
    173, 177, 180; Mekong Legal Advocacy
    Unit (MLAI) 178
Etzioni, Amitai 140
EVN (Electricity of Vietnam) 13, 28, 50, 80,
    82, 175
experts 2, 11, 187, 197, 198; and
    contestation of projects 172, 178, 179,
    183; and EA 115, 126; and law and
    development 36, 48, 60; legal 18, 40;
    legal reform 36–7, 61–2, 65–6, 68, 71,
    75, 77–86; and the MRC 88, 91, 92, 104,
    105, 107; and transparency 137, 140, 144,
    151, 155, 162, 164
'export sophistication' 52

Fisheries Action Coalition Team
    Cambodia 19, 177
Fisher, P. 121
fish/fisheries 9, 100, 107; and the Fisheries
    Programme 100–1; Lao PDR 24, 107;
    and SEA 102–3; Thailand 33
floods 9, 28
Fong, J. 143
Foucault, Michel 53
Fox, C. 93
French Indochina 63–7
'fugitive resources' 3–4

Galison, Peter 198
Gao, Qi 106, 117, 126
Geheb, K. 106, 108, 111
geology 8–9
geopolitical conflict 10
Gillespie, J. 64, 74
global law 47
GMS (Greater Mekong Subregion) 15–16,
    77, 82, 83; and regional energy trade
    78–81
Gooch, G. 92
good governance 15, 41, 62, 86; and finance
    83, 138; and law and development
    54–5; and neoliberal law 72–3; and
    transparency 145
'Go Out' policy (China) 16, 24
governance 36, 37, 51–3, 72; and EA 120–1;
    and RBOs 90–2; and transparency 152;
    see also MRC; transparency
governmentality 53, 87, 108
grassroots community groups 19, 167,
    179–80
'green economy' approach 135–6
Grumbine, R.E. 106
Guttal, S. 163

Harding, A. 77
hard law 45–7, 90, 138–9, 195
Hayek, Friedrich 52, 71, 143–4
Helsinki Rules on the Uses of the Waters of
    International Rivers 8, 40, 69, 96
Heyzer, Noeleen 145
HIAs (health impact assessments) 114
Ho Chi Minh 65
HSAP (Hydropower Sustainability
    Assessment Protocol) 176
human rights 19, 57–9, 72, 144, 167, 191,
    195; Cambodia 75; litigation 181; Office
    of the High Commissioner of Human
    Rights 174
Hun Sen 14, 20, 52, 94, 172

identity 6
ideology 141
IFAD (International Fund for Agricultural Development) 57
IFC (International Finance Corporation) 84, 173–4, 175
IHA (International Hydropower Association) 176
ILC (International Law Commission) 40
IMF (International Monetary Fund) 55, 57, 72, 76, 81
Indicative Basin Plan (1970) 69
informal negotiations 26, 61, 189–90
information, access to 117, 117t, 205; Aarhus Convention 151; and Cambodia 147, 156; and the MRC 149; UNECE 40; and Vietnam 148; and Xayaburi dam project 108
Institut de droit international 40
institutions of dominance 142
integrated water resources management (IWRM) 88, 91–3, 98, 102, 103
international law 47–8, 189, 191–3, 195; firms 18–19; and law and development 56–9; and transboundary river basins 39–41, 89; and transparency 153–8
International Law Association 40
International Law Commission (ILC) 40
International Rivers 21, 28, 105, 173, 178, 180; and the Xayaburi dam 19, 106, 132
interview approach 35–6
Isan (Thailand) 12
IWRM (integrated water resources management) 88, 91–3, 98, 102, 103

Jasanoff, Sheila 53
Jessup, Philip 47
Joint Declaration of Principles for Utilization of the Waters of the Mekong River Basin 69, 70, 96
justice 3, 55, 72, 155–6

Ke, Jian 106, 117, 126
Kennedy, Duncan 42–3
Keohane, Robert 52
Khmer Rouge 66–7
Khong-Chi-Mun irrigation project 12, 33, 70, 96, 110
Kim, Sang-Hyun 53
Kok-Ing-Nan project 35
Kristensen, Joern 150–1

Lancang dams (China) 24
Lao Front for National Construction (LNF) 160–1
Lao PDR 8t, 12–13; and ASEAN 94; and China 17; civil society 20–1, 177; development common sense 146–7; EA laws 117t, 119; geopolitical conflict 10; and the GMS 80; and governance 52; legislation/regulations/decisions/ministerial instructions 48, 49, 50, 200–2; and the Mekong 6; and the Mekong Agreement 8, 15; and the Mekong Committee 68; Nam Theun 2 1–3; and neoliberal law and development 73–4, 83; NMC (National Mekong Committee) 107, 108; and private-sector actors 17, 83, 83–4; and RBOs 91; socialist-style law 65, 66; and Thailand 94; trade in power 82; and transparency 147–8, 163; treaties/standards 203–7; and Vietnam 94; *see also* Don Sahong dam; Xayaburi dam
Lao People's Revolutionary Party (LPRP) 94
Lao Women's Union (LWU) 160–1
Latour, Bruno 193–4
law 36, 193–4, 198–9; approaches to transboundary river basins 39–43, 59–60; defined 44; and development 36, 54–9, 65–7, 70–7, 85; and hydropower decision-making 48–50; legal reform agendas 36, 199; and the MRB 3–6; national 147–8, 189, 196; neoliberal 70–7; and 'non-legal' resistance 182–5; terminology 43–50; *see also* international law; legal reform; legislation
law firms/lawyers 18–19
layering 136
Lazarus, K. 98–9
Lebel, L. 88
legal assistance 70–1, 73–4, 75, 77
legal consciousness 168
legal education 177–80
legal pluralism 43–4
legal reform 36–7, 61–3, 85; and French Indochina 63–7; international water resources management 67–70; neoliberal law and development 70–85; socialist-style law and development 65–7
legal strategies for contestation 165–8, 185–6; and the courts 180–2, 185; legal education 177–80; and non-legal modes of resistance 182–5; and political engagement 168–73; and private actors 173–7

legislation 200–2
letter-writing 172
Li, Tania 92
litigation 180–2, 185, 195
Living Rivers Siam 19, 181
LNF (Lao Front for National Construction)
    160–1
local authorities, and contestation of
    projects 169–70
Lower Sesan 2 dam (Cambodia) 14, 28–9,
    29, 178; contestation against 173, 175;
    and EA 125
Low, S. 74
LPRP (Lao People's Revolutionary
    Party) 94
LWU (Lao Women's Union) 160–1
Lysa, Hong 142

market transparency 143–4
Marxism 140–1
mass mobilization 167, 182–3, 184
Matthews, N. 111
media, and transparency 156
MEE Net (Mekong Energy and Ecology
    Network) 19
Mekokg Water Resources Assistance
    Strategy (MWRAS) 101
Mekon2Rio conference 98
Mekong Agreement 8, 10–11, 15, 48, 69,
    70, 89, 96–7, 108, 110, 190; and EA
    118–19, 123, 128, 134; and the MRC 95,
    98; as soft law 45; and transparency 145,
    149, 153; and the Xayaburi dam 106
Mekong Cascade 69, 70, 99, 105
Mekong Committee 68–9; Joint
    Declaration of Principles 69, 70, 96
Mekong Project 68
Mekong River Basin 6–11, 7, 8t; +and
    development 37–8; and law 3–6
Mekong Secretariat 68–9, 70, 98, 99
Mekong Water Governance Programme 19
Mekong Water Resources Assistance
    Strategy (MWRAS)
Merry, Sally Engle 44, 168
Middleton, Carl 105, 106
ministerial instructions 200–2
MLAI (Mekong Legal Advocacy
    Institute) 178
MLN (Mekong Legal Network) 178
Molle, F. 92
Montini, M. 116, 135–6
MoU (memoranda of understanding)
    79, 206

MRC (Mekong River Commission) 9,
    14–16, 37, 68, 70, 87–8, 95–105, 112–13,
    189, 197; Basin Development Strategy
    10; and China 26; and EA 121, 134, 135;
    and mainstream dams 105–12; and RBOs
    90–2; and science 88–9; and TbEIAs 115;
    and transparency 148–53, 162
Mun river Declaration 183
MWRAS (Mekong Water Resources
    Assistance Strategy) 101
Myanmar 6, 7, 8t, 196; and ASEAN 94; and
    the Mekong Committee 68; and
    RBOs 91

Nam Ngum dam (Laos) 68–9
Nam Theun 1 dam (Lao PDR) *32*
Nam Theun 2 dam (Lao PDR) 30–1, *32*,
    33, 80, 174, 192; and private investors 50,
    83–4; and transparency 158–64; visit by
    authors 1–3, 187
NARBO (Network of Asian River Basin
    Organizations) 91
national constitutional law, and
    transparency 147–8
national interests 14–15, 16
National Mekong Committees 98, 152
nature, and society 193–5
NCPs (national contact points) 176–7
'neoliberal concepts' 71
neoliberal law and development 70–7,
    85–6, 127, 196
'new governance' 51–2
NGOs 19; Cambodia 75; and EA 131–2;
    and legal awareness 178–9; and litigation
    180–1; and the MRC 104–5; and private
    sector actors 173–4; working with
    governments 172
NHRC (National Human Rights
    Commission) (Thailand) 182
Nicholson, P. 74
NIEO (New International Economic
    Order) 56–7, 58
*Niwat v Electricity Generating Authority of
    Thailand* 89, 124, 132, 181
NMC (National Mekong Committee) 98,
    149, 152; Cambodia 126; Lao 107, 108;
    Vietnam 28, 30
norms, legal 45, 89

OECD (Organisation for Economic
    Co-operation and Development) 167,
    176, 209, 210
Oxfam 19, 28, 144

Pak Mun dam (Thailand) 33, *34*, 35, 167, 180, 182–3
PanNature (People and Nature Reconciliation) 19
PAPs (Project Affected Peoples) 50, 159
Parker, Christine 44
participation-management 136
PCPD (Public Consultation, Participation and Disclosure) 161
peace, public 140–1
Peck, J. 71
'People's Constitution' (Thailand) 77
Pham, H.T. 128, 129
pico-hydropower 50
PNPCA (Procedures for Notification, Prior Consultation and Agreement) 24, 69, 97, 103–4, 112, 112–13, 149, 170–1, 181, 197; and contestation of projects 170–1; and the Don Sahong dam 105–6, 107; and the Xayaburi dam 106–11, 131–2
post-Washington Consensus 17, 58, 59
Poumisak, Jit 142
PPA (Power Purchase Agreement), Xayaburi 181, 182
pragmatism 140, 141
private international law 47
private law 35, 167, 173, 188, 190, 195
private sector actors 11, 17–18, 50, 82–3, 110; engaging 173–7; promotion of legal awareness 179
Project Affected Peoples (PAPs) 50, 159
'proto-transparency' 149
public consultations 112, 132, 158, 160, 161, 185; *see also* PNPCA
public international law 47
public participation 36, 38, 142, 188, 192; Cambodia 118; and EA 124–6, 127, 130; and environmental assessment laws 117, 117t; and transparency 138, 142, 154–5, 161; Vietnam 148; and the Xayaburi dam 108, 133
public-private partnerships (PPPs) 17, 52, 82, 145

Ratchaburi Electricity Generating Holding Public Company 17, 81
RBOs (river basin organizations) 89, 90–2, 96, 112; and politics 92–5; *see also* MRC
RCC (Rivers Coalition in Cambodia) 19, 126
regional law 46, 47, 189
regulation 44, 200–2

remedial transparency 138, 140, 155–8
republicanism 140, 141
resettlement/relocation 30, 128–30, 187, 192
Revised Indicative Plan (1987) 105
Reynolds, Craig 142
Rieu-Clarke, Alistair 40–1, 136
rights 3; human 19, 57–9, 72, 75, 167
Rio Declaration on the Environment and Development (1992) 40
rivers 6; and cultural meaning 5–6; as relational phenomena 4; as resources 3–4
Rousseau, Jean-Jacques 140
RPTCC (Regional Power Trade Coordination Committee) 79
rule of law 54, 61, 72–3, 86, 139; and Cambodia 75; and development common sense 145–7; and the MLN 178; and socialist-style law 65; and Thailand 77, 181; and the World Bank 55

Sabel, R. O. 52
'sacred nationalism' 143
Samreth Law Group (Vishnu Law Group) 172, 173, 179
Sarat, Austin 44
Save the Mekong coalition 20
science 88–9, 92, 100–1, 106–7, 188, 198
SEA (strategic environmental assessment) 102–3, 116, 117t, 123; and Xayaburi 119, 133–4
security, state 140–1
sediment 4, 12, 13, 26, 102, 107, 111, 132, 194
Sen, A. 57
Shoemaker, B. 163
SIA (social impact assessment) 115–16, 117t, 130
Sida (Swedish International Development Agency) 79, 80, 81
Simon, Jonathan 43
Sneddon, C. 93
Social Development Plan, and Nam Theun 2 159–60
social environmental standards 73
social impact assessment (SIA) 115–16, 117t, 119
socialist-style law and development 65–7, 85
'social', the 41–3
society, and nature 193–5

socio-legal approach to transboundary river
basins 41–3, 60, 193–5
'socio-technical imaginaries' 53, 60
soft law 36, 39–40, 45–7, 90, 108, 112–13,
131, 195, 197; and transparency 138–9;
see also standards/treaties
sovereignty 14, 59, 188–9
SPSI (science–policy–stakeholder
interface) 92
stakeholder management 159
stakeholders 38, 159, 197; participation
model 92; and transparency 138, 163
Stalnacke, P. 92
standards/treaties 203–10
state actors 11–17, 171–3
state security 140–1
strategic environmental assessment (SEA)
102–3, 116
Suhardiman, D. 103
sustainability 136
sustainable development 5, 15, 72
Sustainable Hydropower Programme 100

Tai Baan programme 101
Tamanaha, Brian 44
TbEIAs (transboundary environmental
impact assessments) 114, 115–16, 117t,
122, 123, 151, 153; Cambodia 117t, 118,
119; Laos 119; Vietnam 128
Tennessee Valley Authority (TVA) 67,
68, 69
TERRA (Towards Ecological Recovery
and Regional Alliance) 19, 105
Thailand 7, 8t, 12, 94; and civil society 20,
165; and contestation of projects 167,
171, 179, 181, 182–5; EA laws 117t,
119; Enhancement and Conservation of
National Quality Act (1992) 45; human
rights 168; and Joint Declaration of
Principles 96; and Lao PDR 94; and law
and development 67; laws/policies and
decision-making 48, 50, 203–7; lawsuit
against developers of Xayaburi dam
89; legislation/regulations/decisions/
ministerial instructions 200–2; and
the Mekong Agreement 8, 15; and the
Mekong Committee 68, 70; National
Human Rights Commission 59;
neoliberal law and development 76–7;
opposition to dams 10; private-sector
actors 17, 83, 174–5; and RBOs 91;
'sacred nationalism' 143; and trade in
power 80, 81; and transparency 145–6,
147, 155–6, 163; *see also* Pak Mun dam

Theun Hinboun dam 82, 174
Theun Hinboun dam (Lao PDR) *32*, 80,
82, 174
Thomas, C. 145
Tolentino, A.S. 190
Tonle Sap Lake (Cambodia) 9
trade in electricity 78–82
transitional justice 55
transparency 38, 59, 72, 137–9, 188, 195–6;
as commonality-for-development 164;
global 153–8; history of 139–48; and
the MRC 148–53; and Nam Theun 2
158–64, 187
Tran, V.H. 130
treaties/standards 203–10
TVA (Tennessee Valley Authority) 67,
68, 69

UNCTAD (United Nations Conference
on Trade and Development) 57
UNDP (United Nations Development
Programme) 57, 128–9, 174
UNECE (United Nations Economic
Commission for Europe) 40
UNESCAP (United Nations Economic
and Social Commission for Asia and the
Pacific) 145
UNIDO (United Nations Industrial
Development Organization) 57
United States 67, 68–9, 70–1, 103
Universal Declaration of Human
Rights 56
UNOHCHR (United Nations Office of
the High Commissioner for Human
Rights) 174
UNTAC (United Nations Transitional
Authority on Cambodia) 74–5
UN (United Nations): Charter 55;
Convention on the Law of the
Non-Navigational Uses of International
Water Courses 8, 89, 97, 98, 205;
Declaration of Human Rights 56;
Development Programme (UNDP) 57,
128–9, 174; Economic Commission
for Asia and the Far East (ECAFE)
68; Human Rights Committee
139–40; Industrial Development
Organization (UNIDO) 57; and law
and development 55, 56–7; Office of
the High Commissioner of Human
Rights (UNOHCHR) 174; Transitional
Authority on Cambodia (UNTAC)
74–5; Watercourses Convention 40
Urban, F. 16

USAID (United States Agency for International Development) 69, 103
utilitarianism 140, 141

Vietnam 7, 8t, 13, 68; and ASEAN 94; and Cambodia 94–5; civil society 20–1; and contestation of projects 171, 179; and Convention on the Law of the Non-Navigational Uses of International Water Courses 8, 89–90; and EA 117, 117t, 120; as French Indochina 63–7; and human rights 167; and Joint Declaration of Principles 96; and Lao PDR 94; laws/policies/treaties/decision-making 48, 49, 50, 203–7; legislation/regulations/decisions/ministerial instructions 200–2; and the Mekong Agreement 8, 15; and the Mekong Committee 68, 70; NMC (National Mekong Committee) 28, 30; private-sector actors 18, 83; and RBOs 91; and SEA 116; socialist-style law 65–6; TbEIAs (transboundary environmental impact assessments) 128; trade in power 81–2; and transparency 148, 163; VRN 19; War 10; and the Xayaburi dam 109, 135; *see also* A Luoi dam; Yali Falls dam
Village of the Poor 185
Viraponh Viravong 189
VRN (Vietnamese Rivers Network) 19

'Washington Consensus' 72
Water Utilization Programme (WUP) 101
well-being, and transparency 140–1
Wheeler, Raymond A 68
World Bank 11; and development 58, 79, 82–3; environmental policies 45; governance 72; Nam Theun 2 2, 31, 160; on transparency 150, 151
World Commission on Dams 33, 45–6, 153, 154, 159, 174, 175, 192–3
World Trade Organization 55, 74
WTO (World Trade Organization) 55, 57, 74, 76
WUP (Water Utilization Programme) 101, 103
WWF (World Wide Fund for Nature) 176

Xayaburi dam (Lao PDR) 12, 21–4, *23*, 37, 171; and contestation of 173, 181; and EA 24, 109, 119, 130–5, 135, 136; PPA 181, 182; and private investors 50, 83; and public participation 125–6, 161; and science 106–7; Thai lawsuit 89; and transparency 164
Xu, J. 106

Yali Falls dam (Vietnam) 26, *27*, 28, 104; and contestation of 179; EIA 132